# The Sedimentary Record of Sea-Level Change

This unique textbook describes how past changes in sea-level can be detected through analysis of the sedimentary record. In particular, it concentrates on the theory of sequence stratigraphy, which provides a framework for how entire sedimentary systems evolve through geological time. This theory is widely used to understand the genesis of the sedimentary record, to examine the global synchroneity of sedimentary cycles and in the exploration for hydrocarbon reserves. The book explains the current sequence stratigraphy model from basics and shows how the model can be applied to both siliciclastic and carbonate successions. It also covers stratigraphical techniques, mechanisms for sea-level change and forward modelling of stratigraphical geometries. A variety of case studies are presented including a deltaic siliciclastic succession from the Gulf of Mexico; an in-depth study of the fluvial to open-marine siliciclastic deposits exposed in the Book Cliffs, USA; a shallow-marine carbonate platform succession from Mallorca; and a carbonate ramp succession from Spain.

Designed for undergraduate and graduate courses, as well as professionals, this textbook includes numerous features that will aid tutors and students alike including full colour diagrams, case studies, set-aside focus boxes and bulleted questions and answers. The book is supported by a website hosting sample pages, selected illustrations to download, and worked exercises: **http://publishing.cambridge.org/resources/0521831113**

The authors of this book have been closely involved in the testing, development and application of the sequence stratigraphy model since the 1980s. They are all field geologists with much experience of collecting and interpreting data from the sedimentary record. Each of them teaches sedimentology and sequence stratigraphy at their respective institution, and they have combined this experience to write a clear and readable book.

The project was initiated by **Angela Coe** (Senior Lecturer at The Open University, Milton Keynes, UK) and **Chris Wilson** (Professor of Earth Sciences at The Open University) as part of the development of a new Open University Course, and the material has been tried and tested by Open University students. Dr Coe's research includes the identification and interpretation of sedimentary cycles within Jurassic and Miocene deposits, particularly mudrocks, together with developing new stratigraphical techniques. Professor Wilson's interests include sedimentology and, based on work he completed as part of two ODP cruises, the tectonic evolution of the Atlantic margin. **Kevin Church** is an Associate Lecturer at The Open University with expertise on sequence stratigraphy of the Carboniferous of England, an example of which has been used to illustrate the theory presented in Chapters 3 to 5. **Stephen Flint** (Professor of Stratigraphy and Petroleum Geology, Director of the Stratigraphy Research Group, University of Liverpool, UK) and **John Howell** (formerly at Liverpool University, now Professor of Production Geology, University of Bergen, Norway) contributed the Book Cliffs case study, which is based on their extensive work and that of other members of the Liverpool University Stratigraphy Research Group. Professors Flint and Howell continue to work on stratigraphical architecture of sedimentary systems and methods for modelling hydrocarbon reservoirs. **Dan Bosence** (Professor of Carbonate Sedimentology at Royal Holloway University of London, UK) has broad experience of both modern and ancient carbonate deposits. His current research projects include high-frequency Jurassic carbonate cycles in Europe and North Africa and numerical modelling of carbonate platform stratigraphy. His expertise was combined with that of Professor Wilson to produce Part 4 of this book.

**Cover photograph**   Foreshore near Santa Barbara, California, USA. The cliffs expose the Monterey Formation, a thick, laterally extensive organic carbon-rich mudrock succession (weathered to a pale yellow). This is interpreted to have been deposited due to upwelling associated with global cooling and sea-level fall in the mid-Miocene. It has been suggested that deposition of these organic-rich mudrocks enhanced global cooling by a positive feedback mechanism, i.e. that as the mudrock was deposited, it led to drawdown of $CO_2$ from the atmosphere, which in turn led to further cooling (the Monterey hypothesis). The Monterey Formation is an important source rock for California's oil reserves. The large Pacific waves cause movement of considerable amounts of foreshore and shoreface sediments along this coast as well as making it popular for surfing. *(Angela Coe, Open University.)*

# The Sedimentary Record of Sea-Level Change

Edited by Angela L. Coe

Authors:

Angela L. Coe

Dan W. J. Bosence

Kevin D. Church

Stephen S. Flint

John A. Howell

R. Chris L. Wilson

The Open University

CAMBRIDGE
UNIVERSITY PRESS

PUBLISHED BY THE PRESS SYNDICATE OF THE UNIVERSITY OF CAMBRIDGE
The Pitt Building, Trumpington Street, Cambridge, United Kingdom

CAMBRIDGE UNIVERSITY PRESS
The Edinburgh Building, Cambridge CB2 2RU, UK
40 West 20th Street, New York, NY 10011-4211, USA
477 Williamstown Road, Port Melbourne, VIC 3207, Australia
Ruiz de Alarcón 13, 28014 Madrid, Spain
Dock House, The Waterfront, Cape Town 8001, South Africa

http://www.cambridge.org

First published 2002

This co-published edition first published 2003

Printed in the United Kingdom by Bath Press, Blantyre Industrial Estate, Glasgow G72 0ND, UK

*Typefaces*: text in Times 11/13pt; headings in Futura

*System*: Adobe Pagemaker

*A catalogue record for this book is available from the British Library.*

ISBN 0 521 53842 4  paperback       ISBN 0 521 83111 3  hardback

This publication forms part of an Open University course, S369 *The Geological Record of Environmental Change*. Details of this and other Open University courses can be obtained from the Call Centre, PO Box 724, The Open University, Milton Keynes MK7 6ZS, United Kingdom. Tel. +44 (0)1908 653231, e-mail ces-gen@open.ac.uk Alternatively, you may visit the Open University website at http://www.open.ac.uk where you can learn more about the wide range of courses and packs offered at all levels by The Open University.

2.1

# Contents

# Preface

*The Sedimentary Record of Sea-Level Change* is about how we can detect past changes in sea-level from an analysis of the sedimentary record. In particular, it concentrates on the revolutionary new concept of sequence stratigraphy. This concept has changed the way in which many sedimentologists and stratigraphers examine sedimentary rocks because it provides a framework for how entire sedimentary systems evolve through geological time and places emphasis not only on the sediments themselves, but also on gaps between sedimentary units. In addition, sequence stratigraphy incorporates two long-standing observations: first, that the sedimentary record shows repetitions or cycles; and secondly, that some sedimentary units can be traced over long distances.

Sequence stratigraphy is a method of dividing the sedimentary record into discrete packages, where each package is interpreted to represent a cyclic change in sea-level and/or sediment supply. Evidence in the sedimentary record shows that rises and falls in sea-level together with changes in sediment supply have occurred repeatedly throughout geological time. These fluctuations are of various magnitudes and took place on a number of different time-scales. The constituent parts of these sedimentary packages map out the position of sea-level and/or the change in sediment supply within a cycle and enable their inter-relationship through time to be analysed.

There are several reasons why sequence stratigraphy is useful. First, it is a more holistic way of examining the sedimentary record because it considers 'what is missing' as well as 'what is present' in the sedimentary record; and secondly, it places emphasis on examination of the sedimentary record over discrete time periods. Therefore, evolving palaeogeographies are studied together with how different sedimentary environments, from fluvial systems to coastal environments and into the deep sea, have interacted and affected each other. The technique has been widely used in the prediction of hydrocarbon reserves and also to examine the global synchroneity of sedimentary cycles; however, their synchroneity still remains unproved as does the mechanism controlling sea-level change for many parts of the geological record. The aim of this book is to explain the theory of sequence stratigraphy and to illustrate this with several case studies. It does not discuss the global synchroneity (or not) of sequence cycles or cover how the sequence stratigraphy model has evolved; instead, the book attempts to present the current model in a pragmatic, clear and concise fashion.

The book is divided into four parts. The first part examines what it is about sedimentary rocks that makes them important and useful for determining sea-level change, how we can determine geological time, and introduces sea-level change. It also includes an examination of some possible evidence for Noah's flood. Part 2 covers the theory of sequence stratigraphy and mechanisms for sea-level change. It ends with a short case study from the Gulf of Mexico. Part 3 is a longer case study of the famous siliciclastic succession from the Book Cliffs, Utah, USA. The succession comprises fluvial to open marine deposits that are well exposed in a spectacular cliff face about 300 km long and up to 300 m high. Part 4 examines how carbonate sediments respond to sea-level change and shows how these responses can be forward modelled to predict different stratigraphical geometries and successions. Both modelling and exposure information are presented in case studies from Mallorca and north-east Spain. The Mallorcan example is particularly impressive as the present-day topographical height of the crest of the reef approximates to an ancient sea-level curve.

This book is aimed at advanced undergraduates and graduates. Readers are expected to have a basic understanding of sedimentary rocks, processes and depositional environments (though further information can be obtained from the references). It is designed as a teaching text for independent study in a linear fashion, and includes short bulleted questions followed directly by answers. These are designed to make the reader pause for a *moment* and think about a fundamental point concerning the subject matter. Supplementary and/or background information, which some readers may already be familiar with, is placed in boxes.

This book forms a part of an advanced undergraduate course offered by The Open University, UK, entitled S369 *The Geological Record of Environmental Change*.

Angela Coe, February 2003

# PART 1 INTRODUCTION

# 1 Sedimentary rocks as a record of Earth processes

*Angela L. Coe*

*Imagine standing on the banks of a large meandering river during a period of continuous heavy rain in the nearby mountains and think about what you would observe; the higher water level would lead to greater water flow and increased current strength, allowing more sediment and larger clasts to be transported. The base and sides of the river channel would be eroded as sediment is hurtled along by the currents. Periodically, parts of the river bank would collapse and more and more sediment would become entrained in the fast-flowing current. As the water level became very high, the river would 'break its banks', bursting through its levées, carrying medium- to coarse-grained sediment and plant debris across the flood plain and rapidly dumping it. Elsewhere, water would pour over the banks and cover the flood plain. In places, the river would erode through old deposits and change its course by taking a short cut across the neck of a meander loop.*

*When the high water levels eventually subsided, a record of this dynamic flooding process would be left behind. On the river bank there would be crevasse splays, composed of cross-stratified sand (Figure 1.1), spreading out from point sources along the river bank. Where the water covered the flood plains, mud and plant debris, which was carried in suspension in the water, would be deposited in a blanket as the water drained away and evaporated (Figure 1.1). Where the river changed its course, cutting off a meander loop, an ox-bow lake would form.*

This scenario is just one example of how dynamic processes that occur on the Earth's surface might be recorded in the *sedimentary* record.

○ What about the igneous and metamorphic rock records; do they record processes at the Earth's surface and in the crust too? If so, how?

● Yes, they do. For instance, a volcanic eruption on the Earth's surface would be recorded as a succession of lavas or pyroclastic deposits, and mountain-building episodes are recorded by metamorphic rocks.

**Figure 1.1** Sharp-based, cross-stratified sandstones (yellow-brown) interpreted as crevasse splay deposits. These are encased in overbank mudstones (grey) that were deposited across a flood plain. The succession is from the Middle Jurassic exposed in Yorkshire, England. The cliff is c. 6 m high. (Chris Wilson, Open University.)

Igneous and metamorphic rocks also record processes within the Earth's mantle and core. However, there are particular advantages of the sedimentary record as a means of recording processes at the Earth's surface, compared with the igneous and metamorphic records. These are their lateral extent (sediments cover almost the entire globe) and their more continuous coverage through time. If we consider the river bank example described above, the river is almost continuously flowing and thus constantly eroding and depositing sediment. Volcanic eruptions on the other hand are often interrupted by long periods of dormancy. Metamorphic rocks can preserve an amazing history of deformation and recrystallization, but again the events tend to be sporadic. Even so, the constant erosion in a river system and the frequent lack of new space to put the sediments into does mean that not every event is recorded, so that even the sedimentary record is not totally complete. Of all the sedimentary environments, the sediments deposited in the seas and oceans generally provide the most continuous record of Earth surface processes because: (i) there is nearly always 'room' to put the sediment; (ii) sediment is always being transported from the land to the sea; and (iii) further sediments are being produced biologically and chemically in the ocean waters. In fact, the majority of sediments are either dumped by rivers into the sea or produced in shallow-marine areas.

Geological processes are a natural series of continuous actions, changes or movements. Processes can be thought of as either internal or external to the sedimentary environment. For instance, the river described above taking a short cut across the meander neck, and the entrances to the meander becoming plugged with sediment to form an ox-bow lake, is an example of a process internal to the environment. External processes can be thought of as those acting on a wider scale, over longer time periods and of a larger magnitude which then affect individual environments and cause them to change. External processes include movement on faults and changes in sea-level through the thawing and freezing of ice sheets. Often there is positive or negative feedback between external and internal processes. External processes are ultimately controlled by major Earth processes such as continental break-up and changes in the Earth's orbit around the Sun. The most common factor linking the sedimentary record to both external and internal processes is sea-level. This is because sea-level is relatively easily changed by external processes, and the preservation of many sedimentary deposits is ultimately linked to sea-level. So, by analysing the sedimentary record for changes in sea-level, we can ultimately learn something about many Earth surface processes and even some acting within and external to the Earth.

Sections 1.1 to 1.3 introduce three important deductions that we can make by using the sedimentary record, and they are all illustrated with a single example of a sedimentary succession shown in Figure 1.2. Section 1.1 presents an example of how time can be deduced from the sedimentary record. It is important to be able to interpret the amount of time represented by sedimentary units and the amount represented by boundaries between such units, and to understand the order, duration and frequency of Earth surface processes. Section 1.2 describes how an external process is recorded in the sedimentary record and introduces the dominance of sea-level in sediment preservation. Section 1.3 indicates how sediments record a repetition of processes over different time-scales.

# 1.1  Interpreting time from the sedimentary record

In the previous Section, we mentioned the fact that the sedimentary record represents the most complete record of changing Earth surface processes through geological time.

○  How are gaps often recorded in the sedimentary record?

●  They are recorded as unconformities.

One of the most visually spectacular unconformities in the British Isles is that at Jedburgh which was recorded by James Hutton in the latter part of the 18th century. Vertically dipping, strongly folded Silurian greywackes are overlain by a breccia and a succession of Devonian sandstones (Figure 1.2).

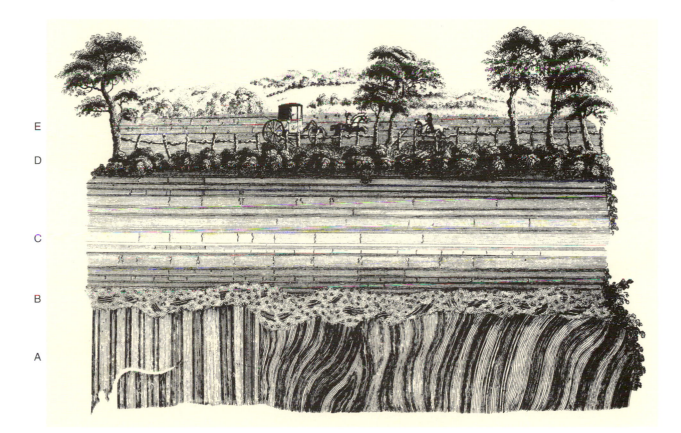

**Figure 1.2**  Engraving of the unconformity between greywackes and a breccia overlain by sandstones at Jedburgh, UK. *(Craig et al., 1978.)* This section is now almost completely obscured by trees and undergrowth. It lies half a mile south of the town, at NT 652198, and is best seen from the west bank of the River Jed. The engraving is by D.B. Pyet and is based on a drawing by John Clerk of Eldin made in 1787 which forms Plate III in Volume 1 of James Hutton's *Theory of the Earth* (1795). The original drawing included the following notes:

'Section of a bank of mineral strata in the River near Jedburgh.  A The Shistus standing upright  B The bed of Pudding-stone composed of the wreck [?] of the Shistus  C Beds of red and marley sandstone, deposited above it  D The Line or Level of the Road cut out of the Bank  E The beds of red marle Sandstone seen above the Road'.

This unconformity, which in fact is better termed an angular unconformity because the beds above and below have a different dip and strike, represents a series of events:

1  Deposition of greywackes as horizontal beds in the sea.

2  Tilting and folding of greywackes to form almost vertical beds together with uplift out of the sea during the Caledonian mountain-building episode.

3  Subaerial exposure and significant erosion of the top surface of the greywackes.

4  Deposition of a breccia composed mainly of clasts of greywacke derived from weathering and erosion of the uplifted landmass.

5  Deposition of sandstones and mudstones by rivers traversing the area.

○  How would you determine the length of time the unconformity represents?

●  By dating the sediments on either side of the unconformity.

The dating has been done using fossils and shows that the greywackes were deposited during the early Silurian and the breccia and sandstones during the late Devonian. Thus, the unconformity surface represents the passage of about 50 Ma. However, despite the fact that there are no sediments actually recording this 50 Ma time period, we can deduce what has happened by noting the presence of the unconformity and examining the unconformity surface together with the sedimentary deposits above and below it. One of the aims of this book is to demonstrate how much more subtle unconformities can be recognized in the rock record, and the different ways in which we might interpret them. The unconformity at Jedburgh is of the same age as the famous unconformity, also described by James Hutton in 1788, exposed on the Scottish coast at Siccar Point some 100 km to the north-east of Jedburgh.

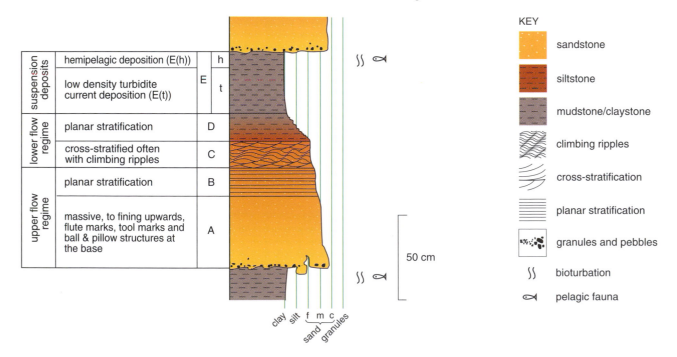

**Figure 1.3**  A full Bouma Sequence showing the different facies deposited from a turbidity current and capping of hemipelagic deposits overlain by the deposits of the next turbidity current.

Let us consider the succession of greywackes underneath the unconformity at Jedburgh (Figure 1.2). These have been interpreted as a succession of Bouma Sequences (Figure 1.3) deposited by turbidity currents during closure of the Iapetus Ocean that once occupied this area.

○ Explain in general whether the sedimentary record is continuous or discontinuous between individual turbidites.

● It is more than likely that the record is discontinuous, each turbidite representing deposition from an individual turbidity flow. Between the action of each of these turbidity flows, an interval of anything from a few hours to tens of thousands of years might have elapsed. In addition, erosion may have occurred at the head of each turbidity flow, removing part of the underlying unit. Flute marks and tool marks at the base of the Bouma Sequence are good indicators of erosion.

○ How might we determine from the sedimentology whether the time that has elapsed between turbidity flows is a few hours or years, or more than thousands of years?

● The presence of the uppermost division of a Bouma turbidite sequence, the E(h) unit, indicates that there has been a gap of much more than a few hours between individual flows. This is because the E(h) division does not represent deposition from the actual turbidity current itself, but sedimentation from suspension of hemipelagic deposits during the quiet period between turbidity currents.

○ Describe four problems concerning interpretation of the amount of time that has elapsed between turbidity flows from the thickness of the E(h) unit.

● First, there might have been erosion of all or part of the E(h) unit by the next turbidity flow. Secondly, not all the time that has elapsed may be represented by sediment. Thirdly, the E(h) unit is sometimes difficult to distinguish from the E(t) unit. Fourthly, the rate of deposition may vary for the E(h) unit.

○ So does the thickness of the E(h) unit give a minimum, average or maximum estimate of the time elapsed?

● A minimum estimate.

Figure 1.4 schematically illustrates the difference between time elapsed and thickness of the stratigraphical succession; it shows a graphic log of a succession of turbidites against the usual vertical thickness scale, together with a 'chronostratigraphical' plot which gives an indication of how much time each part of the turbidite might represent. Note how the claystones, mostly E(h), represent much more time than the sandstones. The amount of time unrepresented because of erosion at the base of turbidites 1, 3, 4, 5 and 7 is shown. Sometimes *more* time is represented by the gap than by the sediment. It has been estimated that the sedimentary record as a whole is on average only 60–70% complete because of gaps between units and the fact that many units only represent part of the event that led to their deposition.

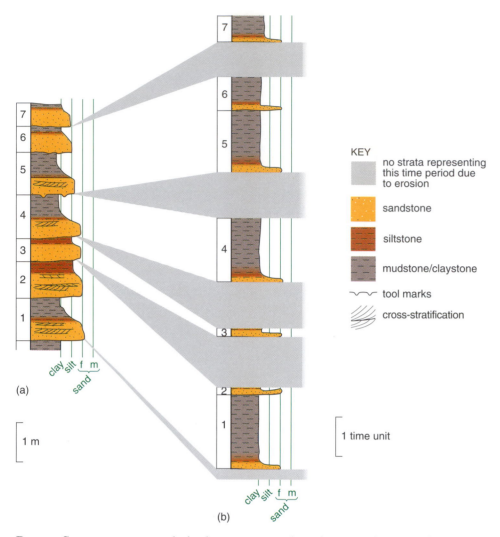

**Figure 1.4** (a) Thickness and (b) chronostratigraphical representation of a succession of turbidites indicating how much time might be represented by gaps in the succession.

Bouma Sequences are a relatively easy example to interpret because they represent two simple processes: (i) turbidity current deposition and (ii) hemipelagic deposition. There are many other examples of different types of gap in the sedimentary record, many of which appear exceedingly subtle at first.

○ Explain if Walther's Law * applies if there is a gap in the sedimentary record.

● If there is a gap in the succession, Walther's Law does not apply because one of its important points is that facies that were originally laid down laterally adjacent to each other should be preserved in a vertical succession.

For instance, Figure 1.5 shows a succession of strata deposited in a storm-dominated strandplain setting. Using the principle of Walther's Law, it is apparent that there is a sedimentary gap at X–X′ because there are no shoreface sediments. This surface X–X′ is also a type of unconformity. In order to ascertain how much time is represented by the unconformity here, we would have either to date the foreshore and offshore transition sediments or determine from a nearby (stratigraphically complete) locality how much sediment is missing and what the sedimentation rate is. We will consider Walther's Law again in Box 2.1.

* Walther's Law states that: 'The various deposits of the same [environmental] area and, similarly, the sum of the rocks of different [environmental] areas were formed beside each other in time and space, but in crustal profile we can see them lying on top of each other . . . it is a basic statement of far-reaching significance that only those [environmental] areas can be superimposed, primarily, that can be observed side by side at the present time.'

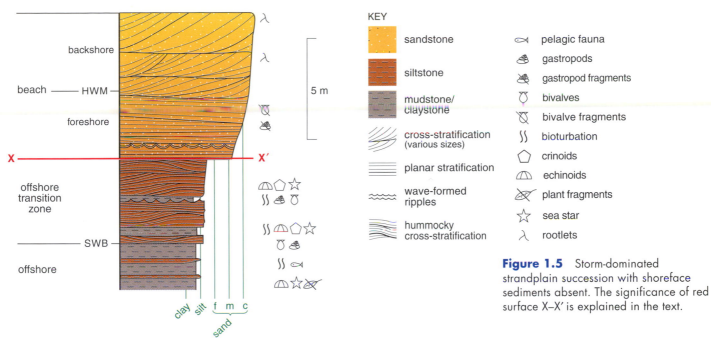

**Figure 1.5** Storm-dominated strandplain succession with shoreface sediments absent. The significance of red surface X–X′ is explained in the text.

Let us now consider what other gaps might be present in the sedimentary succession shown in Figure 1.5.

○ What do the bed boundaries between the mudstones and siltstones represent in the offshore transition zone?

● Small gaps in the sedimentary succession, formed through either non-deposition or erosion or a combination of the two before the next storm brought more sediment into the offshore transition zone.

In a similar manner, each of the planar stratification surfaces within the foreshore sediments also represent small gaps in the sedimentary record. In fact, there is a whole range of scales of gaps in the sedimentary record from individual laminations, to bed boundaries, to boundaries between unconnected facies, to angular unconformities. Thus, it is important that in the analysis of any succession of sedimentary rocks we take into account how much time the rock units and the boundaries between them represent. Chapter 2 covers the basics of how we can detect the passage of geological time in sedimentary rocks using a wide range of techniques.

## 1.2 Processes

The succession shown in Figure 1.2 illustrates a second important observation and deduction that we can make using the sedimentary record; that is, the actions of processes external to those in the particular sedimentary system. The sedimentary deposits below the unconformity were deposited under marine conditions whereas those above the unconformity were deposited under non-marine conditions. The sea-bed on which the greywackes were deposited has moved relative to sea-level; this could either be due to local tectonic processes uplifting the area or the lowering of sea-level. Sea-level can change due to a number of different processes, including regional tectonic movement, the freezing and thawing of ice sheets and thermal expansion in the volume of seawater.

○ Local tectonic movement is favoured as the simplest single reason for the change in position of sea-level relative to the land in Figure 1.2. Why is this so?

● Because, in addition to the change from marine to non-marine sediments across the unconformity, the marine sediments underneath the unconformity have been tilted and folded — an observation that can only be explained by the action of tectonic processes.

Although tectonic movement is the simplest single process to explain the observations in Figure 1.2, it is quite possible that a combination of processes are responsible. Tectonic processes are also important in controlling the amount and type of sediment.

Climatic processes leave their mark in the sedimentary record as they also control the sediment type and supply. For instance, climate affects biological productivity in the surface waters and hence the production of most carbonate sediment; climate also affects the amount of rainfall on land, and therefore how much siliciclastic material is transported into the ocean. Climatic processes also influence the position of sea-level, and hence the sedimentary record, through such mechanisms as the melting and freezing of ice sheets.

The sedimentary record is full of evidence for changes in sea-level both on a dramatic scale as shown in Figure 1.2 and on a much more subtle scale. In fact, it has been demonstrated that deposition, erosion and the resulting preservation of both marine and non-marine sediment is largely a function of sea-level changes. This is because sea-level is usually the most significant datum for all sedimentary systems as they try to achieve a balance between erosion and deposition so that energy is conserved. Why sea-level is so important and how we can analyse the evidence for sea-level change in the sedimentary record is the main theme of this book.

## 1.3  Repetition in the sedimentary record

The third important feature of sedimentary successions, which can be seen in the lower part of the example succession shown in Figure 1.2, is the repetition of similar beds or groups of beds. The greywackes in Figure 1.2 are the result of the *repeated* action of turbidity currents (Figure 1.4a) which, by comparison with the same process occurring on the sea-floor today, we know were most likely deposited over a few thousand years. Repetition of processes can occur on many different time-scales. The decimetre-scale interbedded limestones and mudstones shown in Figure 1.6a indicate repeated switches from mudstone to limestone deposition. Research has shown that each limestone/mudstone pair probably represents subtle climate changes with each pair (or cycle) lasting a few tens of thousands of years. Figure 1.6b shows cycles on an even longer time-scale, with the whole photograph representing a few million years. In this case, each cycle is made up of 1–4 m of dark-grey organic-rich mudstones and 0.1–1.5 m of pale-grey siliceous sediment composed of millions of microfossils called diatoms. The changes in the facies are interpreted to represent fluctuations in the amount of nutrients and temperature of the seawater at the time of deposition. Why processes repeat themselves, over what time-scales, and how this relates to sea-level and sediment supply will be considered later in this book.

(a) Cliff *c.* 15 m high                                              (b) Cliff *c.* 10 m high

**Figure 1.6**   (a) Interbedded limestones (pale-grey) and mudstones (dark-grey) reflecting compositional changes in carbonate, siliciclastics and organic carbon, which were ultimately controlled by short-term cyclic changes in climate during the Jurassic. From the Blue Lias, Lyme Regis, Dorset . (b) Long-term sedimentary cycles from the Monterey Formation, California. Dark-grey units are organic-rich mudstones and pale-grey units are siliceous deposits composed almost entirely of diatoms. *((a) Anthony Cohen, Open University; (b) Angela Coe, Open University.)*

## 1.4   Summary and conclusion

The sedimentary record provides us with a means of examining the interaction, duration and magnitude of surface processes on the Earth, but in order to get a full picture we also need to consider the boundaries between each of the rock units in terms of what they represent and what might be missing. Charles Darwin spent much time examining geology, as well as botany and zoology, on his various expeditions. In 1859, his *Origin of Species* was published in which he wrote:

> 'Some of the formations, which are represented in England by thin beds, are thousands of feet in thickness on the Continent. Moreover, between each successive formation, we have, in the opinion of most geologists, enormously long blank periods. So that the lofty pile of sedimentary rocks in Britain gives but an inadequate idea of the time which has elapsed during their accumulation; yet what time this must have consumed!'

This quotation summarizes well the importance of unconformities and also the very wide variation that there is in sedimentation rate. Chapter 2 reviews the different methods of dividing the stratigraphical record and interpreting time from it. Chapter 3 introduces the concept of sea-level change.

## 1.5   References

### Further reading

TUCKER, M. E. (2001) *Sedimentary Petrology* (3rd edn), Blackwell Science, 272pp. [An excellent introduction to sedimentology. This is the most up-to-date edition; previous editions (e.g. 2nd edition, 1991) are equally good.]

### Other references

DARWIN, C. (1859) *The Origin of Species*, R. & R. Clarke Ltd, Edinburgh, 454pp.

HUTTON, J. (1795) *Theory of the Earth, with proofs and illustrations*, **1–2**, London: Cadell and Davies; Edinburgh: William Creech.

CRAIG, G. Y., McINTYRE, D. B. AND WATERSTON, C. D. (1978) *James Hutton's Theory of the Earth: The Lost Drawings*, Scottish Academic Press, in association with the Royal Society of Edinburgh and the Geological Society of London, 67pp., and accompanying folio facsimiles.

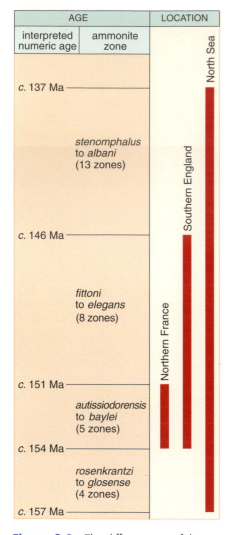

| AGE | | LOCATION |
|---|---|---|
| interpreted numeric age | ammonite zone | |
| | | North Sea |
| c. 137 Ma | | |
| | stenomphalus to albani (13 zones) | Southern England |
| c. 146 Ma | | |
| | fittoni to elegans (8 zones) | Northern France |
| c. 151 Ma | | |
| | autissiodorensis to baylei (5 zones) | |
| c. 154 Ma | | |
| | rosenkrantzi to glosense (4 zones) | |
| c. 157 Ma | | |

**Figure 2.1** The differing age of the Kimmeridge Clay lithostratigraphical unit in Northern France, Southern England and the North Sea.

# 2 Division of the stratigraphical record and geological time

## Angela L. Coe

Stratigraphy is the study of rocks in time and space. It deals with lateral correlation of rock units, and interpreting and understanding geological time from the rock record. The early geologists relied mainly on two techniques to divide the stratigraphical record and correlate it laterally. The first of these is *lithostratigraphy*: rocks of the same or similar lithology were grouped together into a unit that could be recognized over a wide area. Some of these 'lithological units' are still in use today, e.g. the Chalk, a Cretaceous deposit that is widespread across Northern Europe, or the Kimmeridge Clay, a Jurassic mudstone unit rich in organic carbon and found in Southern England, Northern France and the North Sea, where it is the major source rock for North Sea oil. However, as we also know from Walther's Law (see Box 2.1), it is often the case that a particular lithology or facies does not necessarily represent the same period of time everywhere, and such units can be described as *diachronous*. This is certainly the case with the Kimmeridge Clay; whilst the lower part is of a similar age in Northern France, the North Sea and Southern England, the top of this lithostratigraphical unit is far younger in the North Sea than it is in Southern England, and in both of these places it represents much more time than it does in France (Figure 2.1).

This difference in age of the Kimmeridge Clay can be demonstrated by the second method that the pioneering geologists used, namely *biostratigraphy*. This relies on the fact that over time certain animals and plants evolve relatively rapidly, so that any one point in time is characterized by the presence of certain species. These species may then be found as fossils in strata of differing lithology demonstrating that they were laid down synchronously. In the case of the Kimmeridge Clay, ammonites can be used to demonstrate the age differences (Figure 2.1). Biostratigraphy is a type of *chronostratigraphy*, that is, a method of dividing the stratigraphical record into time-significant rock units (Box 2.1). However, biostratigraphy is a relative dating method because it can only be used to ascertain the approximate time represented by one rock unit relative to another rock unit. Absolute dating on the other hand is a means of determining the numeric age of the formation of a rock.

Today, we have a wide variety of stratigraphical techniques available to us and the geological time-scale is constructed by integrating most of these techniques. This has the advantage that it allows us to: (i) use whatever dating technique(s) is most appropriate on a particular set of rocks to date it; and (ii) to compare its age to another set of rocks. In addition, because the different techniques are so closely integrated, we can, by comparison, interpret the absolute age of a rock even if the rock itself has only yielded data suitable for relative dating.

The disadvantage of these compilations of data that give an integrated time-scale is that it is often hard to estimate the uncertainties associated with each data set and thus the correlation between different types of data might appear much better than it is. In Section 2.3, we will consider how an integrated geological time-scale is constructed. Throughout this book, we will use a number of stratigraphical techniques, so a brief introduction to each of them is given here.

## Box 2.1    Walther's Law, diachronous sedimentation, chronostratigraphy and lithostratigraphy

The *Principle of Superposition* tells us that the beds on top were deposited *after* those below (assuming no overturning by later tectonic activity).

The situation is very simple when the same facies is lain down over a wide area during exactly the same time interval. An example of this is the hemipelagic limestones and mudstones shown in Figure 1.6a. In these situations, the lithostratigraphical boundaries between different rock types are coincident with chronostratigraphical boundaries, which subdivide sediments deposited at (arbitrarily) different times.

However, some sedimentary facies are *diachronous*, i.e. the same facies differs in age from place to place. Diachronous sedimentary facies were described by Johannes Walther in 1894, and are dealt with in Walther's Law.

This states that:

'The various deposits of the same [environmental] area and, similarly, the sum of the rocks of different [environmental] areas were formed beside each other in time and space, but in crustal profile we can see them lying on top of each other … it is a basic statement of far-reaching significance that only those [environmental] areas can be superimposed, primarily, that can be observed side by side at the present time.'

It is easiest to illustrate this part of the Law by looking at an example. Consider again a storm-dominated strandplain vertical succession (Figure 2.2). At locality A, we may be fortunate enough to see and interpret the entire succession (from bottom to top): offshore–offshore transition zone–shoreface–foreshore–backshore.

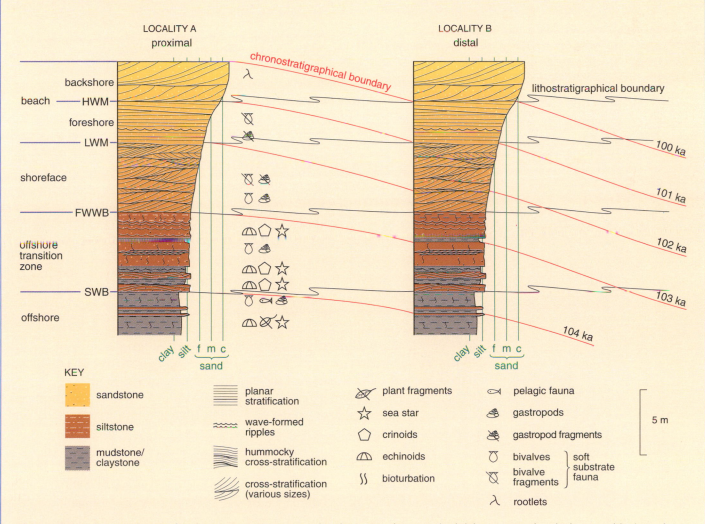

**Figure 2.2**   The correlation of sedimentary successions at localities A and B using both lithostratigraphical (shown in black) and chronostratigraphical (shown in red) boundaries.

At locality B, the same vertical succession is exposed, but let us assume that its position is further away from the land, i.e. is more *distal*. (Conversely, locality A, being closer to the ancient continent, is the more *proximal* succession.) Each depositional zone has been correlated between the two successions using black lines; such a correlation is based on the similarities of the different facies, so these boundaries are lithostratigraphical boundaries.

Whilst being a useful way of linking similar environments together, these boundaries do not take into account the time when these facies were deposited. If you were standing on a beach today and walked out to sea (preferably with a wetsuit!), you would soon leave the beach and encounter more distal environments in the order shoreface–offshore transition zone–offshore. In fact, you would have been walking on a time boundary, representing the time that you walked along it. Such boundaries are therefore chronostratigraphical boundaries.

○ In this example, what is the relationship between lithostratigraphical and chronostratigraphical boundaries? Are they parallel or do they intersect one another?

● Clearly, as you walked from the beach to the offshore zone, you kept to a chronostratigraphical boundary but cut across lithostratigraphical boundaries (i.e. those between the foreshore and shoreface, the shoreface and offshore transition zone, etc.). The two surfaces must therefore cross-cut one another. This relationship is illustrated in Figure 2.2.

As we consider sequence stratigraphy in more detail in later Chapters, you will see that it is important to appreciate the difference between chronostratigraphical and lithostratigraphical boundaries. Whilst chronostratigraphical boundaries are the most difficult to correlate between vertical sections (you could not have drawn those shown in Figure 2.2 without some age information), they are fundamentally more important than lithostratigraphical boundaries when it comes to subdividing a sedimentary succession into depositional units representing a particular time period. How chronostratigraphical boundaries are identified in ancient sedimentary rocks is another theme that we will explore in this book.

The other principle that Walther's Law deals with is the fact that sediments that were not deposited side by side (e.g. foreshore and offshore transition zone sediments) will be separated by an unconformity when they are found in vertical succession. We used this principle in Section 1.1.

## 2.1 Stratigraphical techniques

### 2.1.1 Lithostratigraphy

Lithostratigraphy is the subdivision of the rock record according to lithology. The basic lithostratigraphical division of the rock record is the *formation*. A formation is a body of rock which can be identified by its lithological characteristics and stratigraphical position. It is the smallest mappable unit over a wide area. Ideally, formations should have some degree of lithological homogeneity but their defining characteristics may include chemical and mineralogical composition, fossil content and sedimentary structures. Formations can be divided into *members* which have limited lateral extent. The smallest formal lithostratigraphical unit is a *bed*. This term is used for a distinctive layer within either a member or a formation. Two or more formations with similar characteristics may be put together to form *groups* which, in turn, may form part of a supergroup. As discussed at the start of this Chapter, lithostratigraphical units are often diachronous.

### 2.1.2 Biostratigraphy

William Smith, an 18th–19th century surveyor and canal engineer by trade, made the observation that many sedimentary rocks contained fossils and that the assemblage of fossils varied from layer to layer, producing a characteristic succession. He also discovered that the same succession of fossils could be found at different places and thus proposed his *Principle of Faunal Succession*. This principle is the basis of modern biostratigraphy which relies on the fact that some of the fossils found in rocks (Figure 2.3) can be used to obtain a relative age by comparing the fossils to a known standard succession or 'zonation'.

○ What are the three main reasons that only some biota are useful for biostratigraphy?

● (i) Not all animal and plant groups evolve fast enough to make them temporarily significant; (ii) some fossils are not found in sufficient quantity over a very wide area; and (iii) only some fossils have a good preservation potential.

Over time, animals and plants have evolved and changed their form. Some evolved much faster than others; for instance, ammonites, graptolites and pollen evolved rapidly, whereas most gastropods and fish did not. A good biostratigraphical zonation also depends on how cosmopolitan and mobile the biota are. For this reason, planktonic biota that move around freely in the sea, and pollen that is transported great distances by wind and water, are particularly useful. Although some soft-bodied animals may have evolved rapidly, they will not be useful zonal fossils because their preservation potential is low. Microfossils, such as pollen, have only become important zone fossils in the last few decades since the improvements in microscope technology and rock extraction techniques.

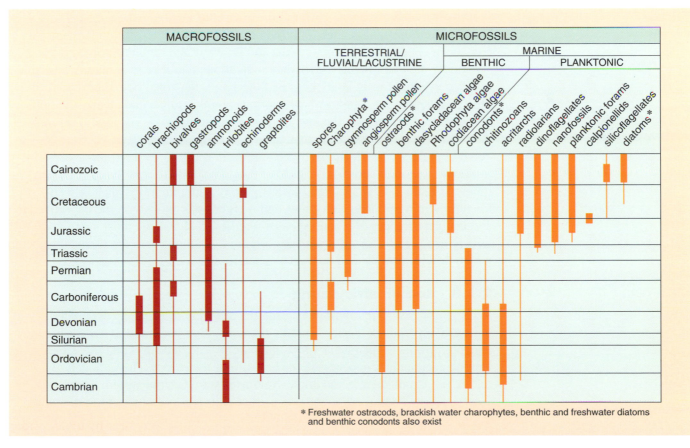

\* Freshwater ostracods, brackish water charophytes, benthic and freshwater diatoms and benthic conodonts also exist

**Figure 2.3** Biostratigraphically useful groups of organisms preserved as macrofossils and microfossils in the stratigraphical record and their age ranges. The thick part of the line indicates the interval where the fossil group is used most extensively. Benthic organisms live on the sea-floor or in the sediments, whereas planktonic organisms live in the water column and have limited ability to move through the water column. *(Left: Nichols, 1999; right: Emery and Myers, 1996.)*

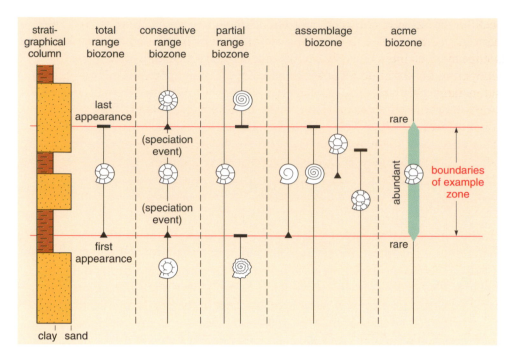

**Figure 2.4** The common zonation schemes used in biostratigraphical correlation. *(Nichols, 1999.)*

For example, ammonites (a type of ammonoid) remain one of the most useful biostratigraphical indicators for the Mesozoic (Figure 2.3). This is because of their rapid evolution into different forms, their high preservation potential and their widespread occurrence. Ammonite zones are typically about 1 Ma long. Despite their usefulness, there still remain problems; some ammonites seem to have been restricted to certain oceans and seaways and to have evolved separately. In addition, there are some indications that they may not always have evolved synchronously in all areas and therefore that zonal boundaries are not always synchronous. Much debate continues on this topic, and clearly the notion of synchrony depends on the level of time resolution that is required. Differences in working practice also mean that, at first, it is not always clear where exactly a zonal boundary is, as biostratigraphers have, in some cases, used the nearest lithological marker beds for the base of the zone, and different types of fossil ranges (Figure 2.4). In general, biostratigraphy using microfossils has been treated more rigorously, with zones and marker beds defined on the basis of first and last appearances of certain species. Figure 2.3 shows the most useful and widely used fossils for different ages. Figure 2.4 summarizes the various ways in which different fossils can be used as zone fossils.

## 2.1.3 Radiometric dating

Radiometric dating involves measuring the amounts of naturally occurring radioactive parent isotopes of certain elements, together with the corresponding daughter product. The rate at which the parent isotope decays and the daughter isotope accumulates depends on the half-life of the parent. The radioactive 'clock' in parent–daughter isotopic systems is set when the system becomes closed (i.e. locked in so that there is neither gain nor loss of parent or daughter except by radiocative decay). Thus, we can calculate the age of the rock or crystal if we can measure the amount of parent and daughter, and if we know the half-life of the parent. An analogous system, commonly used for biological remains, is carbon-dating that involves measurement of $^{14}C$ (half-life 5700 years) which makes it useful for dating recent events as far back as *c*. 50 ka.

Luckily, the half-lives of some radioactive elements are much longer, making them exceedingly valuable for the dating of events throughout Earth history. Isotope systems include potassium-argon, rubidium-strontium, samarium-neodynium and uranium-lead. Most radiometric dating is carried out on igneous and metamorphic rocks to establish the time of crystallization. Direct radiometric dating of sedimentary rocks is much rarer because most of the grains are derived from a mixture of older rocks, and most isotope systems are not reset at the time of sediment deposition. A minimum age for sedimentary rocks can be obtained by radiometrically dating igneous bodies such as dykes and sills that cross-cut them. Clearly, however, there could be large differences between the minimum age determined from the cross-cutting igneous body and the actual age of the sediments. The most successful and widely used method of radiometrically dating sedimentary successions is to use minerals from intercalated volcanic ash layers, but such layers are relatively rare and often poorly preserved. Recent research has shown that a few isotope systems can be used to directly date waterlain sediments because these isotope systems become fixed in the sedimentary rock at the time of deposition. These methods include uranium series and uranium–lead dating of limestones, together with the rhenium–osmium (members of the platinum group of elements) dating of organic-rich mudrocks. Uncertainties in radiometric dating mostly result from analytical uncertainty associated with isotopic analysis, from geological disturbance to the isotope systems, and from uncertainties in the knowledge of their half-lives.

## 2.1.4  Magnetostratigraphy

In the late 19th century, it was discovered that some rocks are magnetized in the opposite direction to that expected, i.e. towards the south magnetic pole rather than the north. Thus, it was suggested that the polarity of the Earth's magnetic field had changed through time. The full significance of this observation was not realized until much later when the theory of plate tectonics was formulated. One of the key observations in the development of the theory of plate tectonics was that the oceanic crust between the continents exhibits a pattern of stripes of high and low magnetic field strength. These stripes vary in width from a few to tens of kilometres and extend for hundreds of kilometres parallel to the axis of the mid-ocean ridges and perpendicular to the direction of movement of the plates. This striped magnetic pattern is attributed to the fact that as magma is emplaced at mid-ocean ridges it becomes magnetized in the direction of the Earth's prevailing magnetic field as it cools through a particular temperature called its Curie point. The formation of new oceanic crust along ocean ridges became known as sea-floor spreading — a process that acts as a kind of tape recorder of the change in polarity of the Earth's magnetic field. High magnetic field strength shows that the magnetic field is in the same orientation as today, i.e. it is of normal polarity, and low magnetic field strength shows that it was the reverse of today (reversed polarity). Figure 2.5 shows how the Earth's magnetic field has alternated over part of the Jurassic and Cretaceous between reversed polarity (marked as white stripes) and normal polarity (black stripes).

○  Using Figure 2.5, describe, in general, how often the Earth reverses its magnetic field.

**Figure 2.5**  Changes in the Earth's magnetic field through part of the Jurassic and Cretaceous. (Ogg, 1995.)

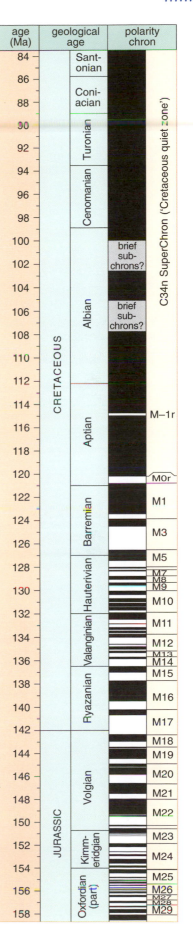

● This is in fact very irregular: sometimes it changes as often as every few hundred thousand years (e.g. the Oxfordian), whereas at other times it can stay normal or reversed for long periods (e.g. tens of millions of years spanning much of the Cretaceous).

As sediments are deposited and become compacted and cemented, the magnetic minerals in them (chiefly iron oxides, but also clay minerals) align themselves with the magnetic field. Provided these rocks are not then metamorphosed and heated above their Curie point, their magnetic field can be measured and their reversed or normal polarity determined. The polarity pattern from both the sea-floor basalts and exposed or drilled sedimentary deposits can be correlated within a framework of other dating techniques completed on the same rocks, e.g. biostratigraphy and radiometric dates, enabling a polarity time-scale like that shown in Figure 2.5 to be constructed. An additional important feature of magnetostratigraphy is that because the rate of sea-floor spreading is practically constant over 5–10 Ma periods, and we can radiometrically date the sea-floor basalts every 5–10 Ma to obtain key tie points, we can then extrapolate these ages using the width of the polarity stripes to determine the time represented by each stripe. This provides a fairly good indication of the amount of absolute time that has elapsed during each phase of normal or reversed polarity, which for periods of Earth history where there have been frequent reversals provides a much higher-resolution time-scale than the radiometric dates.

### 2.1.5  Chemostratigraphy

Chemostratigraphy involves the utilization of chemical and, particularly, isotopic signals that have changed through geological time as a means of both relative dating and correlating different successions by comparing the chemical signature from each. Relative dating of a succession involves comparison of chemical data from rocks of an unknown age with a reference data set of known age. Correlation between sections using chemical signals is of two types: (i) utilization of a particularly well-marked, and usually relatively short lived, chemical signal that relates to an event or series of events in Earth history; and (ii) utilization of a chemical signal representing changes over a long period of time; this is particularly used for astrochronology (see Section 2.1.6).

A variety of elements have been used for chemostratigraphy, and new techniques with other elements are being developed, making chemostratigraphy an expanding field of research. There are three isotopic systems that have received particular attention: oxygen, carbon and strontium. In each case, it is a ratio between two stable isotopes of different atomic weights that is measured. Unlike the radiogenic isotopes discussed earlier (Section 2.1.3), stable isotopes do not decay. Instead, the amounts of the heavier isotope compared to the lighter isotope (usually expressed as a ratio) change as a function of various Earth processes.

Strontium isotopes are soluble in water, and at any one time in seawater they are at a certain ratio. There are two inputs of strontium isotopes into seawater: a high $^{87}Sr/^{86}Sr$ isotope ratio from weathering of the continental crust which is carried to the ocean in river waters; and a low $^{87}Sr/^{86}Sr$ isotope ratio from hydrothermal vents at mid-ocean ridges or weathering of oceanic or basaltic crust. Throughout most of the Neogene and Quaternary, the $^{87}Sr/^{86}Sr$ isotope ratio has been rising (Figure 2.6). This is interpreted to be related to the dominant input of the high $^{87}Sr/^{86}Sr$ isotope ratio from weathering of the uplifting Himalayas. At other times in the past, the $^{87}Sr/^{86}Sr$ isotope ratio has decreased. These decreasing ratios are equated with periods of increased sea-floor spreading, e.g. at the beginning of the Jurassic when the supercontinent Pangea was starting to rift. It has been found that calcitic fossils such as belemnites (an extinct type of mollusc) and foraminifers are particularly good at preserving this ocean Sr-isotope signature and a standard curve for comparison has been compiled (Figure 2.6).

○ If the $^{87}$Sr/$^{86}$Sr isotope ratio in a rock was 0.7086, what are the possible ages and how would you distinguish between them?

● The possible ages are early Neogene, early Devonian, late Silurian, late Ordovician. In the absence of any contextual clues, one would need a further independent dating method such as biostratigraphy to distinguish between them.

This example may lead you to believe that the method is very crude but the following more geologically realistic example shows how powerful the technique can be. Take for example the rocks shown in Figure 1.6a, which we know were deposited in the Jurassic. This knowledge considerably narrows down which part of the curve we need to look at, so if calcitic fossils from the beds near the base of the photograph yield a $^{87}$Sr/$^{86}$Sr ratio of 0.7077, using the compiled strontium curve we would be able to demonstrate that these beds were deposited near the middle of the Hettangian Age in the Early Jurassic (note that Figure 2.6 does not show the position of ages within the Jurassic).

Oxygen-isotope stratigraphy was pioneered in the 1970s and involves measurement of the ratio between the light stable isotope $^{16}$O and the heavier stable isotope $^{18}$O, from calcareous fossils, and, in some instances, from calcareous rocks. The factors that control which isotope is more abundant are more complex than for strontium, and ratios in marine fossils vary with salinity, temperature and the volume of ice-caps and glaciers. Ratios can also be easily reset during diagenesis. Oxygen isotopes are discussed further in Section 3.2.1.

Carbon-isotope stratigraphy is often completed in conjunction with oxygen-isotope measurements and involves measurement of the ratio of the light stable isotope $^{12}$C and the heavier stable isotope $^{13}$C. The experiments are typically completed on carbonate fossils and rocks or on marine or terrestrial organic

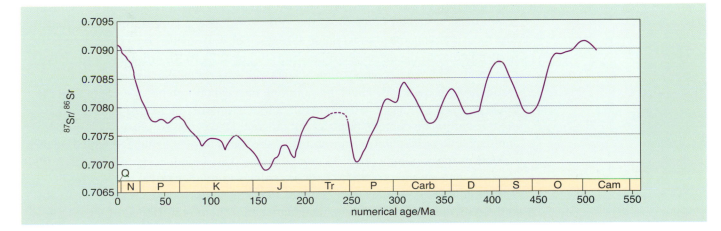

**Figure 2.6**   Sr-isotope curve for the Cambrian to Quaternary. Q = Quaternary, N = Neogene, P = Palaeogene, K = Cretaceous, J = Jurassic, Tr = Triassic, P = Permian, Carb = Carboniferous, D = Devonian, S = Silurian, O = Ordovician, Cam = Cambrian. Note that this curve is generalized and does not show all the detail. (McArthur, 2001.)

carbon. Carbon-isotope ratios track changes in the carbon cycle; the amount of fractionation is strong between the atmosphere, ocean, and plants, thus deposition and weathering of organic carbon, particularly in marine sediments, has a strong effect on the oceanic and atmospheric carbon-isotope composition.

Carbon and oxygen stable isotope analyses have been used to address the following problems:

1 Temperature changes in the oceans and volume of the polar ice-caps through time, mainly based on oxygen isotopes; this work is linked to astrochronology (see Section 2.1.6).

2 Changes in the vertical water-mass structure of the oceans through time.

3 Extreme events in the world's oceans. For instance, major events such as burial or release of large amounts of organic carbon can be detected in the carbon stable-isotope record, often all over the world. These events are useful for correlation as well as understanding changes in climate. An example is given in Section 2.2.

4 Changes in the carbon budget, estimating biological productivity and global organic carbon burial. Carbon isotopes have also been used to try to understand changes in the amount of atmospheric carbon dioxide through time.

Other elements, compounds and isotopes commonly used in chemostratigraphy include calcium carbonate, organic-carbon content, magnesium/calcium and strontium/calcium ratios, neodymium-, sulphur-, nitrogen- and osmium-isotopes, and some platinum group elements including iridium which can be concentrated in the sedimentary record through particular types of meteorite impact.

## 2.1.6 Astrochronology

Notable advances over the last 20 years or so, in both celestial mechanics (study of the motion of the planets and stars and the effects of their motions on each other) and the collection and analysis of stratigraphical data, have led to the development of one of the most high-resolution dating methods: astrochronology or astronomical calibration. This method utilizes the regular periodicities of Milankovich cycles (Box 2.2) recorded in the rocks to erect a chronostratigraphical time-scale. There are several stages to this method. First, it is necessary to determine whether or not the cycles present in the sedimentary deposits under consideration (e.g. those cycles shown in Figure 1.6) are Milankovich cycles. This involves measuring a chemical or physical parameter in the rocks that reflects primary compositional variations. The measurements need to be made at regular, closely spaced intervals perpendicular to the direction of bedding (Figure 2.7a,b). In order that a record of the gradual change in composition is recorded, several measurements need to be made through each bed. Thus, the sampling interval chosen has to be less than the thickness of the thinnest bed (Figure 2.7a,b). Commonly measured parameters include carbonate content, organic-carbon content, oxygen isotopes (Section 2.1.5) and magnetic susceptibility (the ability of the rocks to be magnetized which, in turn, reflects changes in mineralogical composition). This length-series of chemical or physical data are then tested by Fourier analysis (a mathematical method for determining the components of complex periodic signals) to see if there are cycles in the rock composition of the same length and if there is sufficient of them to be statistically valid. Cycles of the same length in a succession are said to be regular and are interpreted to have been deposited by the repetition of a process over a fixed period of time which changed the composition of the rock. The mathematical analysis enables cycles of very similar but not exactly the same length to be grouped together (Figure 2.7c). Small variations in sedimentation rate and measurement error commonly result in not every cycle being of exactly the same length. Most successions contain more than one regular cycle: this makes the sedimentary signal complex because the different cycles interfere with each other. Mathematical treatment of the data enables these different cycles to be separated (e.g. the two cycles shown in Figure 2.7c).

Once the presence of regular cycles is established, the ratio of the length of the cycles to each other, and/or approximate estimates of the time represented by the section from other stratigraphical methods, are then used to determine the likely cycle periodicity. These ratios or periodicities can be directly compared to the Milankovich cycle ratios and periodicities. Assuming that regular cycles of particular frequencies are found, the data can be filtered and 'tuned' to eliminate variations in sedimentation rate, measurement error, etc. Tuning involves 'fixing' all of the cycles to the regular periodicity (Figure 2.7d). Thus, it can be established that the rock unit contains a fixed number of cycles of known duration. The number of cycles multiplied by their duration yields the length of time taken to deposit that particular rock unit.

○  Is this length of time a minimum, maximum or average estimate?

●  It is a minimum estimate of the amount of time because there may be cycles missing.

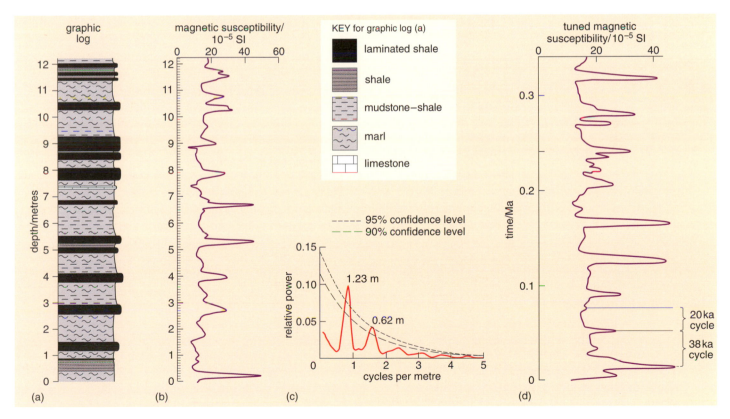

**Figure 2.7**  Diagram to show the principles of astrochronology using an example from the Jurassic Kimmeridge Clay Formation, Dorset. (a) Graphic log through the mudrock succession under consideration. (b) Magnetic susceptibility measurements made every 10 cm (shown by tick marks on the scale bar) to ensure that measurements are made in even the thinnest bed. Note how the laminated shales which have a high organic carbon content correspond to the higher magnetic susceptibility values whereas the marls and limestone have lower values. (c) Power spectra showing two regular cycles at 1.23 m and 0.62 m. (d) The tuned magnetic susceptibility data against the time-scale deduced from the Fourier analysis. The 1.23 m cycle is interpreted as the 38 ka obliquity cycle and the 0.62 m cycle as the 20 ka precession cycle (note that both the obliquity and the precession cycles were of shorter duration in the Jurassic). ((b)–(d) Weedon, 1999.)

For the more recent part of geological time, the time-scale can be refined even further because of advances in celestial mechanics. Researchers have computed a mathematical solution for the orbital motion of the Earth back from the present day, taking into account all the different interactions from other planets and the Moon; this is called the orbital solution, and from it the amount of insolation through time (insolation target) can be plotted. The orbital solution has been calculated back to about 35 Ma ago (Eocene/Oligocene boundary). A solution for the orbital obliquity and precession of the Earth's orbit is unlikely to be extended much further back than this because mathematically it is intrinsically limited due to the planets' chaotic behaviour and the imprecise knowledge of some variables. However, it is possible that the longer-term 413 ka eccentricity cycle (Box 2.2) could be modelled and used to calibrate astronomically the geological time-scale further back in time.

The insolation target for the past 35 Ma can be directly compared with tuned data from the rock record, and correlated using the amplitude and symmetry of the cycles. If there are gaps in the tuned data (i.e. missed cycles), the record will not match the orbital solution exactly. However, which cycles are missing can usually be deciphered by matching the cycles above and below the gap. Thus, for the past 35 Ma of the sedimentary record, it is possible to ascertain the age of rocks to a few thousand years provided cycles within the rock succession can be matched to the orbital solution. Other dating techniques such as biostratigraphical and chemostratigraphical correlation points can be directly tied to the astronomical time-scale because fossils or chemical data can be collected from the same sedimentary deposits. The compilation of this astronomical solution has also provided some interesting feedback to radiometric dating methods (Section 2.1.3) where it has been found that some parts of the astronomical time-scale and individual radiometric dates from particular sections did not agree and that the half-life constant used for particular radiometric dating techniques needed revising.

## Box 2.2    Milankovich theory of climate change

Milankovich theory states that perturbations in the Earth's tilt and orbit influence the amount of incoming solar radiation or insolation at different latitudes. The resultant regular temperature variations in the atmosphere promote changes in climate, with the resultant waxing and waning of ice-caps, and the thermal expansion and contraction of the oceans, both of which in turn instigate global sea-level fluctuations. Three parameters are recognized (with present-day periodicities):

1   Changes in the eccentricity of the Earth's orbit (Figure 2.8a). This is the gradual change in the shape of the Earth's orbit from an ellipse to a circle and back again. The resultant periodicities of this motion include the 95 ka, 123 ka (which form the *c.* 100 ka cycle) and the *c.* 413 ka cycle.

2   Changes in angle of tilt (obliquity; Figure 2.8b) of the Earth's axis with respect to the plane in which it orbits the Sun from 21.8° to 24.4°. This occurs every 41 ka at the present time.

3   Changes in the precession (wobble; Figure 2.8c). There are two components to the precession: that related to its axis of rotation and that relating to the elliptical orbit of the Earth. The Earth's axis sweeps out the shape of a cone in space through time. This is similar to the gyrations of a spinning top. The elliptical orbit of the Earth around the Sun also precesses. The main resultant periodicities of both these motions are 19 ka and 23 ka at the present time, giving a mean cycle of 21 ka.

The eccentricity cycles mainly control the amplitude of the precession cycle and have a small control on the total yearly radiative heating of the Earth. The obliquity and precession affect the distribution of the solar radiation over the Earth's surface. The obliquity component determines the degree of seasonality. Precession controls the timing of the seasons relative to the perihelion (point where the Earth is at its shortest distance away from the Sun in each orbit or year) and aphelion (where the Earth is at its furthest distance away from the Sun in each orbit or year) and hence the total radiative heating in each season.

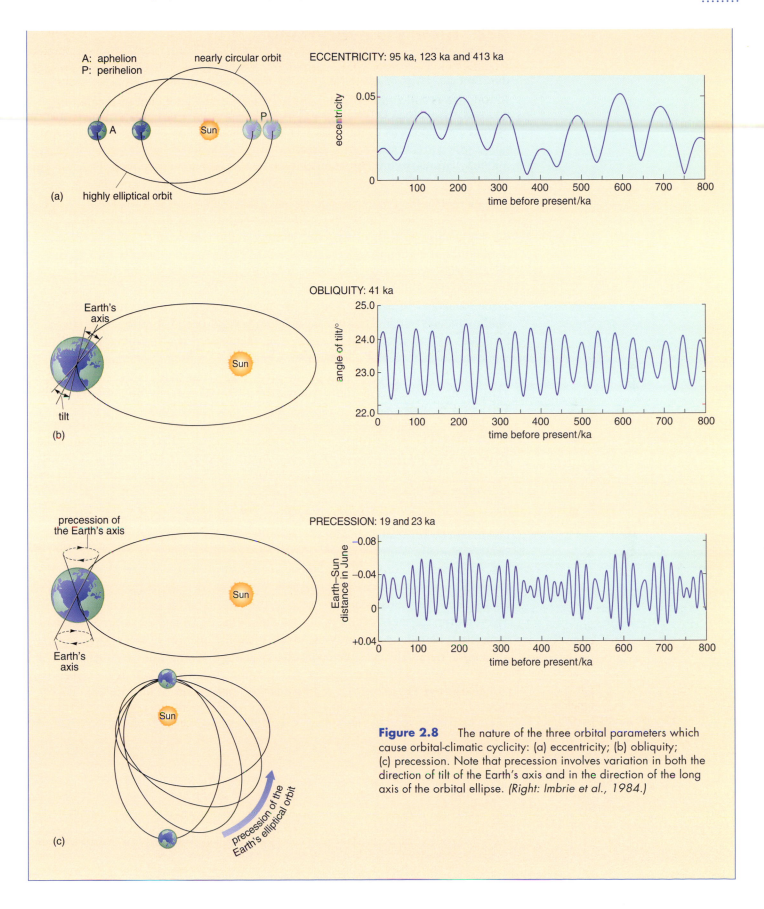

**Figure 2.8**   The nature of the three orbital parameters which cause orbital-climatic cyclicity: (a) eccentricity; (b) obliquity; (c) precession. Note that precession involves variation in both the direction of tilt of the Earth's axis and in the direction of the long axis of the orbital ellipse. *(Right: Imbrie et al., 1984.)*

## 2.2  Timing is everything

Refinement of the geological time-scale is the key to understanding the rate, duration and order of geological events. Research has shown that close to the Palaeocene/Eocene boundary (55 Ma), there was a marked change in the marine and terrestrial biota; between 35% and 50% of benthic foraminifers became extinct and there was a large turnover in land mammal assemblages. Associated with these extinctions, there is sedimentological evidence indicating a change to a much warmer, wetter climate. Both the terrestrial and marine records across this boundary show anomalous carbon-isotope values (termed a carbon-isotope excursion). In the marine record, the carbon-isotope excursion is −3‰ *. All of these effects at the Palaeocene/Eocene boundary have been attributed to decomposition of methane hydrate† leading to the release of c. 1200–2000 gigatonnes of carbon into the atmosphere. Clearly, in order to fully understand the consequences, causes and order of events, a well-defined time-scale is important. Prior to a recent study by Norris and Röhl, carbon-isotope excursions at the Palaeocene/Eocene boundary had been reported from terrestrial soils and mammal teeth, but uncertainties in the number and lateral correlation led to estimates of their age ranging between 54.88 and 55.5 Ma. However, the recent work shows that most of this methane was released into the ocean and atmosphere in one pulse that lasted no more than a few thousand years, making it a truly catastrophic event. Furthermore, the evidence indicates that it took about 120 ka for the carbon cycle to recover and for this major perturbation to be removed by carbon burial.

### Dating the Palaeocene/Eocene methane hydrate release event

The recent high-resolution dating of this event was obtained by astronomical calibration (Section 2.1.6) of some sedimentary deposits obtained by drilling in the North Atlantic. This drill site went through the thickest-known succession of this magnetic reversal zone and was assumed by the researchers to be stratigraphically complete. They carried out Fourier analysis on magnetic susceptibility measurements and the iron content of these deposits (Figure 2.9). From these analyses and palaeomagnetic data, they determined that depending on the stratigraphical position in the section there were one or two regular cycles. They interpreted these cycles to correspond to the 19 ka and 23 ka precession cycles or where there was only one cycle to be the average of these; the c. 21 ka precession cycle. Because the cores also contained appropriate biostratigraphical markers (Figure 2.9) and a magnetostratigraphical signal as well as the actual carbon-isotope excursion, they could use this stratigraphical information to

---

* The symbol ‰ represents per mil or parts per thousand. See Section 3.2.1 for further explanation.

† A clathrate is a solid compound in which molecules of one substance, commonly a gas, are enclosed within the crystal structure of another substance, as if in a cage. Clathrates include methane hydrate which is a crystalline solid that looks much like water ice, in which methane gas molecules are enclosed within a cage-like lattice of water molecules. The main factors controlling the formation and stability of methane hydrates are cold temperatures, high pressures, adequate supplies of methane and water, and suitable geochemical conditions. They occur in oceanic sediments below about 300 m water depth and beneath permafrost. The present-day global amount of carbon bound in gas hydrates is estimated to be more than twice the amount of carbon locked up in all other known fossil fuels (coal, oil, natural gas etc.).

correlate directly to the extinction and climatic events in other records. By counting the number of cycles that correspond to the initial sharp shift to relatively light marine carbon-isotope values, and the slower recovery back to post-excursion steady-state ratios, the researchers calculated the duration of each part of the event (Figure 2.9). They pointed out that there are some possible errors associated with these data including several thousand years missing at the chalk clast bed. Nevertheless, the major conclusions described remain the same.

What is important about this event is not only that it was a cause of a change in climate and the evolution of life 55 Ma ago, but also that the rate of release of extra carbon by these natural means is similar to modern rates of anthropogenic carbon input into the atmosphere. It is also interesting to note that the estimated 14 000 gigatonnes of carbon locked up as hydrate in the world's oceans today could have a large impact on the rest of the total exchangeable carbon reservoir of 42 000 gigatonnes. This fact is often not included in current carbon-cycle models that are used to predict warming and future climate change. It is also clear from this example that the time taken for steady-state to be reached again as the carbon was reburied was about 120 ka. None of these conclusions could have been reached without an accurate time-scale.

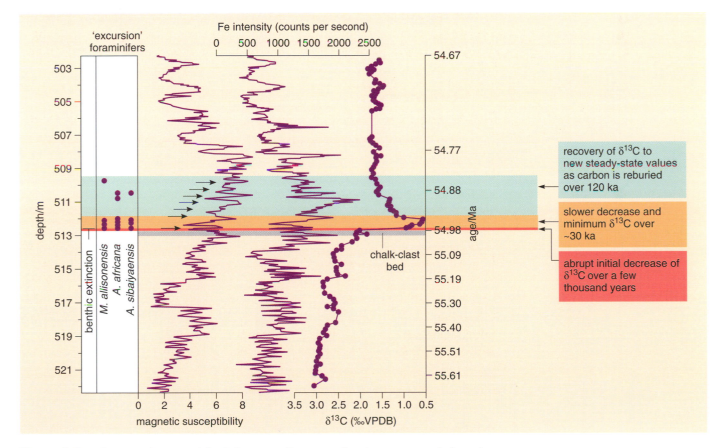

**Figure 2.9**   Stratigraphy around the Palaeocene/Eocene carbon-isotope anomaly from the Ocean Drilling Program Site 1051 in the western Atlantic. This carbon-isotope anomaly is interpreted to be the result of a large amount of methane hydrate being released from oceanic sediments. Small arrows indicate the approximate position of the 21 ka precession cycles. The event was synchronous in the oceans and on land, and began between 54.93 and 54.98 Ma ago. The steady-state carbon-isotope ratio prior to release of the methane hydrate capsule was c. 2.75‰ whereas after release the steady state was c. 1.75‰. (Norris and Röhl, 1999.)

## 2.3  The geological time-scale

The variety of stratigraphical techniques covered in Section 2.1 can be integrated to produce a geological time-scale. Integration is completed in a variety of ways. The ideal situation is that all the techniques are applied to one set of stratigraphically complete rocks so that any lateral correlation problems are eliminated. Clearly, in practice this is rarely possible since not many successions lend themselves to all the techniques, and it is also unusual for one individual section to be stratigraphically complete. However, it is often possible to combine techniques such as biostratigraphy, chemostratigraphy and magnetostratigraphy on favourable sections of sedimentary rocks and to compensate for gaps in the succession by using several sections and correlating them. Problems often arise when tying in the very important, but generally much rarer, radiometric dates. Magnetostratigraphy is one of the common methods by which radiometric dates completed on, for instance, sea-floor basalts are used to produce an integrated time-scale, because magnetostratigraphy can be applied to both igneous and sedimentary rocks.

The geological time-scale is a summary of a huge amount of stratigraphical data from all over the world, and is constantly being revised; this is why you will find that the numeric ages on different time-scales do not always agree. It follows that it is also more precise to use periods, epochs, ages or the actual specific type of data used for dating the rock under consideration rather than radiometric dates (e.g. Magnetochron 22r, or the *Pavlovia fittoni* Biozone).

## 2.4  Summary

- Chronostratigraphy is the subdivision of the rock record into time-significant units. Chronostratigraphical correlation lines are time lines. Lithostratigraphy is the division of the sedimentary record into lithologically similar units. Chronostratigraphical and lithological correlation lines may (and often do) cross each other.

- Stratigraphical techniques include the following relative dating techniques: biostratigraphy, which uses rapidly evolving, abundant and commonly preserved biota; magnetostratigraphy, which uses the geologically frequent flips in the Earth's magnetic field; and chemostratigraphy, which uses, for example, changes in the stable isotope ratio of certain elements through a succession. Absolute (radiometric) dating utilizes the decay of the radioactive isotopes of various elements to other isotopes over a known time-span. Astrochronology is a relative dating technique based on counting the number of regular cycles of known duration and/or matching these cycles to an orbital solution. Where there are fixed datums and the orbital solution is well established (i.e. from present day back as far as the Eocene), astrochronology can be classed as an absolute dating technique.

- A high-resolution integrated geological time-scale allows the ages, duration and sequence of Earth processes to be determined.

- At the Palaeocene/Eocene boundary, the decomposition of methane hydrates released up to 2000 gigatonnes of methane into the ocean and atmosphere in a few thousand years but it took over 120 ka for the carbon cycle to recover. Accurate timing and duration of the different parts of the event was determined from astrochronology. It is postulated this event was the cause of a marked change in the marine and terrestrial biota, including several extinctions.

- The geological time-scale is a compilation of a large body of diverse data. In order to avoid confusion or miscorrelation, it is best to state the age of a rock in line with the technique that was used to date it.

# 2.5  References

## *Further reading*

MIALL, A. D. (2000) *Principles of Sedimentary Basin Analysis* (3rd edn), Springer, 616pp. [Contains advice on collecting and interpreting sedimentological and stratigraphical data together with information on sedimentary basin formation and facies analysis.]

NICHOLS, G. (1999) *Sedimentology and Stratigraphy*, Blackwell Science, 355pp. [An undergraduate textbook which describes the basics of sedimentology and stratigraphy.]

NORRIS, R. D. AND RÖHL, U. (1999) 'Carbon cycling and chronology of climate warming during the Palaeocene/Eocene transition', *Nature*, **401**, 775–778. [Recent study described in Section 2.2 constraining the duration and timing of the stages of the Palaeocene/Eocene methane hydrate release.]

## *Other references*

HOUSE, M. R. AND GALE, A. S. (eds) (1995) 'Orbital forcing timescales and cyclostratigraphy', *Geological Society Special Publication* No. 85, 207pp., Geological Society of London.

JENKYNS, H. C., JONES, C. E., GRÖCKE, D. R., HESSELBO, S. P. AND PARKINSON, D. N. (2002) 'Chemostratigraphy in the Jurassic: applications, limitations and implications for palaeoceanography', *Journal of the Geological Society, London*, **159**, 351–378.

PARRISH, J. T. (1998) *Interpreting Pre-Quaternary Climate from the Geological Record*, Columbia University Press, 338pp.

TARLING, D. H. AND TARLING, M. P. (1977) *Continental Drift: A study of the Earth's moving surface*, a Pelican Book, Penguin Books, 154pp.

WEEDON, G. P. (1993) 'The recognition and stratigraphic implications of orbital forcing of climate and sedimentary cycles', in V. P. WRIGHT (ed.) *Sedimentary Review*, **1**, 31–50, Blackwell Scientific Publications.

WEEDON, G. P. (2003) *Time Series Analysis and Cyclostratigraphy: examining stratigraphic records of environmental cycles*, Cambridge University Press, 259pp.

# 3 Sea-level change

## Angela L. Coe and Kevin D. Church

'Time and tide wait for no man.'          Anon., 16th century proverb.

In Chapter 1, the fact that sedimentary deposits record changes in sea-level was introduced. In the first part of this Chapter, this scenario will be turned around. We know that sea-level has changed within the lifetime of humans, because such changes have been recorded, and in some cases directly measured with instruments. Understanding the effects of these recent sea-level changes on the sedimentary record gives us clues for interpreting sea-level changes from the geological record. This Chapter ends by using some of these clues to show how an example of cyclic sediments from the Carboniferous can be interpreted in terms of cycles of sea-level change. This example is revisited in Chapters 4 and 5, after further techniques have been introduced.

## 3.1 Noah's flood: a record of sea-level change

Noah's flood is one of the most widely reported changes in sea-level in the historical record. It is described most famously in the Old Testament of the Bible but similar accounts exist in ancient Greek and Middle Eastern literature. However, neither the timing nor geographical location of the flood is clear and several explanations for the flood have been proposed. Traditionally, the area of Noah's flood has been proposed as being in lower Mesopotamia (Figure 3.1) on top of the recently built-up delta area of the Euphrates, Tigris and Kurun rivers. For many thousands of years, this area has been a marshy, flat, low-lying plain a few feet above sea-level in the Arabian Gulf. The tale of the flood is supposed to have been told by the inhabitants of the area (Sumerians) from at least 4000 BC. Many historians believe that it was the same event that was later recorded on clay tablets engraved around 2000 BC that recount the tale of Gilgamesh, an ancient Babylonian king. It has been postulated that flooding of lower Mesopotamia by the sea occurred because of an earthquake in the area, which generated a large tsunami (tidal wave). The narrowness of the Arabian Gulf would have increased the height and therefore the destructive power of such a wave. However, it is also possible that the delta could have been flooded by: (i) a storm surge from the sea; (ii) flooding of the rivers due to heavy rainfall; or (iii) a rise in sea-level when glaciers melted, although the latter would have caused only a gradual inundation of the area because the Arabian Gulf is open to the ocean. The problem with all these theories is that because Mesopotamia is such a low-lying area they are all events that are likely to have happened a number of times and thus do not necessarily stand out as particularly dramatic and worthy of being recorded.

More recently, two American scientists, Bill Ryan and Walter Pitman, have uncovered and published (Ryan and Pitman, 1998) new evidence that the water level in the Black Sea (Figure 3.1) rose dramatically about 7600 years ago (i.e. about 5600 BC). They postulate that this flooded more than 100 000 km² of land in a matter of months, and most startling of all that this great volume of water could have only poured into the Black Sea through a narrow valley called the Bosporus (Figure 3.1) in a flow that probably reached the equivalent of 200 Niagara Falls! Prior to this flood, there is good evidence from fossils that the Black 'Sea' was actually a freshwater lake isolated from the saline waters of the Mediterranean (Figures 3.1, 3.2). At the time of the flood, sea-level in the Sea of Marmara (Figure 3.1) is interpreted to have gradually risen until it was poised to invade the Bosporus valley. At first, the seawater would have swept through the Bosporus valley carrying with it the soil and debris that once dammed it before plunging 150 m down into the Black Sea. The debris-laden fast-flowing torrent would have rapidly eroded the bedrock of the Bosporus valley: the deeper it cut the faster it flowed, and the faster it flowed the deeper it cut until there was a channel up to 85 m wide and 145 m deep. Ryan and Pitman hypothesize that this was the flood that was later recounted in Mesopotamia by the Sumerians and recorded in the tales of Gilgamesh. Certainly, it seems probable that such a dramatic event would be told and retold and could have eventually evolved into the biblical story.

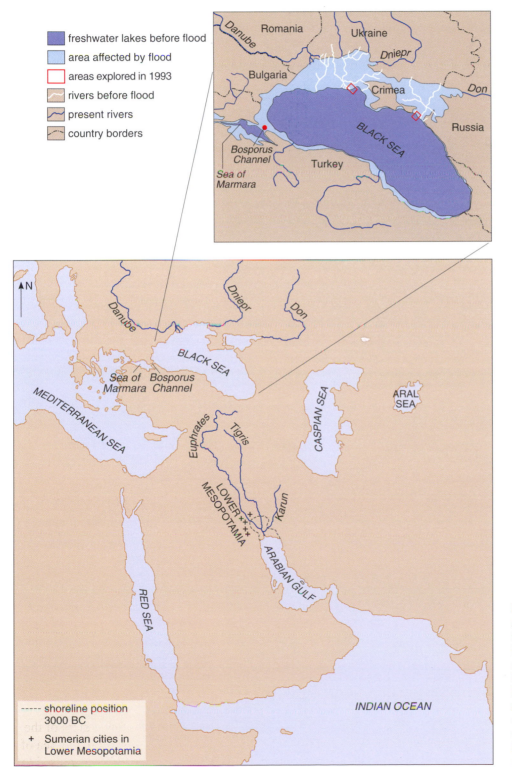

**Figure 3.1**  Present-day map of the Middle East and Eastern Europe showing the position of lower Mesopotamia at the head of the Arabian Gulf. Detailed map shows the extent of the two freshwater lakes (dark blue) and rivers around the Black Sea area prior to the flood compared to the area supposedly affected by it. Note that the water level after the flood in the Black Sea was higher than the present day. (Detailed map: Mestel, 1997.)

○  So, what evidence could you look for to establish if there was a rapid rise in water level in the Black Sea thousands of years ago?

●  Several different lines of evidence could be used: a change in the sedimentary record at the bottom of the Black Sea; buried shorelines nearer the middle of the Black Sea than the current-day shoreline; a change in salinity from freshwater to brackish; and flooded human settlements.

This is exactly what Ryan and Pitman looked for in the summer of 1993. They joined a Russian ship completing other scientific research in the Black Sea. They drilled many shallow boreholes (1–3 m deep) to recover sedimentary deposits on the sea-floor and they used an instrument that sends sound waves down through the water and sedimentary deposits to map out the sea-bed. A summary of some of the key results indicating there was a sudden flood in the Black Sea 7600 years ago is shown in Figure 3.2.

○  What common feature marks the flood event?

●  A ubiquitous layer of olive-grey homogeneous mud that drapes over all the previous deposits.

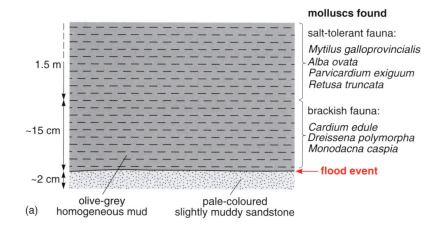

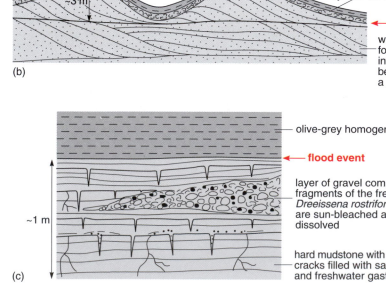

**Figure 3.2**  Sketches of some of the features found in different parts of the Black Sea demonstrating a flood and change from freshwater to marine conditions 7600 years ago based on information presented in Ryan and Pitman (1998). (a) 68 m water depth near the palaeomouth of the Don river (Figure 3.1); (b) 76 m water depth; (c) 100 m water depth near the mouth of the Dneister (Danube) and Dniepr rivers (Figure 3.1). The lowermost sedimentary deposits and fossils recovered indicate the shores of a freshwater lake whereas the uppermost sedimentary deposits and fossils indicate marine conditions. Carbon-14 ($^{14}$C) dating of mollusc shells from the very base of the olive-grey homogeneous clay indicates an age of 7600 years.

○ What is the evidence that there was a change in salinity in the lake from freshwater to marine?

● The different types of mollusc found.

○ How did the scientists determine the timing of the flood?

● By dating the mollusc shells using $^{14}C$ radiometric dating.

The results of all their coring and surveys demonstrated that prior to the flooding the level of the Black Sea shoreline was somewhere between 160 and 170 m below its present sea-level. The rapid rise in water level is estimated to have been up to 15 cm per day, and to have moved inland by as much as a mile a day; this would have had a devastating effect on any local communities that lived close to the shores of the lake. However, Ryan and Pitman also discuss evidence to show that this was not all 'doom and gloom' and that the flood probably led to the migration of people and spread of farming from the Middle East and Eastern Europe into Central and Western Europe, Asia, Egypt and Mesopotamia, thus marking an important event in the history of civilization.

Two key questions remain: (i) what was the primary cause of a sea-level rise in the Mediterranean at 7600 years ago? and (ii) was there a documented global event at the time? Climate proxies indicate the flooding event occurred about 200 years after the initiation of a particularly marked increase in global temperature (Figure 3.3), which culminated in the Holocene thermal maximum (i.e. the warmest part of the Holocene). Thus, what appears to have happened is that during the Holocene thermal maximum it was sufficiently warm to cause global sea-level to rise (through melting of ice caps) to such a height that the sea in the Mediterranean flooded over the top of the Bosporus valley threshold, created a channel and poured into the Black Sea. During this and other warm interglacial periods, the climate was both warmer and wetter. This may well be the reason that the legend mentions 40 nights and days of rain, though clearly, as in many legends, the time-scale seems to have been somewhat foreshortened.

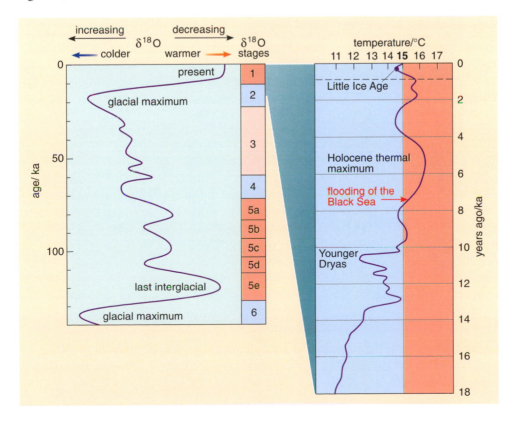

**Figure 3.3** Relationship between flooding of the Black Sea, oxygen isotopes over the past 140 ka (left), and average global temperature interpreted from climate proxy records for the past 18 ka (right). Global mean surface temperature today is 15 °C. Oxygen-isotope stages 1 and 5 were warm periods, stage 3 was intermediate, and the rest were cool periods. Note that the maximum and minimum oxygen-isotope values define the isotope stages and that the boundaries between them are taken at the midpoint between maximum and minimum. Technically, this means that the boundary between isotope stages can vary between different records because the amplitude of the signal varies from record to record. See Section 3.2.1 for further discussion of oxygen isotopes. (Left: Lowe and Walker, 1997 based on Martinson et al., 1987; right: Duff, 1994.)

Ryan and Pitman end their book on the flood by reflecting that although the scientific evidence for Black Sea flooding is good, as yet no direct evidence has been found for Neolithic settlements and human occupation in the area, mainly because the area has been flooded. However, the *Sunday Times* of 17 June 2001 reported that beams, fragments of walls, stone tools and a rubbish dump had been found about 100 m below the water level in the Black Sea, indicating the area was probably inhabited at the time of the flood. This evidence was found by the same team who located the wreck of the *Titanic*. They are now searching for further evidence that significant numbers of people lived in fairly permanent dwellings on the shores of the Black Sea before the water level increased so dramatically.

Whether or not the Black Sea was the site of Noah's flood, there is clear sedimentary and palaeontological evidence that 7600 years ago the water level in the Black Sea increased dramatically and it became saline.

*Clues for interpreting sea-level and salinity change from the sedimentary record:*

- *Water-level rise can lead to deposition of lithologically uniform sediments over a wide geographical area (olive-grey mud, Figure 3.2a–c);*
- *Fossils can indicate changing conditions from freshwater to brackish to marine, and are useful for dating (Figure 3.2a);*
- *Rapid water-level rise may lead to deposition of a winnowed lag of previously deposited material (layer of abraded shell debris, Figure 3.2b);*
- *Rapid water-level rise does not always erode previously deposited sediments, even where they have a topography (sand dunes, Figure 3.2b).*

## 3.2 Measurement of sea-level change

### 3.2.1 Oxygen isotopes

For much of the Neogene and all the Quaternary, the oxygen-isotope record of marine fossils can provide a valuable insight into the volume of polar ice sheets which in turn can be directly linked to sea-level change. This is because the more water that is locked up in polar ice-caps and glaciers, the less water there is in the oceans.

The ratio of the stable isotopes $^{16}O$ and $^{18}O$ in water varies with salinity, temperature and the volume of ice. As water evaporates, molecules with the lighter oxygen isotope $^{16}O$ evaporate more readily (because of their higher vapour pressure) so that the co-existing water vapour becomes enriched in this isotope. The enrichment of $^{16}O$ in water vapour in the atmosphere is enhanced by the fact that when water condenses, and falls as either rain or snow, the heavier oxygen isotope $^{18}O$ is preferentially precipitated back into the water. The continued preferential removal of $^{16}O$ from the water through repeated evaporation and precipitation results in the water vapour becoming more and more depleted in $^{18}O$. This process is limited: first, by an oppositely directed isotope exchange which causes the amount of $^{18}O$ in the water to peak when the liquid is 15–25% of its original mass, and secondly by the possible addition of freshwater during evaporation. Towards the Earth's poles, the water vapour becomes even more depleted in $^{18}O$ because of the selective removal of $^{18}O$ in initial condensates for the reasons described above, as well as for the following additional reasons:

1   enrichment of the lighter isotope in water vapour increases with decreasing air temperature;

2   evaporation into undersaturated air leads to additional kinetic fractionation;

3   re-evaporation of water from rain droplets and from surface water; and

4   evapotranspiration from plants which favour the lighter isotope.

Thus, when the water vapour is precipitated as snow in the polar regions, it is much depleted in the $^{18}O$ isotope relative to the oceans. The larger the ice-caps, the higher the relative proportion of $^{18}O$ in seawater and the lower the relative proportion of $^{18}O$ in ice-caps.

The proportion of $^{18}O$ isotope to the $^{16}O$ isotope is determined with respect to a standard and so is noted as the change relative to the standard in parts per thousand (per mil or ‰) as:

$$\delta^{18}O = \frac{(^{18}O/^{16}O)_{sample} - (^{18}O/^{16}O)_{standard}}{(^{18}O/^{16}O)_{standard}} \times 1000$$

Because it is the isotopic change that is measured, the $\delta^{18}O$ value (i.e. the oxygen-isotope composition) can be either positive or negative. The standards used are generally either, for water, silicates and ice records, seawater today (standard mean ocean water, SMOW) or, for carbonates and organic matter, a particular belemnite fossil (a type of mollusc) known as the Pee Dee Belemnite (PDB).

Changes in the oxygen-isotope composition of seawater through geological time as the ice-caps have changed in size are preserved in calcium carbonate that precipitated in the seawater. The seawater signal is only well preserved if the calcium carbonate does not suffer later diagenesis. For this reason, most studies involving oxygen isotopes have concentrated on the recent geological past, particularly the Quaternary (1.81 Ma to the present day). The calcareous skeletons of microfossils, particularly benthic and planktonic foraminifers which are found throughout marine Cainozoic sediments, have been favoured as they are abundant in both time and space, and any diagenetic alteration is relatively easy to detect. The exchange of oxygen isotopes between water and dissolved inorganic species such as carbonate is rapid. Hence, the $\delta^{18}O$ value of the water will determine the $\delta^{18}O$ value of the precipitated minerals. Carbonate is enriched in $^{18}O$ isotope compared to the water it precipitated from, due to mineral–water fractionation factors and the fact that $\delta^{18}O$ value of the carbonate is also dependent on temperature. As temperature rises, the activity of the $^{16}O$ isotope increases relative to that of $^{18}O$ isotope resulting in carbonate that forms at a higher temperature containing a higher proportion of $^{16}O$ than if it formed at a lower temperature. Because the $\delta^{18}O$ values of marine fossils are affected by both temperature and ice volume, more reliable estimates of ice volume can be obtained by using $\delta^{18}O$ values from benthic organisms rather than from planktonic organisms. The reason for this is that there is less temperature variation in the bottom waters of the oceans than there is in the surface waters. From this discussion, it follows that the higher the $\delta^{18}O$ values in benthic marine fossils, the greater the enrichment of $^{18}O$ isotope in seawater and the larger the ice-caps are. On average, a 1 °C change in temperature causes a 0.2‰ change in $\delta^{18}O$ value whereas typical changes in ice volume between glacials and interglacials result in a 1.2‰ shift. This indicates that ice volume has a more significant effect on $\delta^{18}O$ values than changes in temperature. Thus, by measuring the $\delta^{18}O$ value of benthic organisms during periods when there are polar ice-caps, we can obtain a good proxy for changes in ice volume and hence global sea-level. It has been calculated that a 1‰ change in $\delta^{18}O$ value is equivalent to about a 100 m change in global sea-level.

The oxygen-isotope composition also varies with salinity. More saline waters are enriched in the heavier isotope because, as already described, the $^{16}O$ isotope evaporates more easily. Thus, river water has a lower $\delta^{18}O$ value than seawater. Such factors need to be taken into account in oxygen-isotope studies where there is potentially any salinity variation, or in arid climatic conditions where there is increased evaporation, or where there is a large river water flux into the ocean, for instance near deltas.

Studies of Cainozoic marine sedimentary deposits have been completed in hundreds of cores from the world's oceans, as part of a large international project (the Deep Sea Drilling Program which was later replaced by the Ocean Drilling Program). The oxygen-isotope composition of foraminifers within the deposits has been measured and this has lead to the construction of a precise oxygen-isotope curve for much of the last 20 Ma (small parts of this are shown in Figures 3.3 and 3.4). This curve has been used to understand the timing and duration of glacials/interglacials and as an indication of sea-level change.

Additional oxygen-isotope data for the last 400 ka come from ice cores that were drilled from present-day ice-caps. One project was based in Greenland (Greenland Ice Core Project, or GRIP) and one in Antarctica at the Russian Vostok research station. The oxygen-isotope composition of the ice was measured at regular intervals, and a curve was obtained. As larger ice-caps preferentially store more of the lighter $^{16}O$ isotope, the glacial and interglacial episodes can be detected. In addition, the ice cores contain dust incorporated into the ice during colder periods when the wind was stronger. These ice-core data have been correlated with the data from sedimentary deposits using a variety of stratigraphical techniques.

Figure 3.4 shows an example of part of a calibrated benthic oxygen-isotope curve and its correlation to shallow-marine sedimentary cycles in New Zealand. The lowest $\delta^{18}O$ values which infer high sea-level correspond to the deepest-water facies — the shelf siltstones. Increases in $\delta^{18}O$ values inferring decreasing sea-level correspond to the change from shelf siltstones to shoreline sandstones, indicating a shallowing-upward succession that is entirely consistent with decreasing sea-level. The highest $\delta^{18}O$ values inferring lowest sea-level correspond to erosion surfaces interpreted to have formed at the lowest sea-level. Decreasing $\delta^{18}O$ values inferring relative sea-level rise correspond to shell beds deposited during transgression. Note also that the highest $\delta^{18}O$ values at oxygen-isotope stages 78 and 82 (oxygen-isotope stages are explained in the captions to Figures 3.3 and 3.4) correspond to the preservation of less shoreline sandstone facies (probably because of more erosion during the relative sea-level fall) than the smaller-magnitude isotopic maximum at stages 80 and 84.

In Chapter 6, we will consider oxygen-isotope and sedimentological data from sediments deposited during the Quaternary on the edge of the Mississippi delta.

*Clues for interpreting sea-level change from the sedimentary record:*

- *Oxygen-isotope records from fossils record temperature and ice-volume changes and can therefore also yield information about sea-level change.*
- *Sedimentary facies can indicate changes in sea-level.*

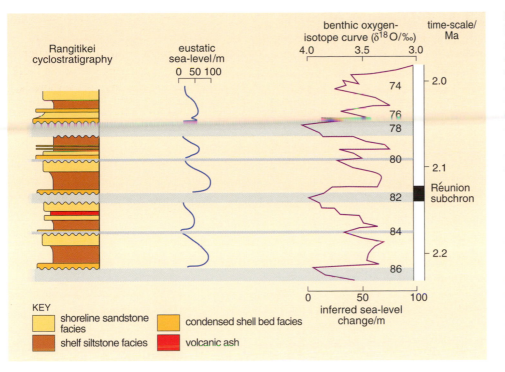

**Figure 3.4** Example of the correlation between the benthic oxygen-isotope record for c. 2–2.2 Ma from which changes in sea-level related to ice-cap and glacier volume can be inferred, and sea-level change interpreted from the sedimentary record in New Zealand. The column on the left is a graphic log of the sedimentary facies with erosion surfaces in the succession shown as grey shaded areas. The eustatic (global) sea-level curve to the right of the graphic log is interpreted directly from these sediments and their fossils. On the right is a benthic oxygen-isotope curve. The sea-level that can be inferred directly from this benthic oxygen-isotope curve is shown on the horizontal scale. The maximum and minimum oxygen-isotope values are known as isotope stages ($\delta^{18}O$ stages): maxima (corresponding to cool periods) have even numbers and the minima (corresponding to warm periods) odd numbers. *(Naish, 1997.)*

## 3.2.2 Coastline maps and coastal sediments

We can easily see that sea-level has changed over the last few decades by comparing coastline maps plotted at different times. A worldwide plot of the coastline at particular points in time would allow changes in global sea-level to be deduced. This principle of plotting the coastline can be applied throughout geological time by identifying sediments that were deposited along the coast and thereby reconstructing a palaeocoastline map. The problem with this is that the data set is often incomplete because of later erosion but nevertheless the technique has been used to construct some sea-level curves.

Sediments deposited along the coastline and in shallow-marine areas are the most sensitive to changes in sea-level. As the coastline sediments change position as sea-level rises and falls, so do all the other adjacent alluvial and marine facies belts. Sedimentary facies and the amount of encroachment of shoreline facies along cross-sections perpendicular to continental margins have been examined and used to interpret sea-level change. How the sedimentary record may be analysed in detail to determine sea-level changes is covered in Chapter 4.

*Clues for interpreting sea-level change from the sedimentary record:*

- *Maps of the distribution of coastal sediments can show how sea-level has changed through time.*
- *The vertical and horizontal movement of sedimentary facies in space, particularly shallow-marine sediments, indicates how sea-level has changed.*

### 3.2.3 Tide-gauge records

If you spend a day on the beach, you may witness clear evidence for sea-level change. Twice a day, the tides force sea-level to rise, driving you up the beach before the subsequent fall. This is sea-level change on a very short (diurnal) time-scale and it is clear that the sea is rising and falling with respect to the land. Tide gauges can be used to measure this short-term tidal variation in sea-level, but they can also show the longer-term variation in sea-level if used over an extended period. The results from some such instruments are shown in Figure 3.5. It is the longer-term sea-level variations that we are interested in extracting.

Figure 3.5 clearly shows that sea-level has changed between 1890 and 1970. However, the records also show that whilst at one locality sea-level is nearly constant, it is falling significantly at others.

○ What might cause sea-level changes to apparently differ at various localities on the same continent?

● The local tectonic setting.

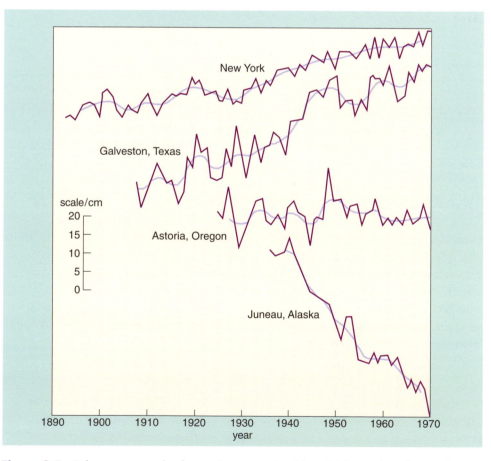

**Figure 3.5** Tide-gauge records of annual mean sea-level (purple) from selected coastal localities in the USA. The large annual range is caused by meteorological changes (e.g. low pressure causing higher sea-levels, and onshore winds, which can drive water up against coasts). The smoother pale-blue curves indicate a five-year average. (Davis, 1994.)

The sea-level measured at one particular location on the Earth is not necessarily going to be a global signal. It may be affected by glacial rebound, as is the case for Juneau, Alaska, or by tectonic uplift, e.g. Astoria in Oregon. The curve for New York actually corresponds more or less to the global average because it is on a stable continental crust. The similar but slightly higher rate of sea-level rise at Galveston, Texas, is interpreted to be due to the added effect of subsidence because of extraction of oil and/or groundwater.

This discussion poses three questions: (i) What are the processes causing sea-level change? (ii) How do we separate the different processes that are contributing to a change in sea-level at any one location? And (iii) how do we obtain a global signal?

The answers to these questions are complex issues which will be briefly introduced in Section 3.3 and then discussed more fully in the rest of the book.

*Clue for interpreting sea-level change from the sedimentary record:*

- *The data from any one geographical location do not give a global signal of sea-level change as this may be influenced by local processes.*

## 3.3   Why does sea-level change?

The example shown in Figure 3.5 indicates that long-term sea-level changes may result either from the level of the sea or the elevation of the land changing. This leads to the conclusion that processes controlling sea-level change are of two types: those which require the volume of seawater to change and those which require the total volume of basins containing seawater to change (see Box 3.1 for an explanation of different types of basin).

### Changes in the volume of seawater

Whilst water can be added to the oceans through volcanism, and removed during the subduction of oceanic crust at destructive plate margins, these processes only cause minor fluctuations on the thousand to million-years time-scale that we are interested in. The principal cause of variation in the volume of seawater is through the formation and melting of ice-caps and glaciers. Complete melting of the present-day Antarctic ice-cap and Greenland ice sheet would lead to a sea-level rise of *c.* 60–80 m, sufficient to reduce the global land area by about 20%! Another mechanism for changing the volume of seawater is thermal expansion; an increase in average global ocean water temperature of 10 °C would cause sea-level to rise by about 10 m.

It should be clear from this brief introduction that there is a close link between climatic processes and sea-level. When global temperatures increase, the volume of seawater expands and polar ice-caps melt, both of which in turn cause global sea-level to rise. Conversely, when global temperatures decrease or polar ice-caps grow, the volume of seawater decreases which causes global sea-level to fall.

### Changes in the volume of basins containing seawater

The second mechanism we need to consider is how the size and shape of basins (Box 3.1) containing seawater can change. These can either be ocean basins formed on oceanic crust or continental basins formed on continental crust. The major changes in volume are due to ocean basins changing in size and shape.

Complete desiccation of a basin containing seawater can result in redistribution of the seawater into other basins.

○ What is the main process that controls changes in volume of the seawater-filled basins?

● Plate tectonics.

During continental break-up, an increased length of mid-ocean ridge spreading centres or an increase in the rate of sea-floor spreading will decrease the volume of the ocean basins. This is because the hot rising magma causes the lithosphere to be uplifted along the mid-ocean ridges, thus decreasing the volume that the ocean basins occupy, causing the oceans to flood the continental margins. Conversely, during continental collision, orogenic (mountain-building) episodes will decrease the area taken up by the newly formed continent (relative to the two old continents from which it was formed). This will increase the volume of the ocean basin, causing a fall in sea-level. Furthermore, the respective thickening and thinning of the continental crust that this entails will lead to changes in the elevation of coastal regions. This process is counteracted by the subsequent erosion of the newly formed mountain chain. As sediment is deposited, it will decrease the volume of the basin.

We will not consider the reasons for the change in continental elevation in any detail, though you should be aware of the following mechanisms:

• Fault movement along coastal regions may cause rapid tectonic uplift or depression of part of a continent's margin.

• Loading of the continents by glaciers causing them to subside, and isostatic rebound (uplift) following glacial melting.

• Sediment loading within a sedimentary basin: as sediment is deposited within a basin, the added weight causes the basin and its margins to subside.

These mechanisms all act on different time-scales; in addition, some are global whereas others influence sea-level in just one location. It is clearly important to try to discriminate between global sea-level changes, due to 'true' sea-level fluctuation, and local or regional changes, due to the vertical motion of a particular continent.

In Chapter 5, we will return to the issues briefly outlined above in an attempt to understand the causes of sea-level change and the time-scale on which each of these causes acts.

## 3.4 What do we mean by 'sea-level'? Definitions of eustasy, relative sea-level and water depth

Following on from the above discussion, it is clear that it is useful to define different types of sea-level change. Eustatic sea-level, relative sea-level and water depth all have specific meanings. *Eustatic sea-level* (or eustasy) is global sea-level and is a measure of the distance between the sea-surface and a fixed datum, usually taken as the centre of the Earth (Figure 3.6). Variations in eustasy

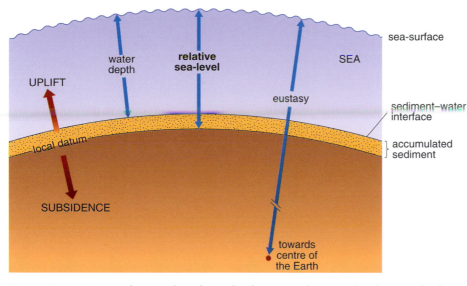

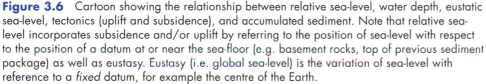

**Figure 3.6**   Cartoon showing the relationship between relative sea-level, water depth, eustatic sea-level, tectonics (uplift and subsidence), and accumulated sediment. Note that relative sea-level incorporates subsidence and/or uplift by referring to the position of sea-level with respect to the position of a datum at or near the sea-floor (e.g. basement rocks, top of previous sediment package) as well as eustasy. Eustasy (i.e. global sea-level) is the variation of sea-level with reference to a *fixed* datum, for example the centre of the Earth.

are controlled by changes in ocean-water volume (e.g. by varying the volume of water locked up in glaciers, causing glacio-eustasy) and by changing the volume of the ocean basins (e.g. by varying the total length or rate of sea-floor spreading and therefore the volume of the world's ocean ridge systems, causing tectono-eustasy). *Relative sea-level* on the other hand is the distance between the sea-surface and a local datum, for example the top of the basement rocks in a sedimentary basin (Figure 3.6). Relative sea-level change is therefore influenced not only by eustasy, but also by changes in the elevation of continents and the sea-floor. Relative sea-level change is a useful term as it does not imply that a particular mechanism is responsible for the sea-level change or that it is global in extent. For instance, relative sea-level change rather than eustasy is a better term to use when considering sea-level change in a local area, as it accounts for both local subsidence (or uplift) and eustatic changes in sea-level.

○   Describe a situation in a sedimentary basin whereby eustatic sea-level is falling during a relative sea-level rise.

●   If eustatic sea-level is falling, then the only way that relative sea-level can rise is if the basin is subsiding at a higher rate than the eustatic sea-level fall. This will increase the distance between the sea-surface and the base of the basin.

Water depth is not the same as relative sea-level. Whilst the latter is the distance between the sea-surface and a fixed local datum, *water depth* is the distance between the sea-bed, i.e. the top of the sediments (Figure 3.6), and the sea-surface or water level. So even if basin subsidence and eustatic sea-level are static, water depth will be reduced as sediment fills the basin.

## Box 3.1  Sedimentary basins

Sedimentary basins are generally regions of prolonged subsidence of the Earth's surface. They are the places where the majority of sedimentary rocks are found because they are essentially a 'hole' that gets filled. Sedimentary basins may be filled with either marine and/or non-marine sedimentary deposits and, of course, seawater and freshwater. Processes associated with plate tectonic motions within the cool, relatively rigid lithosphere are responsible for the formation of sedimentary basins. There are three basin-forming processes that may occur either independently or may change during the history of the basin. They are:

1  *Purely thermal mechanisms:* examples are the cooling and subsidence of ocean lithosphere as it moves away from mid-ocean ridge spreading centres, or thermal contraction causing sagging.

2  *Changes in crustal/lithospheric thickness:* thinning of the crust by mechanical stretching causes fault-controlled subsidence, whereas thinning of the lithosphere produces thermal uplift.

3  *Loading:* the lithosphere will be deflected or flexurally deformed as it is loaded. Examples include subsidence that occurs adjacent to mountain belts as they are forming, and loading of the crust by a volcano or seamount.

These mechanisms produce a range of different types of basin. The ones considered in this book include:

- *Ocean basins:* these are formed on oceanic lithosphere by purely thermal mechanisms. As the new oceanic lithosphere moves away from the mid-ocean ridge spreading centre, it cools and subsides, creating a basin either side of the spreading ridge over the entire ocean. The Atlantic Ocean is a modern-day example.

- *Rift basins:* these form as the lithosphere is thinned by mechanical stretching. Rift basins form part of an evolutionary sequence from the initial stages when stretching of the continental crust causes faulting and subsidence, to their possible evolution into oceanic spreading centres and an ocean basin if the stretching is strong and continuous enough. If new sea-floor does form, the faulted and subsided edge of the continent forms a passive continental margin or passive margin. The North Sea is an example of a failed rift basin, that is, it rifted but the stretching forces were never sufficient to form oceanic lithosphere. The formation and trapping of hydrocarbons has occurred in the North Sea because it is composed of a number of rift basins which filled with a thick pile of sediments having a range of compositions.

- *Strike–slip basins:* these are formed by local stretching in intricate fault zones. These basins tend to be complex; they form linked structural systems, in that each area of extension is balanced by compression. There are many such basins along the San Andreas fault system in California.

- *Foreland basins:* these form through flexure of the lithosphere as it is loaded in a continental collision zone during the formation of mountains. Modern examples of flexural basins can be found near the Alps and Himalayas. The sedimentary rocks of Book Cliffs covered in Part 3 were deposited in a foreland basin.

- *Fore-arc basins and deep oceanic trenches:* these are formed through flexure as the lithosphere approaches a subduction zone. The driving force is the loading of the lithosphere by the mass of the magmatic arc and/or the gravitational forces exerted by the downgoing slab.

For further information see Allen and Allen (1990) which provides a good summary of sedimentary basin formation and evolution.

## 3.5  An example of sea-level change from the Carboniferous sedimentary record

This Section presents an example of sedimentary cycles from the Namurian (part of the Carboniferous) of northern England. It will raise questions which will be answered more fully when we reconsider this example in Section 4.4.

During the late Carboniferous, deltas built out (or prograded) southwards from what is today Scandinavia and the North Sea to fill deep marine rift basins (see Box 3.1) that occupied the area that is now Britain (Figure 3.7). Progradation of these deltas was occasionally interrupted by regionally extensive, marine transgressions which pushed the delta shoreline back to the north. This was followed by a renewed phase of progradation in which the deltas again advanced southward.

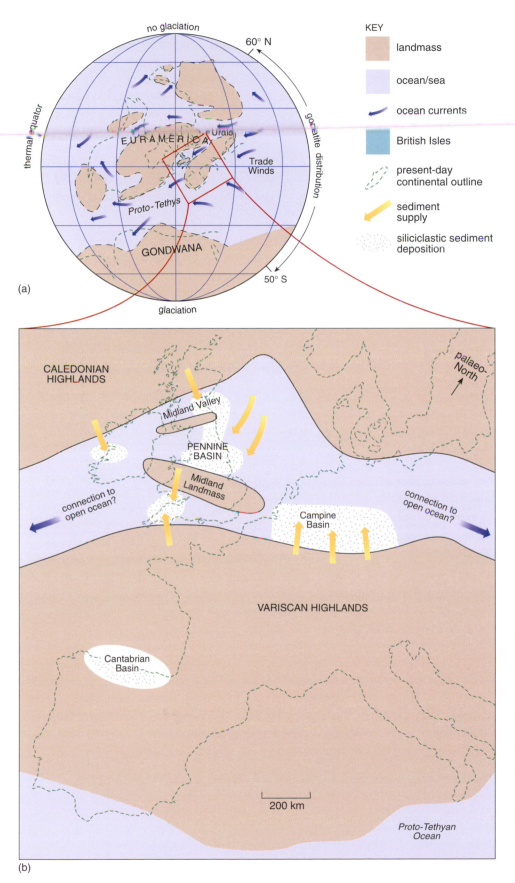

(a)

(b)

**Figure 3.7**  (a) The global distribution of continents and oceans during the Namurian. There is evidence of glaciation over the South Pole during the Namurian. The evidence is preserved in parts of the present-day continents (South America, Antarctica, Africa, Arabia, India and Australia) that made up the supercontinent Gondwana in the Namurian. The North Pole is not thought to have been glaciated at this time. This imbalance between north and south would have shifted the thermal equator some 10° to 20° north of the geographical equator.
(b) Palaeogeographical map of north-western Europe during the Namurian indicating continents, seaways, oceans, marine sedimentary basins and areas of active sediment deposition.
((a) Ramsbottom, 1971; (b) Collinson, 1988 based on Anderton et al., 1979 and Ziegler, 1982.)

From a global perspective, the supercontinents of Gondwana (comprising South America, Antarctica, Africa, Arabia, India and Australia) and Euramerica (comprising North America, North Asia and Europe) were separated by a 'Proto-Tethys' Ocean during the late Carboniferous (Figure 3.7a). Evidence of boulder clays, glacial tills and glacial striations in rocks of this age suggests that Gondwana was glaciated to varying degrees during the late Palaeozoic and underwent a longer episode of glaciation in the Namurian when most of Gondwana was covered in ice. Eustatic sea-level change induced by the waxing and waning of the Gondwanan ice sheet is therefore a mechanism that we need to consider to explain the periods of delta advance and marine transgression.

Figure 3.7b is a palaeogeographical map of north-western Europe during the Namurian. It was produced by mapping those areas of known sediment erosion (interpreted as continental) and those of sediment deposition (interpreted as marine). It shows an east–west elongate seaway, exceeding 3000 km in length, sandwiched between the Variscan and Caledonian highlands (each the product of two earlier ocean closures and orogenies). We know the seaway extended eastwards as far as the Urals in Russia (another 2200 km) because rocks of similar age, and suggesting similar deltaic conditions, are known to exist between there and Britain.

○ Why are we not sure about the westward extension of the seaway?

● The westward extension leads into what is now the Atlantic Ocean, an ocean that did not exist in the late Carboniferous. Rifting associated with its opening in the Jurassic and Cretaceous would have destroyed much of the evidence for such a seaway.

Further evidence for this seaway emerges on the other side of the Atlantic in the Appalachian Mountains.

The deltaic successions now preserved in central and northern England prograded into the Pennine Basin from the north and north-east (Figure 3.7b). These deltaic sediments are built up of many cycles. An idealized deltaic sedimentary cycle for this area is in the order of 30–50 m thick and comprises dark-grey, marine mudstones containing goniatites (free-swimming, coiled-shell marine organisms closely related to ammonites). These pass upwards through prodelta mudstones, interbedded delta-front mudstones and sandstones, and mouth-bar sandstones, into delta-plain fluvial distributary channels associated with coals and palaeosols (Figure 3.8). It is important to note that the thickness of each interval often bears little relation to the time taken to deposit it. The marine mudstones containing goniatites are examples of condensed* deposits. So, although they may total less than a metre in thickness, they may have been deposited over thousands of years, whereas the rest of the deltaic cycle may have been deposited in only a few hundred years.

Exposure across northern England is often so poor that trying to match up sandstones (which often have similar characteristics) between exposures is problematical. Fortunately, goniatites greatly aid correlation of the lithologies. They evolved rapidly through time to give us many different species which are principally recognized by the different ornament on their shells. Consequently, where the same goniatites are found at different localities, we know that they represent the same marine transgression. This is an example of the use of biostratigraphy in the correlation of sediments (Section 2.1.2, Figure 3.8).

* Condensed is defined here as a long period of time being represented by a small thickness of sediment.

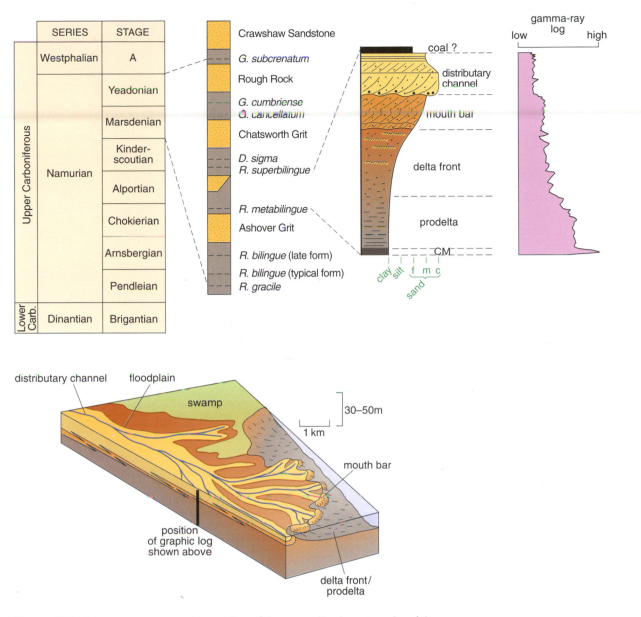

**Figure 3.8**   Namurian stages and an outline of the generalized stratigraphy of the Marsdenian and Yeadonian stages for central England. Each successive delta cycle has its own named sandstone, and each is terminated by a unique goniatite species. Also illustrated is a graphic log of a typical deltaic cycle showing the component facies and the typical corresponding gamma-ray log (Box 3.2). The lower part of the diagram shows a block diagram of a prograding delta. The position of the graphic log is marked. G = *Gastrioceras*; D = *Donetzoceras*; R = *Reticuloceras*; CM = condensed marine mudstones with goniatites.

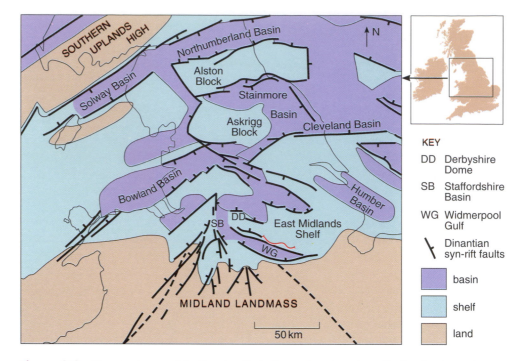

**Figure 3.9** Structural map of the Pennine Basin (central and northern England) prior to the deposition of the Namurian deltas showing land areas and areas covered by sea which are divided into deeper fault-bounded basins or troughs and shallower shelf areas (shelf areas are also variously termed blocks, platforms, domes, highs). Note that the deltas prograded into the Pennine Basin from the north and north-east. Red line shows position of cross-section in Figure 3.10. *(Guion et al., 2000.)*

Figure 3.9 shows the structure of the Pennine Basin just prior to the deposition of the Namurian deltas. Far from being a flat surface in the middle of an ocean, there were a series of predominantly north-west–south-east oriented shelf areas and basins, each bounded by faults, with some of the basins showing marked asymmetry. It is known that these faults ceased to be active before the deltas prograded into the area. However, the difference in water depth (or palaeobathymetry) between the basins and shelf areas had a marked effect on the distribution of the earliest deltas which were restricted to the basins or only thinly developed over the shelf areas. We will focus our attention on the deltas deposited in the Widmerpool Gulf and on the East Midlands Shelf (Figure 3.9). The Derbyshire Dome (Figure 3.9) acted as a barrier to the southward advance of the deltas into the Staffordshire Basin and Widmerpool Gulf until the late Kinderscoutian (Figure 3.8).

Figure 3.10 is an interpreted cross-section through the Marsdenian and Yeadonian intervals of the Namurian and lowermost Westphalian (Figure 3.8). The cross-section is made up of data from nine boreholes, and consists of graphic logs constructed from cores and a natural gamma-ray log response (Box 3.2). Ignoring for a moment the sandstones marked in orange, Figure 3.10 shows ten goniatite-bearing marine mudstone layers, each of which is overlain by a new phase of delta progradation into the marine basin. Most of these progradation phases are shown as the colour change from grey below to yellow above, representing the coarsening-upward transition from prodelta mudstone/delta front sandstone (grey) to mouth bar/delta front/delta top sandstone (yellow). The most clearly defined parts of the progradational trends are shown by the black triangles labelled 1 to 9 on Figure 3.10. Some of these delta progradation intervals are very thin, being represented by a few metres of prodelta mudstone/delta front

> ## Box 3.2 The gamma-ray log
>
> Boreholes are drilled to find water, oil or gas. They also allow data to be gathered on the lithologies that they penetrate. Whilst this is most effectively done by extracting sections of core, it is also the most expensive way, and so coring is usually only undertaken for rocks of considerable interest. For oil companies, this might mean sandstones thought to be good hydrocarbon reservoirs. As an alternative, instruments may be lowered into the borehole after (or during) drilling to investigate remotely the properties of the surrounding rocks. One property routinely measured is the natural radioactivity of the rocks.
>
> The gamma-ray log measures gamma rays emitted by small concentrations of naturally occurring radioactive elements, principally the isotopes of thorium (Th), potassium (K) and uranium (U). The logs in Figure 3.10 do not discriminate between the three isotopes though this is possible to do with spectral gamma-ray tools. Values are measured as a ratio against a standard sample held by the American Petroleum Institute (API) and are quoted in 'API' units.
>
> In general terms, the radioactive isotopes are concentrated in the micas and clay minerals in mudstones, so mudstones record high values of gamma rays, whilst sandstones record low values. The gamma-ray log thus records sandstones and mudstones as 'end-member' low and high values respectively. Intermediate values result from a mixture of lithologies, for example when sandstones and mudstones are interbedded. Though the log may appear to resemble a grain-size profile, its shape is in fact determined by changes in mineralogy. For this reason, mudstones of different clay minerals give different values. Organic carbon-rich mudstones often give particularly high gamma-ray values as they also tend to contain high concentrations of uranium because it is fixed in the sediment under anoxic conditions. The gamma-ray log is widely used as a correlation tool in the petroleum and coal industry.

sandstone sandwiched between goniatite-bearing marine mudstones (e.g. between the *R. gracile* and *R. bilingue* marine layers). These progradational intervals are nevertheless recognizable from the gamma-ray signal. Above most progradational intervals lies a transition back to prodelta mudstone and delta front sandstone (this is shown as a colour change from yellow below to grey above, e.g. above the *D. sigma* marine layer). These indicate an end to progradation and the gradual onset of transgression as the sea flooded (or transgressed) the top of the delta plain to deposit finer-grained sediments on what was formerly a terrestrial environment. In each case, the acme of the transgression is marked by the deposition of goniatite-bearing marine mudstones. In the Carboniferous, these marine mudstones are prominent on the gamma-ray log because they are enriched in uranium and so have very high gamma-ray log responses (Figures 3.8, 3.10). The pairing of delta progradational successions and marine deposits represents one depositional cycle.

○ How might we explain the development of these depositional cycles in terms of relative sea-level change (i.e. subsidence and eustatic sea-level) and/or sediment compaction?

● Clearly, we need a mechanism to allow the sea to periodically transgress the land. This can be done either by allowing eustatic sea-level to rise in stages, or by keeping eustatic sea-level constant and allowing subsidence of the basin in stages, both cases equate to pulses of relative sea-level rise. Alternatively, relative sea-level could have remained constant and compaction of newly deposited sediment within the delta may have caused the delta to subside beneath the sea-surface.

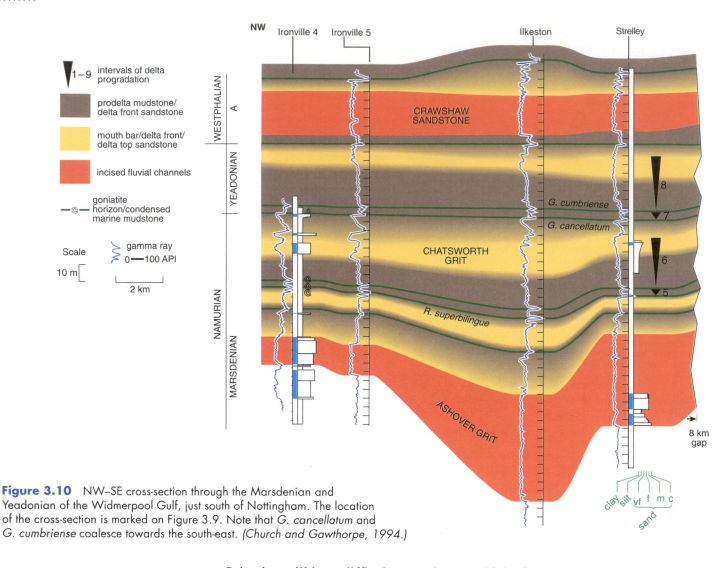

**Figure 3.10** NW–SE cross-section through the Marsdenian and Yeadonian of the Widmerpool Gulf, just south of Nottingham. The location of the cross-section is marked on Figure 3.9. Note that *G. cancellatum* and *G. cumbriense* coalesce towards the south-east. (*Church and Gawthorpe, 1994.*)

Scientists still have difficulty agreeing on which of these mechanisms was dominant during the deposition of the Namurian in the Pennine Basin. Most would concede that as the South Pole was glaciated, the first option is more likely, simply because there is a mechanism in place that is capable of accounting for these observations. However, it may also be true that both tectonic subsidence and compaction of the sediment played some part in causing each of the transgressions.

Now consider the sandstones marked in orange on Figure 3.10. They contain sedimentary structures indicative of deposition within fluvial distributary channels. However, these sandstones appear to overlie erosively the delta deposits (which of course also contain their own distributary channel sandstones). This suggests that they cannot have been deposited concurrently with the underlying delta. Instead, they must have been deposited following a period of erosion which occurred after the delta prograded into the area.

○ Bearing in mind our conclusion regarding the fact that each depositional cycle represents progradation of the delta followed by flooding by the sea, what might have caused the erosion of its top?

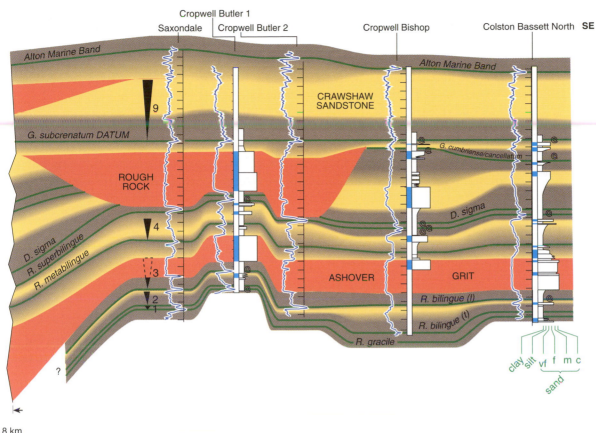

- It could be argued that a fall in relative sea-level would leave the old delta plain stranded high above the sea-surface. As a consequence, rivers would incise down into the delta plain as they endeavoured to adjust to a new gradient consistent with a lowering of relative sea-level. As they did this, they would carve out new valleys, which would subsequently fill with fluvial sand, recreating what we see in Figure 3.10. This can be done simply by allowing eustatic sea-level to drop. Although tectonic uplift of the basin is another plausible mechanism, it is unlikely because it would require compression of a basin that we know to have been in extension at the time.

From this simple example, we have identified sedimentary cycles with varying thickness and facies development, and whose deposition was probably controlled by fluctuating relative sea-level. Each cycle is composed of a progradational phase (deposition of deltaic sediments), a transgressive phase (the transition back to prodelta and delta front sediments and the capping by a goniatite-bearing marine mudstone) and some have an erosive phase (fluvial channel sandstones lying within incised valleys), caused by a drop in relative sea-level. In the next Chapter, we will look at sedimentary cycles in general to see how they develop during one cycle of relative sea-level rise and fall, using the principles of sequence stratigraphy. In Section 4.4, we will return to our Carboniferous case study to apply these principles to the sedimentary cycles discussed in this Section.

## 3.6  Conclusion

As we move further back in time, we obviously lose the ability to measure sea-level change directly. However, as we have seen from the Black Sea example in Section 3.1, analysis of sedimentary successions themselves can indicate changes in sea-level. Interpretation of the sedimentary record in terms of sea-level change involves combining evidence from unconformities and sedimentary deposits so that we can obtain a holistic record (Chapter 1) together with integrating the appropriate part of the geological time-scale (Chapter 2). In Chapter 4, we will present the sequence stratigraphy model for interpreting changes in sea-level from the sedimentary record. In the remaining Chapters, we examine many case studies from the sedimentary record deposited over the past 350 Ma, in a variety of tectonic settings and depositional environments.

## 3.7  Summary

- There is sedimentological, palaeontological and archaeological evidence to suggest Noah's flood recorded in the Bible might well be related to flooding of the Black Sea 7600 years ago when seawater from the Mediterranean entered the area through the Bosporus valley. This event can be correlated with eustatic sea-level rise associated with the Holocene thermal maximum.

- Oxygen-isotope data, coastline maps, maps of the distribution of coastal sediments and tide gauge records can all provide information about sea-level change over long time periods.

- Data on sea-level change from any one particular point does not usually provide a eustatic sea-level signal. This is because local sea-level changes may be governed by a combination of eustatic changes, local tectonic movement and sediment compaction.

- Sedimentary basins are generally regions of prolonged subsidence of the Earth's surface; there are a variety of different types depending on tectonic conditions.

- Sea-level change is caused either by variations in the volume of basins filled by seawater or by alteration of the volume of seawater. The former is controlled by tectonic processes and the latter mainly by ice-cap and glacier growth and melting together with thermal expansion and contraction of seawater.

- Relative sea-level is the distance between the sea-surface and a local datum.

- Eustatic sea-level is the distance between the sea-surface and a fixed datum (usually the centre of the Earth).

- Water depth is the distance between the sea-surface and the top of the sediments.

- Cyclic deposition of deltaic, incised fluvial, and open-marine sediments which can be correlated across northern England using biostratigraphy, indicate that relative sea-level changed during the deposition of the Namurian in central and northern England. Evidence from continental areas that made up the supercontinent of Gondwana during the Namurian suggests that the South Pole was glaciated at this time. Fluctuation in the volume of the ice-cap due to climate change is the most likely mechanism of relative sea-level change during this period of Earth history.

- Gamma-ray logs provide additional information on the composition of sedimentary succession. They can often be used as a proxy for grain size (clay minerals give a high value whereas quartz gives a low value).

# 3.8 References

## Further reading

ALLEN, P. A. AND ALLEN, J. R. (1990) *Basin Analysis; principles and applications*, Blackwell Scientific Publications, 451pp. [A clear account of sedimentary basin formation and evolution.]

EINSELE, G. (2000) *Sedimentary Basins Evolution Facies and Sediment Budget* (2nd edn), Springer, 792pp. [Contains short sections on the types and evolution of sedimentary basins and more extensive sections on depositional systems and facies together with a quantitative approach to denudation and accumulation of sediment.]

MIALL, A. D. (2000) (see further reading list for Chapter 2).

NAISH, T. R. (1997) 'Constraints on the amplitude of late Pliocene eustatic sea-level fluctuations: new evidence from the New Zealand shallow marine sedimentary record', *Geology*, **25**, 1139–1142. [Concise research paper describing the case study from New Zealand used in Section 3.2.1.]

RYAN, W. B. F. AND PITMAN, W. C. (1998) *Noah's Flood: the new scientific discoveries about the event that changed history*, Simon & Schuster, USA, 319pp. [Account of the data found in the Black Sea. Interesting read but somewhat rambling.]

## Other references

DAMIANOVA, A. (2001) '*Titanic* explorer hunts for signs of Noah's ark', *Sunday Times* 17 June.

FAIRCHILD, I. J., HENDRY, G., QUEST, M. AND TUCKER, M. E. (1988) 'Chemical analysis of sedimentary rocks', in TUCKER, M. E. (ed.) *Techniques in Sedimentology*, Blackwell Scientific Publications, 274–354.

FAURE, G. (1977) *Principles of Isotope Geology*, John Wiley and Sons, 589pp.

LEEDER, M. AND STRUDWICK, A. E. (1987) 'Delta-marine interactions: a discussion of sedimentary models for Yoredale-type cyclicity', in MILLER, J., ADAMS, A. E. AND WRIGHT, V. P. (eds) *The Dinantian of Northern England, European Dinantian Environments*, John Wiley and Sons, 115–130.

MCCABE, P. J. (1977) 'Deep distributary channels and giant bedforms in the Upper Carboniferous of the central Pennines, northern England', *Sedimentology*, **24**, 271–290.

MESTEL, R. (1997) 'Noah's Flood', *New Scientist*, 4 October.

SANDARS, N. K. (1972) *The Epic of Gilgamesh* (1972) Penguin Classics, Penguin Books, 127pp.

WALKER, R. G. (1966) 'Shale Grit and Grindslow Shales: Transition from turbidite to shallow-water sediments in the Upper Carboniferous of northern England', *Journal of Sedimentary Petrology*, **36**, 90–114.

# PART 2  SEQUENCE STRATIGRAPHY AND SEA-LEVEL CHANGE

# 4  Sequence stratigraphy

## Angela L. Coe and Kevin D. Church

In this Chapter, we will examine the application of a revolutionary new concept that has now been widely applied to the sedimentary record. This concept, called sequence stratigraphy, has evolved over the past few decades and changed the way in which geologists interpret the development of sedimentary environments in time and space. Rather than being based either on correlation of rocks using lithology, fossils or other stratigraphical techniques (Section 2.1), or on facies analysis to construct past sedimentary environments and systems, sequence stratigraphy combines the two approaches and recognizes packages of strata each of which was deposited during a cycle of relative sea-level change and/or changing sediment supply. This genetic approach means that the packages of strata are bounded by *chronostratigraphical* surfaces (Box 2.1). These surfaces include unconformities formed during relative sea-level fall and flooding surfaces formed during relative sea-level rise.

Although the sequence stratigraphical technique was initially developed within the oil industry to help correlate rock units and hence predict new hydrocarbon reserves, sequence stratigraphy is now much more widely used for a number of reasons:

- To try to understand and predict gaps (unconformities) in the sedimentary record.

- To divide the sedimentary record into time-related genetic units, which are useful for stratigraphical correlation and prediction of facies.

- To obtain a holistic view of the distribution of sedimentary facies in time and space.

- To determine the amplitude and rates of past changes in sea-level, and so aid our understanding of the nature of crustal (e.g. fault movement, isostasy, ocean-floor spreading) and climatic processes operating in the past.

- To help identify, classify and understand the complex hierarchy of sedimentary cycles in the stratigraphical record. Sequence stratigraphy is useful for analysing cycles that range, in duration, from the 10 ka to >50 Ma scale.

Sequence stratigraphy is therefore a tool that allows geologists to draw together many different lines of evidence when analysing the fill of a sedimentary basin (Box 3.1). Geologists tend to use the logical thought progression of observation–process–environment to interpret sedimentary rocks. Using sequence stratigraphy, we will show how this thought progression can be taken one stage further to investigate sedimentary environments in both time and space. The predictive nature of the sequence stratigraphy model can greatly aid in integrating and correlating a range of depositional environments at a number of localities. In this way, a wider, more regional picture of the sedimentary deposits can be built up. By combining evidence from both the sedimentary packages and the unconformities, we can thus use sequence stratigraphy to gain a fuller picture of environmental changes documented in the sedimentary record through time.

## 4.1 Sediment accommodation space — principles and controls

Sequence stratigraphy emphasizes the importance of the space that is made available within a basin for sediment to be deposited and the amount of sediment supplied. In order for either marine or non-marine sediment to be deposited, there has to be space available to put it in; this is termed *accommodation space*. The amount of *marine* accommodation space is governed by changes in *relative* sea-level (Figure 4.1). In Sections 3.3 and 3.4, we discussed the fact that changes in relative sea-level are controlled by both eustatic sea-level fluctuations and tectonic subsidence/uplift (Figure 3.6). We also discussed the fact that eustasy and tectonic subsidence act independently of each other, so a eustatic sea-level fall coupled with a higher rate of tectonic subsidence will still result in a relative sea-level rise and hence an increase in accommodation space. For this reason, relative sea-level is used as an all-encompassing term for eustatic and tectonic changes and can be directly equated to changes in marine accommodation space. How tectonic movements and eustatic sea-level also control *non-marine* accommodation space is discussed later in this Section.

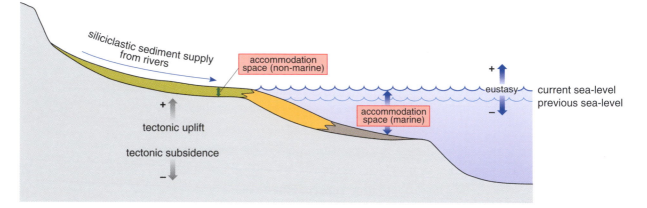

**Figure 4.1** Sediment accommodation space and its relationship to eustatic sea-level and tectonic uplift and subsidence. Marine accommodation space created during a rise in relative sea-level has been partially filled with sediment (yellow and dark-grey), whereas the non-marine accommodation space created during the rise in relative sea-level has been totally filled with sediment (yellowish-green).

If there is zero accommodation space available, the sediments will be transported to an area of (positive) accommodation space where they can be deposited. Thus, areas of zero accommodation space are sites of sediment by-pass. If there is a negative amount of accommodation space, the previously deposited sediments will be eroded and transported to an area of (positive) accommodation space. This is because all sedimentary systems are trying to achieve and then preserve an *equilibrium profile* (or depositional profile) where the available accommodation space is balanced by the amount of sediment supplied.

○ What would happen if the sediment supply increased at a much greater rate than the increase in accommodation space?

● All the accommodation space would be filled with sediment, resulting in a regression and the deposition of a shallowing-upward succession.

Therefore, if either the *rate* of sediment supply or the *rate* of change of accommodation space is altered, the equilibrium will be upset. This will lead to regression (basinward or seaward shift of the shoreline) or transgression (landward shift of the shoreline) in marine areas and re-adjustment of the areas of deposition and erosion in land areas.

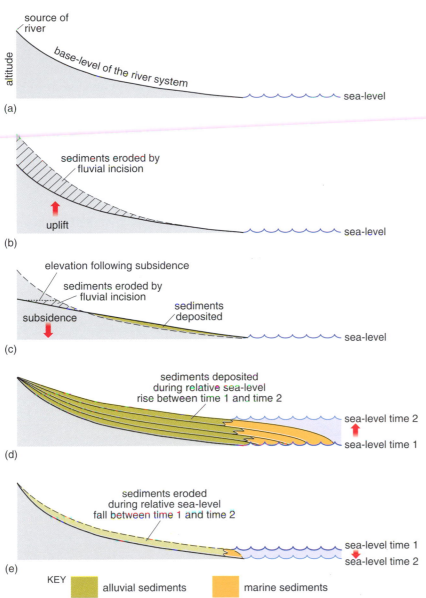

**Figure 4.2**  (a) The equilibrium profile of an alluvial system. In order to maintain the equilibrium profile, erosion or deposition of alluvial sediments will take place if there is a relative sea-level change and/or tectonic movement in the source area. (b) Erosion of sediments due to uplift of the source area. (c) Erosion and deposition of sediments along the alluvial profile due to subsidence of the source area. (d) Deposition of sediments due to a relative sea-level rise. (e) Erosion of sediments due to a relative sea-level fall.

The ideal equilibrium profile of an alluvial system (sometimes referred to as the longitudinal or depositional profile) is an exponentially curved topographical gradient from the source area to sea-level (Figure 4.2a); this curved topographical gradient is a result of the height of the source area and the increase in the rate of discharge downstream. Uplift of the source area will cause the rivers to cut down (or incise) and sediment to be removed (Figure 4.2b). Subsidence of the source area of the river will change the shape of the profile and the river system will re-adjust by eroding and depositing sediment to regain a stable equilibrium profile which will be flatter (Figure 4.2c). Similarly, a change in relative sea-level will also alter the equilibrium profile of the alluvial system. As relative sea-level rises, the position of the coastline with respect to the gradient of the equilibrium profile changes, the river channels infill and hence the gradient flattens out (Figure 4.2d). As relative sea-level falls, alluvial systems will cut a new lower river profile by incision (Figure 4.2e) until the alluvial system profile and sea-level are back in equilibrium. Rivers with an irregularly shaped equilibrium profile, for example due to individual fault movement, will erode and deposit sediment along the profile to establish an equilibrium. The level along the equilibrium profile below which sediment will be deposited and above which sediment will be eroded is referred to as *base level*.

In aeolian systems, the base-level is the water table, so deserts have a fairly flat equilibrium profile. Relative sea-level has a strong influence on the position of the water table, which is a key factor in the preservation of aeolian sediments.

○ How might the position of the water table in deserts affect the aeolian sediments preserved?

● Sediments that are situated below the water table stand a better chance of being preserved because: (i) they are wet and therefore not as easily transported by the wind; and (ii) they are preferentially cemented by minerals in the ground water. Therefore, the sediments situated above the water table are more likely to be eroded and transported.

In the shallow-marine setting, there is also an equilibrium profile governed by the position of high and low tide levels together with fairweather wave-base and storm wave-base (Figure 4.3), each of which defines the specific base-level for the individual coastal zones ranging from the backshore to offshore transition zone. However, when the whole equilibrium profile from the land to the deep sea is considered, these various base-levels are all relatively close to sea-level, so the latter can be effectively taken as base-level in marine environments. This is particularly true for carbonate environments where the greatest supply of carbonate grains and matrix is in shallow-marine areas and all the available accommodation space up to sea-level can be rapidly filled (Chapters 11 and 12). Other base-levels may be present along the continental shelf, for instance, where ocean currents regularly impinge on the sea-floor, but we shall not consider these in any detail here.

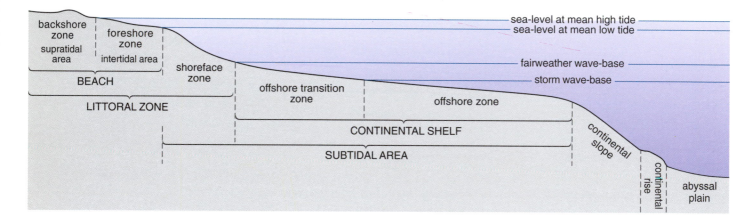

**Figure 4.3** The shallow-marine equilibrium profile and various base levels discussed in the text; for simplicity, sea-level can be taken as the base-level in shallow-marine environments.

Equilibrium profiles in all environments ranging from alluvial to deep sea are perturbed by tectonic and climatic factors which also govern relative sea-level and sediment supply. Sea-level is thus particularly important in affecting deposition and erosion along both non-marine and marine equilibrium profiles. In the majority of circumstances, the amount of accommodation space above sea-level is generally less than it is below. Thus, in siliciclastic environments, sediment is transported from non-marine systems into the nearest accommodation space available, which is the shallow-marine setting. In carbonate environments, the majority of sediment is produced and accumulates in this shallow-marine setting. In the deep sea, below the edge of the continental shelf, there is a large amount of accommodation space but it is mainly underfilled as most of the sediment is deposited in the shallow-marine areas. We can think of shallow-marine environments as the 'sediment traffic jam' area (because it is the carbonate factory where the majority of carbonates are produced, and where the majority of siliciclastics are dumped by rivers). For this reason, the shallow-marine environment is the most sensitive area to changes in accommodation space and sediment supply.

The compaction of sediments over time is an additional mechanism by which there can be an increase in accommodation space. Compaction of sandstones may cause up to 30% reduction in sediment volume, whilst dewatering in mudstones may cause the sediment volume to decrease by up to 80% (compaction is discussed further in Section 5.3). Consequently, even if relative sea-level remains constant and sediment deposition ceases, it may still be possible for accommodation space to increase due to sediment compaction.

From this discussion, it should now be clear that changes in accommodation space and the rate at which it is filled are controlled by the interplay of a number of factors. Climate contributes to eustatic sea-level change through such mechanisms as glacio-eustasy, by controlling the rate of erosion and transportation of sediment from the higher inland areas (or hinterland) to the site of sediment deposition, and by controlling the rate at which carbonate sediment is generated. Tectonic movements determine the relative positions of the sea-bed and land surface and contribute to tectono-eustasy by altering the volume of ocean basins and hence the amount of accommodation space. Tectonic movements also affect sediment supply, and sediment compaction can increase accommodation space.

## 4.2   Filling basins with sediments and the development of parasequences

The simplest way to master sequence stratigraphy is to consider the 'sediment traffic jam' area of the coastal and shallow-marine siliciclastic depositional environments where changes in relative sea-level are easiest to interpret. Relative sea-level and therefore accommodation space, together with sediment supply vary over a number of different time-scales. A 'bottom-up' approach will be used, i.e. the smallest, simplest sequence stratigraphical depositional unit will be considered first. Then we will consider how these small-scale units stack together due to changes in accommodation space and/or sediment supply over longer time periods to form sequences.

The small-scale units are termed *parasequences*, and each of them results from a small-amplitude, short-term oscillation in the balance between sediment supply and accommodation space. To examine how each parasequence forms, we will consider a coastal environment where the rate of increase of accommodation space is less than the rate of sediment supply. As sediment enters the sea, it fills up the most proximal areas first, and then, because there is still more sediment than accommodation space, the sediment will be transported into more distal areas, causing the shoreline to move progressively basinward (or prograde, Figure 4.4a).

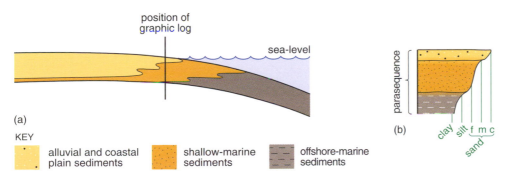

**Figure 4.4**   (a) Cross-section showing progradation of a coastal succession due to the rate of increase in accommodation space being less than the rate of sediment supply. (b) A simplified graphic log of the resultant coarsening-upward succession termed a parasequence.

In keeping with Walther's Law, the facies within this succession will represent conformable deposits that accumulated through time in progressively shallower-water depths, and thus any vertical succession through the parasequence is shallowing-upward and usually coarsening-upward, as higher, younger sediments are deposited under progressively more proximal, higher-energy conditions (Figure 4.4b). If the rate of sediment supply and rate of creation of accommodation space remained constant, the succession would continue to prograde into the basin until the sediment had filled all the available accommodation space.

○ What will happen to the coastal shallowing-upward succession if there is an increase in the rate of creation of accommodation space (i.e. a relative sea-level rise) such that it is greater than the rate of sediment supply?

● The increase in accommodation space will cause transgression of the sea over the succession, terminating its shallowing-upward deposition.

The rock record shows that, in the majority of cases, this transgression event is marked either by a much thinner set of facies representing transgression or, more often, by a distinct surface which caps the succession, termed a *flooding surface* (there are a number of reasons why sediments representing small-scale short-term transgressions are not preserved, and these are described later in this Section). The flooding surfaces represent deepening and they cap each parasequence. Following transgression, a new parasequence will start to build out on top of the first one, utilizing the accommodation space that was newly created as relative sea-level rose.

A parasequence is thus a small-scale succession of relatively conformable beds or bed sets bounded by flooding surfaces. Parasequence thickness is highly variable, ranging from less than a metre to a few tens of metres. The lateral extent of parasequences varies between tens to thousands of square kilometres, depending on the geometry of the depositional area and the characteristics of the particular sedimentary system. They are the smallest bed-scale cycle (rather than laminar-scale) commonly observed in the sedimentary record and the smallest unit usually considered in sequence stratigraphical analysis. Figure 4.5 shows graphic logs of some typical parasequences. For instance, a strandplain succession of conformable beds representing the offshore transition zone, shoreface, foreshore and backshore might form a parasequence (Figure 4.5a), as would the coarsening-upward succession of delta front mouth bar sandstones overlain by the fluvial channel fills of the delta plain (Figure 4.5b) and a conformable succession of intertidal and supratidal carbonates (Figure 4.5c). Most parasequences coarsen upwards, as shallowing produces higher-energy conditions. A few parasequences fine upwards, e.g. in estuarine depositional settings or a muddy tidal flat to subtidal environment (Figure 4.5c); nevertheless these cycles still represent shallowing upwards.

**Figure 4.5 (opposite)** Parasequences from a range of sedimentary environments: (a) a siliciclastic strandplain succession; (b) a deltaic succession; (c) an intertidal to supratidal carbonate ramp succession.

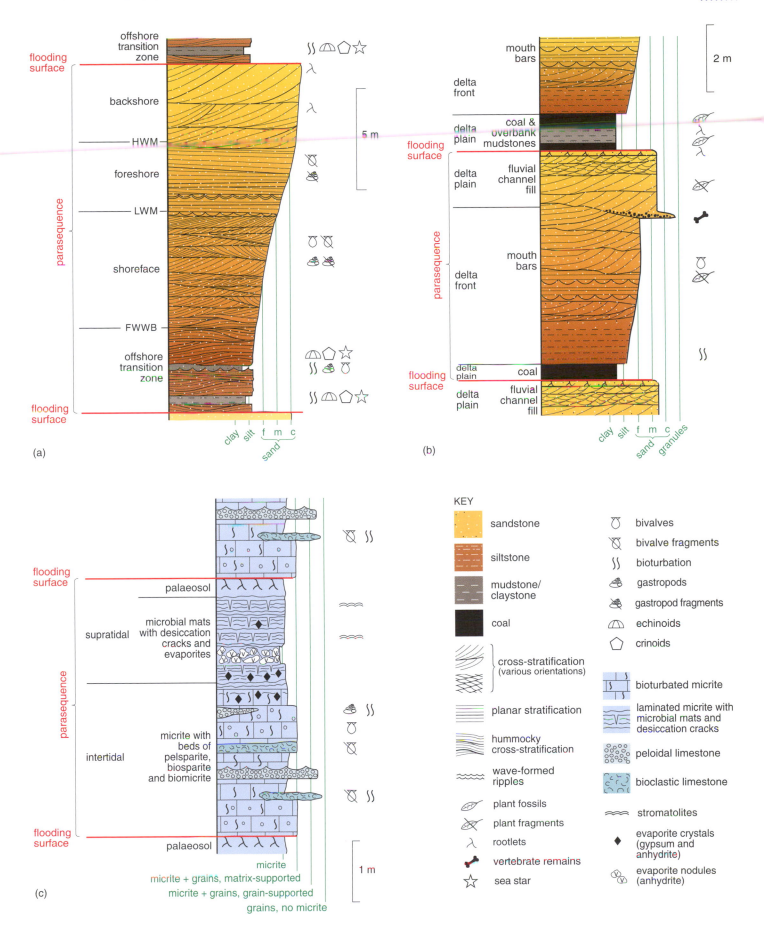

(a)

(b)

(c)

### Reasons why the deepening part of parasequences is not commonly preserved as transgressive sediments

There are several reasons why the deepening part of a parasequence is not represented by a succession of upward-deepening facies. These relate either to the mechanisms by which parasequences may form or the intrinsic conditions produced during deepening. During deepening in shallow-marine areas, wave processes will transport marine sediment landward, thus the sediment will be trapped in more proximal areas and not deposited until the following stillstand or shallowing. In addition, rivers that previously supplied sediment to the area will not incise and produce sediment because of flooding. Parasequences may also be asymmetric if the mechanism producing the accommodation space is subsidence on a fault. This may be an important mechanism in highly tectonically active areas but the movements need to be both repeated and pulsed. Many parasequences are deposited in subsiding areas so the rate of increase in accommodation space due to eustatic sea-level rise or compaction is greatly enhanced by the subsidence. In this case, the sedimentary system does not have a fast enough response time to keep up with rise so no sediments are deposited. Another asymmetric mechanism by which accommodation space is created is through compaction of sediments (Sections 4.1, 5.3). This is because compaction of the sediments can only result in deepening; the newly created accommodation space then progressively fills with sediment and a shallowing-upward succession is deposited.

### 4.2.1 Parasequence sets, progradation, aggradation and retrogradation

Now let us consider the patterns developed when a series of parasequences are superimposed on different longer-term trends in the delicate balance between sediment supply and accommodation space. Each parasequence and its overlying flooding surface represents one cycle of decreasing and increasing accommodation space for a steady sediment supply (Figure 4.4). But if we superimpose these short-term cycles on *longer-term* changes in the rate of sediment supply, or rate of change of accommodation space, it will lead to changes in the stacking pattern of the individual progradational parasequences and cause the next parasequence to start prograding from a different point. If the same longer-term trend continues over several parasequences, then successive parasequences will exhibit similar characteristics in their movement relative to the sediment source area. Successions of parasequences which form distinctive stacking patterns are termed *parasequence sets*. Progradational, retrogradational and aggradational parasequence sets are recognized. These different patterns are explained here with reference to a shallow-marine setting but the same patterns can be observed in other depositional settings.

### Retrogradation

If for each *successive* marine parasequence the increase in accommodation space is *greater* than the constant rate of sediment supply, the deposits of each depositional zone in the successive parasequence will shift landward relative to those in the parasequence below it (Figure 4.6a). Because in this case the shoreline has moved in a landward direction, the parasequences are said to have *retrograded* (or backstepped) and a succession of parasequences showing this pattern is called a *retrogradational parasequence set* (Figure 4.7a). A retrogradational pattern would also result if for each successive parasequence the long-term rate of increase in accommodation space was constant but the rate of sediment supply decreased (Figure 4.7b).

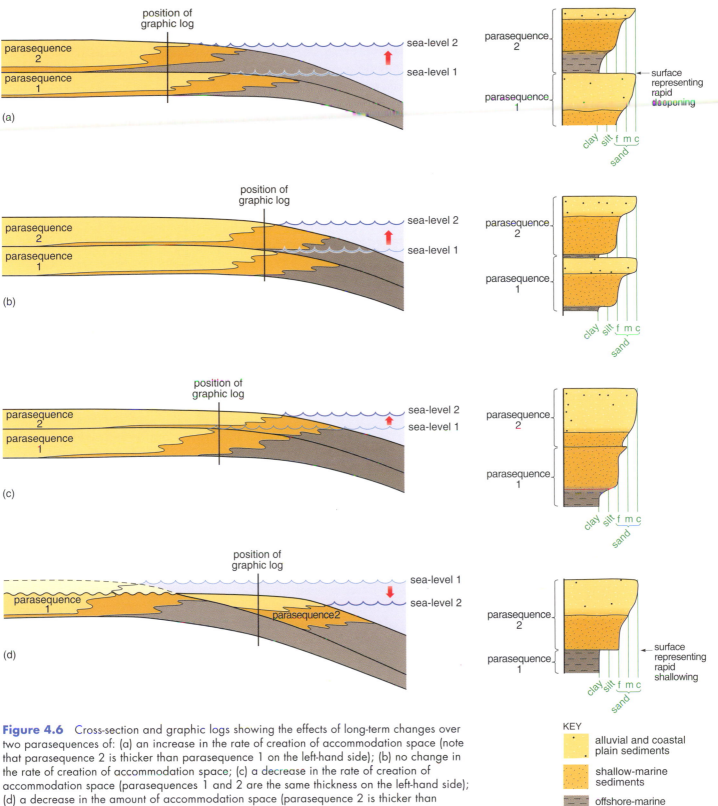

**Figure 4.6** Cross-section and graphic logs showing the effects of long-term changes over two parasequences of: (a) an increase in the rate of creation of accommodation space (note that parasequence 2 is thicker than parasequence 1 on the left-hand side); (b) no change in the rate of creation of accommodation space; (c) a decrease in the rate of creation of accommodation space (parasequences 1 and 2 are the same thickness on the left-hand side); (d) a decrease in the amount of accommodation space (parasequence 2 is thicker than parasequence 1 on the left-hand side). Note that, if the accommodation space continues to decrease, it is unlikely that alluvial and coastal plain sediments will accumulate. The rate of sediment supply is assumed to be constant in each case.

## Aggradation

If for each successive parasequence the increase in accommodation space is *equal* to the rate of sediment supply, the deposits of each depositional zone in the successive parasequence will build out from the same lateral position as the previous parasequence (Figure 4.6b). In this case, the shoreline will have stayed in the same position; the pattern is described as *aggradational* and a group of parasequences showing this pattern form an *aggradational parasequence set* (Figure 4.7c).

## Progradation

If for each successive parasequence the increase in accommodation space is *less* than the constant rate of sediment supply, the deposits of each depositional zone in successive parasequences will shift basinward relative to those in the parasequence below (Figure 4.6c). Because in this case the shoreline has moved in a basinward direction, the parasequences are described to have *prograded* and a succession of parasequences showing this pattern is called a *progradational parasequence set* (Figure 4.7d). Progradation would also result if for each successive parasequence the long-term rate of increase in accommodation space was constant but the rate of sediment supply increased (Figure 4.7e). Depending on exactly how much the rate of increase in accommodation space is less than rate of sediment supply, a spectrum of different types of progradational geometry will result (Figure 4.7d–g). However, because in *every* case the shoreline and facies belts move in a basinward direction, they are all classified as progradational. If there is a decrease in accommodation space between parasequences 1 and 2, but sediment is still being supplied, the more proximal areas will be exposed above sea-level and will therefore be subject to subaerial erosion (Figure 4.6d). Accommodation space will be reduced in all areas, forcing the shoreline to shift rapidly basinward, to a position from which the new marine parasequence will start prograding.

○ What are the relative contributions of accommodation space and sediment supply to the progradational geometry in Figure 4.7e compared to Figure 4.7g?

● In Figure 4.7e, the rate of sediment supply is greater than the rate of increase in accommodation space in order for the parasequence set to prograde and for each parasequence to have the same thickness in the proximal area and a greater thickness in the distal area. However, in Figure 4.7g, there is a decrease in the amount of accommodation space because of a relative sea-level fall. This is independent of the amount of sediment supply.

In fact, Figure 4.7g is the only one to show a relative sea-level fall (a decrease in accommodation space). This leads us on to an important concept, which is the distinction between two types of regression, as explained in Section 4.2.2.

## 4.2.2 Regression and forced regression

In sequence stratigraphy, a distinction is made between '*regression*' and '*forced regression*'. Whilst these are both processes that involve a decrease in accommodation space compared to the rate of sediment supply, they result from different mechanisms. A 'normal' regression (i.e. basinward migration of the shoreline) occurs when the rate of *increase* in accommodation space (through either a relative sea-level rise (Figure 4.7d,e) or a stillstand (Figure 4.7f)) is less than the rate of sediment supply. In this case, all the available accommodation space is filled with sediment and a regressive succession is deposited. Periodic regression and

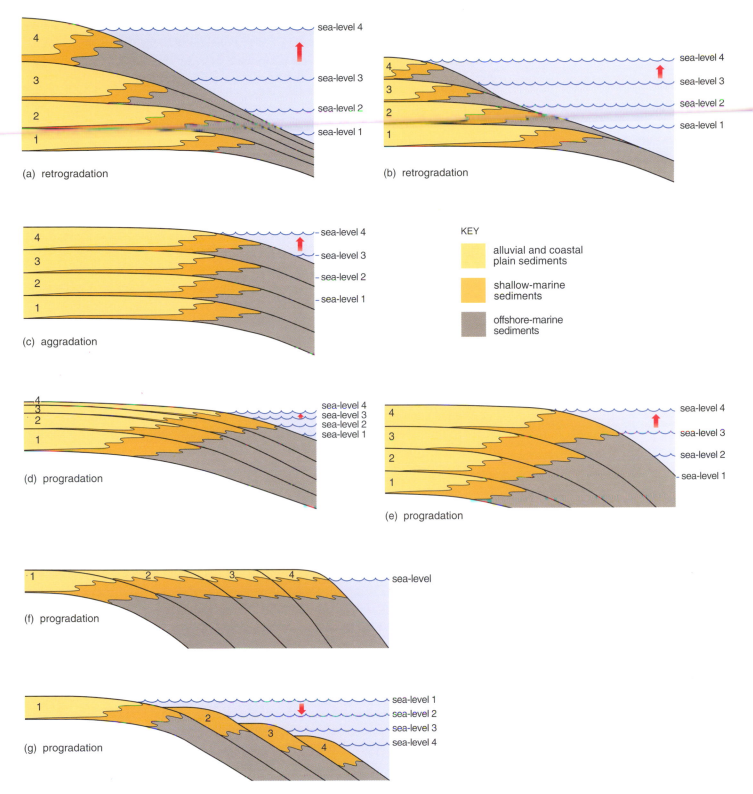

**Figure 4.7** Stacking patterns of parasequences. (a) Retrogradational parasequence set resulting from an increase in the rate of creation of accommodation space for each parasequence that is greater than the constant sediment supply (as shown by the increase in distance between each sea-level). (b) An alternative retrogradational parasequence set resulting from a constant rate of increase in accommodation space (as shown by the equal distance between each rise in sea-level) between parasequences but a decrease in the rate of sediment supply. (c) Aggradational parasequence set resulting from the rate of sediment supply being matched by the rate of increase in accommodation space. (d) Progradational parasequence set resulting from the rate of creation of accommodation space between parasequences being less than the constant rate of sediment supply. (e)–(g) Alternative progradational parasequence sets. (e) Progradational parasequences resulting from a constant rate of increase in accommodation space and an increase in the rate of sediment supply. For (f), there is no long-term increase in accommodation space (i.e. a sea-level stillstand) but there is a continuous sediment supply. See text for explanation of (g).

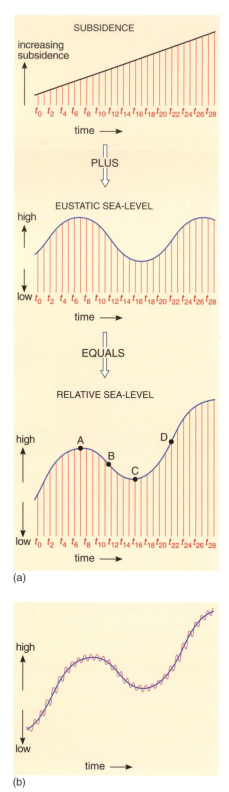

(a)

(b)

transgression may occur, for instance, in a basin with a fairly constant subsidence but fluctuating sediment supply. In contrast, a forced regression is a specific case in which the accommodation space is *decreased* by a relative sea-level fall (which in turn could be due to either the rate of eustatic sea-level fall exceeding the rate of tectonic subsidence or the rate of eustatic sea-level rise being less than the rate of tectonic uplift). Forced regressions are independent of the variations in sediment supply. In every case of forced regression, the shoreline will not only be moved basinward but will also be moved *lower* down the depositional profile than it was previously.

○ Which part(s) of Figure 4.7 results from a forced regression? Give a reason for your answer.

● Figure 4.7g results from a forced regression because this is the only example where the shoreline is lower than it was previously.

Sediment supply, particularly for siliciclastic environments, can be significant during a forced regression, because regardless of any rivers that were already transporting sediment into the basin, additional sediment will be derived by subaerial erosion and fluvial incision into the previously deposited sediments as relative sea-level falls.

Now that we have seen the way in which changes in accommodation space caused by changes in relative sea-level, and changes in sediment supply, control the internal stacking geometries of parasequence sets on the short time-scale, we will step up our scale of observation to investigate how the same controls operate over longer time-scales and at greater amplitudes to control the stacking of successive parasequence sets.

## 4.3 Sequences and systems tracts

A *sequence* or *depositional sequence* is composed of a succession of parasequence sets. Parasequences are the building blocks of sequences. Each sequence represents one cycle of change in the balance between accommodation space and sediment. Some sequences are characterized by forced regressions whereas others show only regression. Sequences generally range in thickness from a few metres to tens or even hundreds of metres in thickness, and they are the next larger (and longer-duration) cycles above parasequences. Similar to parasequences, sequences are the result of either changes in eustatic sea-level, or subsidence/uplift, or changes in sediment supply, or a combination of any of these factors. Every sequence is composed of *up to* four 'systems tracts' each of which represents a specific part in the cyclic change in the balance between accommodation space and sediment supply. Each systems tract is made up of at least one parasequence set. Different conditions may result in one or more of the systems tracts not being developed and/or preserved. This is quite common in the geological record. The concepts presented in Sections 4.2 and 4.3 are used to identify and discuss the evidence for relative sea-level change in the sedimentary record.

**Figure 4.8** Sea-level curves used in Section 4.3: (a) relative sea-level curve derived from addition of a uniform subsidence rate and a sinusoidal change in eustatic sea-level; (b) complex curve (shown in purple) that results from combining the relative sea-level curve in (a) with shorter-term changes in accommodation space associated with the development of parasequences. Equal time divisions are shown by red lines numbered $t_0$, $t_1$, etc. In the example described in this Section, each of these time units is assumed to be the time taken for the deposition of a parasequence. A–D are described in the text, which provides further explanation.

### 4.3.1   The construction of a sequence

The high number of different factors, such as subsidence rate, sediment supply, eustatic sea-level change, climate and lithology type involved in any one geological situation means that the resultant sequences are *highly* variable. However, every sequence has similar genetic components related to changes in the rate of accommodation space creation and sediment supply. We will explore the effect of such changes by considering the idealized stratigraphy resulting from one cycle of relative sea-level change produced by superimposing a sinusoidal change in eustatic sea-level on a uniform rate of subsidence; this is illustrated in Figure 4.8a.

○  Using Figure 4.8a, decide whether or not there will be any forced regressions, and give a reason for your answer.

●  Yes, there will be a forced regression between time 7 ($t_7$) and time 16 ($t_{16}$) because the rate of eustatic sea-level fall shown is greater than the rate of subsidence (note that the lower part of Figure 4.8a shows a decrease in relative sea-level between point $t_7$ and $t_{16}$. This will result in relative sea-level fall, which will cause a forced regression.

For simplification, the changes in accommodation space associated with each parasequence are not shown on the sea-level curves included in later Figures in this Section. In this case, the duration is assumed to be the same for each parasequence. Figure 4.8b shows what the curve would look like if the short-term parasequence-related changes were added. Also for simplification, the rate of sediment supply will be assumed to be constant so that each parasequence contains the same volume of sediment. We will examine two depositional profiles from the land to the sea: (i) a shelf-break margin; that is, a margin with a narrow continental shelf and pronounced change in gradient at the shelf break (Figure 4.9a); and (ii) a ramp which dips at a shallow angle away from the alluvial profile to the deep sea (Figure 4.9b). There are up to seven genetic features formed in the sedimentary record during one sequence cycle which are described in stages in the text and Figures below.

### *Absolute sea-level and rates of sea-level change*

We need to consider both absolute highs and lows in relative sea-level together with *rates* of relative sea-level change. It is important to distinguish between absolute highs and lows of relative sea-level and *rates* of relative sea-level change, because as demonstrated in Sections 4.1 and 4.2 it is the *rate* of change that governs how much new accommodation space is being created (or destroyed) at any one time. To explore these issues, consider the following two questions.

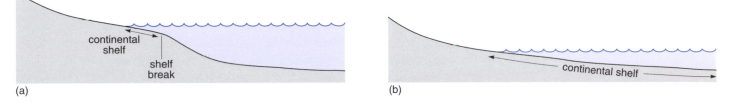

(a)                                                                                      (b)

**Figure 4.9**   Depositional profiles considered in this Section: (a) a shelf-break margin; (b) a ramp which dips at a shallow angle.

○ Where on the sea-level curve in Figure 4.8a is relative sea-level (a) rising at its maximum rate and (b) falling at its maximum rate?

● (a) Although relative sea-level is rising between the trough and the following peak on the sea-level curve, the maximum *rate* of rise is at the mid-point between the two, i.e. at the inflection point (D on Figure 4.8a). It is therefore at this point in time that accommodation space is being created most rapidly. (b) It therefore follows that the maximum *rate* of fall is located at the mid-point between the highest and subsequent lowest relative sea-level (i.e. between the peak and the next trough; marked B on Figure 4.8a).

○ At which points on the sea-level curve is the rate of relative sea-level change zero?

● At the peaks and troughs (i.e. the maximum and minimum, marked A and C on Figure 4.8a) of the curve when relative sea-level is at its highest or lowest respectively. It is here that new accommodation space is being neither created nor destroyed.

## The highstand systems tract (HST)

Consider Figure 4.10: part (b) shows that between $t_0$ and $t_7$, relative sea-level is rising so new accommodation space is being created. Between $t_0$ and $t_2$ for the relative sea-level curve chosen, the amount of new accommodation space created is the same for each of these two parasequences and is balanced exactly by the sediment supply. So, the parasequences deposited from $t_0$ to $t_1$ and $t_1$ to $t_2$ will be stacked aggradationally (note that each individual parasequence will still be progradational). Now consider $t_2$ to $t_3$: relative sea-level is still rising but at a lower rate than between $t_0$ and $t_2$, thus the amount of new accommodation space being created in the proximal areas is less than the amount of sediment being supplied for this parasequence (remember that each parasequence contains the same volume of sediment). Consequently, the sediments deposited between $t_2$ and $t_3$ are transported further into the basin where there is available accommodation space and thus the parasequence progrades relative to the two parasequences below. As relative sea-level approaches its maximum as shown by the curve, the *rate* of relative sea-level rise continues to decrease and thus each parasequence progrades further into the basin (Figure 4.10b). At $t_7$, the rate of relative sea-level rise is zero, and no new accommodation space is created.

This package of sediment deposited between the maximum rate of relative sea-level rise and maximum relative sea-level is termed the *highstand systems tract*, or HST for short. The HST is composed of aggradational to progradational parasequence sets (Figure 4.10c,d). Depending on the conditions, the base of the HST may form later than the maximum rate of relative sea-level rise (see later discussion in this Section).

## The sequence boundary (SB)

Just after $t_7$ (Figures 4.10, 4.11 overleaf), relative sea-level starts to fall; this will result in the sea-level being lower than the top of the coastal sediments deposited at $t_7$ and the equilibrium profile re-equilibrating by eroding into the previously deposited alluvial and coastal facies. Sediment from incised river valleys will be transported out into the sea, where accommodation space is available. Thus, in the alluvial, coastal plain and nearshore areas, no sediment will be deposited and the proximal part of the depositional profile will become an area of sediment by-pass and an unconformity surface will start to form. As relative sea-level continues to

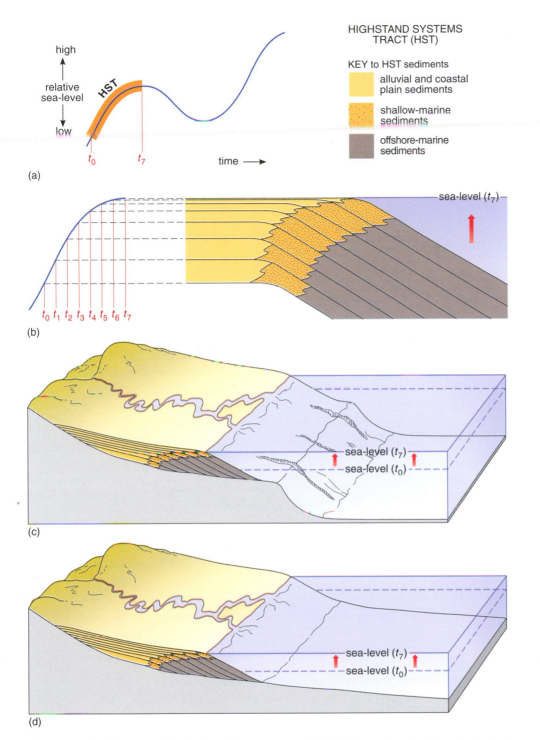

**Figure 4.10** Features of the highstand systems tract (HST). (a) Interval on the theoretical relative sea-level curve shown in Figure 4.8, during which the HST is deposited. (b) Detail of the relative sea-level curve (in blue) and the HST sediments deposited. The curve spans a phase of decreasing relative sea-level rise (i.e. a decrease in rate of creation of accommodation space). The relative sea-level curve is divided into equal time units (red lines $t_1$, $t_2$ etc.). Dashed horizontal black lines in the middle indicate the amount of accommodation space created during each time step in the relative sea-level rise. The right-hand part shows the sediments deposited for each of the equal time intervals, assuming a constant rate of sediment supply. Note how the decrease in accommodation space in the proximal areas results in an aggradational to progradational parasequence stacking pattern. Sediments deposited over the time interval between the maximum rate of sea-level rise and maximum sea-level form the HST. (c) Typical geometry and features of the HST along a margin with a shelf break. Note that in the example illustrated, the HST did not prograde as far as the shelf break. (d) Typical geometry and features of the highstand systems along a ramp margin. (c) and (d) not to scale.

**Figure 4.11** Features of
the sequence boundary (SB).
(a) Interval ($t_7$–$t_{16}$) on the
theoretical relative sea-level
curve where the sequence
boundary forms. $t_7$ = initiation
of sequence boundary
formation and time of
formation of the correlative
conformity. $t_{16}$ = last point
where fluvial incision can take
place. (b) Geometry and
features of the sequence
boundary along a margin
with a shelf break. (b) and (d)
Sediments that may be
deposited simultaneously with
formation of the sequence
boundary are not shown.
(c) Chronostratigraphical
diagram from $t_0$–$t_{22}$ to show
the time represented by the
unconformity surface of the
sequence boundary (shaded
in pink) and its relationship in
time to the correlative
conformity surface. (d)
Geometry and features of the
sequence boundary along a
ramp margin. (e)
Chronostratigraphical
diagram from $t_0$–$t_{22}$ showing
similar features to (c). Note
that for simplicity the
hemipelagic and pelagic
sediments that will be
deposited in the deeper part
of the basin are not shown;
the correlative conformity will
pass through these. Deposits
shown above the correlative
conformity part of the
sequence boundary ($t_7$) in (c)
and (e) are discussed later in
this Section. (b) and (d) are
not to scale.

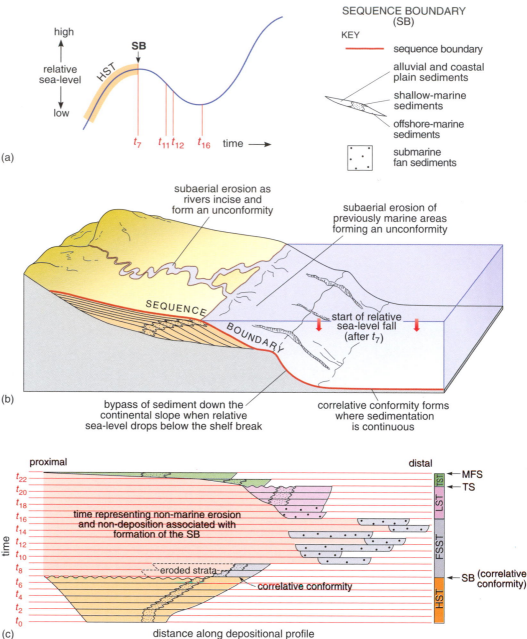

fall, the rivers will continue to incise, and further sediment will be eroded. These
three processes of incision, sediment by-pass and forcing of sediments further and
further into the basin will reach their greatest at the maximum rate of relative sea-
level fall (between $t_{11}$ and $t_{12}$, Figure 4.11a). The three processes, at progressively
reducing rates, will continue until the lowest point in relative sea-level is reached at
$t_{16}$. As relative sea-level starts to rise again, sediment by-pass will continue until the
by-pass area is flooded by the relative sea-level rise, but the downstepping of the
shoreline and incision will cease. What happens to the sediments transported out
into the sea during falling relative sea-level is discussed overleaf (the falling stage
systems tract (FSST)).

In summary, the potential results of this relative sea-level fall will be incised river
valleys, subaerial erosion and exposure of previously deposited marine sediments,
marine erosion of sediments as currents and wave-base impinge lower down on the
sea-floor, and transport and deposition of the eroded sediment to progressively more
basinward positions. In carbonate environments, relative sea-level fall may cause

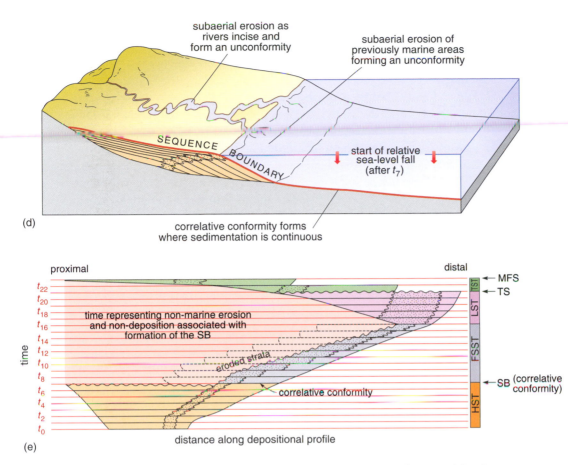

subaerial erosion as rivers incise and form an unconformity

subaerial erosion of previously marine areas forming an unconformity

SEQUENCE BOUNDARY

start of relative sea-level fall (after $t_7$)

(d)

correlative conformity forms where sedimentation is continuous

proximal · · · · · · · distal

time representing non-marine erosion and non-deposition associated with formation of the SB

eroded strata

correlative conformity

time

distance along depositional profile

(e)

MFS — TST
TS — TST
LST
FSST
SB (correlative conformity)
HST

the sediments to dissolve away rather than be redeposited; this is discussed further in Chapter 12. The surface representing this relative sea-level fall, formed during sediment by-pass and erosion, and which will underlie any sediments deposited during the falling relative sea-level, is an unconformity and is termed the *sequence boundary*. The amount of time represented by the sequence boundary will vary along the depositional profile. It will represent the greatest amount of time in the proximal areas where by-passing and erosion occurred earlier, and where the sedimentary deposits that overlie the unconformity were deposited later when the shoreline reached its most proximal location as relative sea-level rose again (see text later in this Section). This chronostratigraphical relationship is illustrated in Figure 4.11c and e. In this representation, the horizontal axis represents distance along the proximal to distal profile and is exactly the same as Figure 4.11b and d respectively. However, the vertical axis represents time instead of thickness. Thus, these chronostratigraphical diagrams * illustrate where and during which time intervals sediment was deposited and preserved and where it was never deposited or was later removed by erosion. This type of diagram is widely used to show the time represented by unconformities or periods of non-deposition and their lateral extent.

○ As the sequence boundary is traced further offshore into the more distal sections, or into an area of higher subsidence, explain if you would still expect it to be marked by an unconformity.

● No, it will not be marked by an unconformity in these areas because this is where most of the sediment will have been transported to. Although accommodation space was also reduced here, it was never zero or negative, so deposition continued unabated during the relative sea-level fall.

* Chronostratigraphical diagrams are also sometimes referred to as Wheeler diagrams, after H. E. Wheeler who wrote a famous scientific paper on them that was published in 1958.

Surfaces that correlate laterally with unconformities, but are not themselves unconformities, are termed *correlative conformities* (Figure 4.11). These can be difficult to identify because there is often no distinct change in facies. For instance, if the water depth at a particular point along the marine depositional profile suddenly changes from 100 m to 90 m (i.e. all below storm wave-base) where mud is being deposited, there is unlikely to be a pronounced change in the composition of the muds above and below the correlative conformity but the features preserved depend on the particular sedimentary system. As the correlative conformity is traced further landward, or into areas of lower subsidence, it may exhibit a slight change in facies either side of the surface (e.g. siltstones gradationally overlying mudstones, or a more subtle change from organic-rich mudstones to organic-poor silty mudstones). The correlative conformity is important for dating the sequence boundary. It lies at the base of any sediments deposited during falling relative sea-level and represents the time at which relative sea-level started to fall.

During a regression, as opposed to a forced regression illustrated here, the sequence boundary would form during the temporary decrease in the *rate* of relative sea-level rise or the stillstand that occurs between the decreasing rate of relative sea-level rise when the highstand systems tract is deposited and the increasing rate of relative sea-level rise when the overlying lowstand systems tract is deposited (see p.77). The sequence boundary in this case would be at the surface where the facies belts reach their most proximal position and the minimal erosion results in the unconformity part being absent or poorly developed.

### The falling stage systems tract (FSST)

The combined effect of relative sea-level fall and hence increased erosion will be to: (i) increase supply of siliciclastics or reworked carbonates into the sea; and (ii) move all the facies belts in a distal direction (i.e. progradation) and, topographically, down the depositional profile (i.e. the shoreline will be at a lower height than in each previous time step (Figure 4.12)). Thus, foreshore sediments may be found unconformably overlying, for instance, offshore transition zone sediments. For siliciclastic environments, sediment will mainly be sourced from rivers that are incising and transporting sediment into the sea in response to the fall in relative sea-level. Carbonate sediments will be produced *in situ* at a lower position on the depositional profile, or derived from erosion of previously deposited carbonates, and under appropriate conditions, downstepping reefs will track falling sea-level (Chapters 12 and 13). Sediment deposited during falling relative sea-level is termed the *falling stage systems tract* (FSST) (Figure 4.12).

**Figure 4.12 (opposite)**   Features of the falling stage systems tract (FSST). (a) Interval on the theoretical relative sea-level curve shown in Figure 4.8 during which the FSST is deposited. (b) Detail of the relative sea-level curve (in blue) and the FSST sediments deposited. The curve spans a phase of increasing and then decreasing rate of relative sea-level fall (i.e. a slow, then rapid, and finally a slow decrease in the amount of accommodation space). The relative sea-level curve is divided into equal time units (red lines $t_7$, $t_8$ etc.). Dashed horizontal black lines in the middle indicate the amount of accommodation space lost during each time step in the relative sea-level fall and hence the potential depth down to which fluvial valleys might incise. The right-hand part shows sediments deposited for each of the equal time intervals assuming a constant rate of sediment supply, that sediment continues to be deposited, and that the depositional profile is gently sloping. The decrease in accommodation space in proximal areas results in a progradational parasequence set. Sediments deposited over the time interval between maximum and minimum relative sea-levels form the FSST. The tops of the previously deposited parasequences in the FSST are eroded as relative sea-level continues to fall. (c) Geometry and features of the FSST along a margin with a shelf break. In this case, the first two parasequences of the FSST are deposited on the continental shelf as downstepping parasequences ($t_7$–$t_8$ and $t_8$–$t_9$). It is assumed that in this case after $t_9$ relative sea-level fell below the shelf break and therefore that all further sediment was transported down the steep continental slope into the deep sea where it was deposited as submarine fans on the basin floor. (d) Geometry and features of the FSST along a ramp margin showing the deposition of an attached FSST. (c) and (d) are not to scale and show a slightly greater relative sea-level fall than (a) and (b) in order to show all the features.

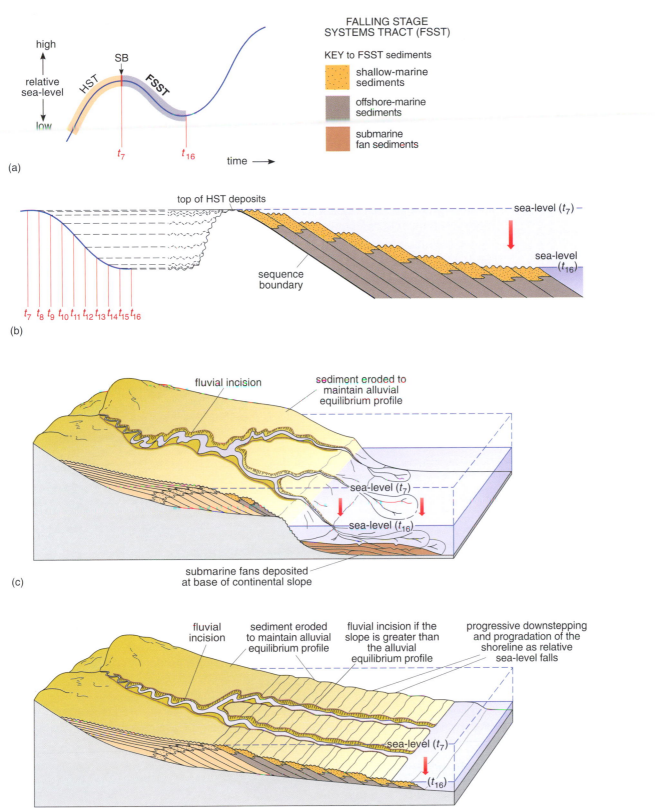

(a)

FALLING STAGE
SYSTEMS TRACT (FSST)

KEY to FSST sediments

- shallow-marine sediments
- offshore-marine sediments
- submarine fan sediments

(b) top of HST deposits — sequence boundary

(c) fluvial incision — sediment eroded to maintain alluvial equilibrium profile — submarine fans deposited at base of continental slope

(d) fluvial incision — sediment eroded to maintain alluvial equilibrium profile — fluvial incision if the slope is greater than the alluvial equilibrium profile — progressive downstepping and progradation of the shoreline as relative sea-level falls

The preservation, geometry and lateral position of the FSST is variable, depending on the shape of the depositional profile, the magnitude and rate of the relative sea-level fall, the rate of sediment supply and the changes of sedimentary process that occur as relative sea-level falls and a larger area is subaerially exposed (Figure 4.13). For shelf-break margins, if the HST sediments have not filled all the space available on the continental shelf, downstepping parasequences will be deposited as shown for the first two parasequences $t_7$–$t_9$ in Figure 4.12c. However, when relative sea-level falls below the shelf edge the relatively steep depositional profile results in sediment from the incised river valleys accumulating at the base of the continental slope as submarine fans (Figure 4.12c, $t_9$–$t_{16}$). In contrast, for a ramp type margin, with its much more shallow gradient, the effect of a similar magnitude sea-level fall can be much more dramatic. On ramp type margins, the whole depositional system can be forced basinward by as much as several kilometres (Figure 4.13a) because the amount of nearby accommodation space lost compared to the total amount of accommodation space available is much greater than on the shelf break margins (Figure 4.14). Because of this wide potential geographical separation of the HST and FSST, many sediments deposited during forced regression on ramp margins have probably gone unrecognized. The relative sea-level fall can alternatively result in the deposition of parasequences all along the depositional profile distal from the last highstand shoreline, each one progressively downstepping as relative sea-level falls (Figures 4.12b,d and 4.13b,c).

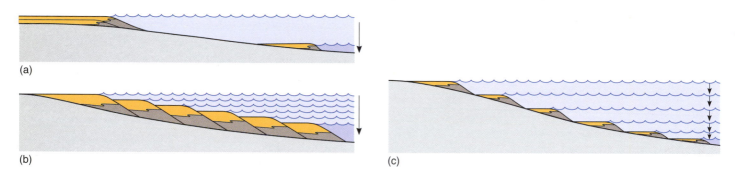

(a)

(b)

(c)

**Figure 4.13** Different types of FSST. The parasequences shown on the left are older than those shown on the right of each profile. (a) A detached systems tract. (b) An attached FSST. (c) Another type of detached FSST. Arrows show relative sea-level falls. See text for further explanation. *(Posamentier and Morris, 2000.)*

○ Figure 4.13b and c shows two possible scenarios for a ramp of the same dip. Which part of the Figure represents a higher sediment supply or a slower rate of relative sea-level fall?

● Figure 4.13b represents a higher rate of sediment supply or a slower rate of relative sea-level fall because the parasequences gradationally downstep whereas in Figure 4.13c the distance that each parasequence downsteps is much greater, resulting in each one being isolated from the next.

The parasequences shown in Figure 4.13a and c are sometimes called stranded parasequences or a *detached* FSST because the shoreline in each case is separated from the previous one. In contrast, the parasequences in Figure 4.13b make up an *attached* FSST.

In situations where, during the forced regression, sediment supply is low to moderate, or the rate of relative sea-level fall is particularly high, or there is continued erosion along the depositional profile to the lowest sea-level, a FSST will not be preserved. The lack of a FSST is quite common in the geological record. In situations where there is a regression rather than a forced regression, there will *never* be a FSST, because this systems tract is solely a consequence of forced regression.

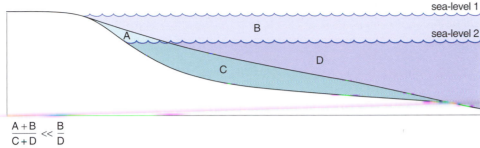

$$\frac{A+B}{C+D} \ll \frac{B}{D}$$

**Figure 4.14**   Sketch to show that the proportion of coastal accommodation space lost during a relative sea-level fall is much greater for a ramp type margin than a shelf break margin. A–D are volumes of water. A + B = volume of water lost after the relative sea-level fall on a shelf break margin; C + D = volume of water remaining after the relative sea-level fall on a shelf break margin; B = volume of water lost after the relative sea-level fall on a ramp type margin; D = volume of water remaining on a ramp type margin after the relative sea-level fall.

### The lowstand systems tract (LST)

At time $t_{16}$ on the theoretical sea-level curve (Figure 4.8), relative sea-level fall reaches its minimum and accommodation space is neither being created nor destroyed. But, by $t_{17}$, a small amount of accommodation space has been created; this allows the shoreline to start building upwards from its lowest position. Progradational marine sediments will be deposited and the fluvial system will cease to incise (Figure 4.15a,b overleaf). This process will continue until there is a much more pronounced increase in the amount of accommodation space which for the theoretical curve shown in Figures 4.8a,b and 4.15 (overleaf) is at $t_{21}$. The package of sediment deposited between the minimum relative sea-level and the pronounced increase in accommodation space is termed the *lowstand systems tract* (LST); this is composed of progradational to aggradational parasequence sets. In the case of the shelf break margin, the LST may comprise one or more submarine fans deposited on the shelf slope on top of the recently formed slope of the FSST submarine fans. These slope fans may be overlain by one or more progradational parasequences as the shoreline progrades again over the top of the gently dipping slope fan (Figure 4.15c). For ramp type margins, progradational to aggradational parasequences will build up above the last FSST shoreline (Figure 4.15b,d).

### The transgressive surface (TS)

As relative sea-level begins to rise at a significant pace, it will reach a point where the long-term rate of creation of accommodation space is greater than the rate of sediment supply and there will be a transgression. The locus of sedimentation will be shifted in a landward direction and there will be deposition of a retrogradational parasequence or parasequence set (the transgressive systems tract, see p.80). The base of these retrogradational parasequences marks the *transgressive surface* (TS). In the scenario that we have used, this will start to form at $t_{21}$ (Figure 4.16a overleaf). The transgressive surface is often particularly well developed in the coastal environments of the shoreface and foreshore where the rise in relative sea-level results in minor erosion and reworking of the sediments from increased wave-, tide-, and storm-induced activity resulting in the formation of a minor unconformity (see erosion surface at $t_{21}$ marked on Figure 4.11c,e). In the proximal (i.e. landward) areas, the transgressive surface may be marked by marine sediments overlying non-marine sediments (Figure 4.16).

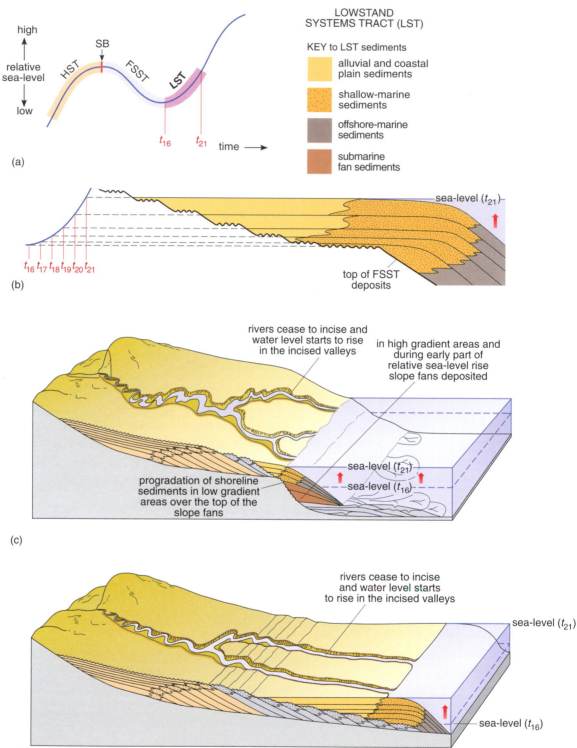

**Figure 4.15** Features of the lowstand systems tract (LST). (a) Interval on the theoretical relative sea-level curve shown in Figure 4.8, during which the LST is deposited. (b) Detail of the relative sea-level curve (in blue) and the LST sediments deposited. The curve is over a phase of slowly rising relative sea-level (i.e. an increase in the rate of creation of accommodation space). The relative sea-level curve is divided into equal time units (red lines $t_{16}$, $t_{17}$ etc.). Dashed horizontal black lines in the middle indicate the amount of accommodation space created during each time step in the relative sea-level rise. The right-hand part shows sediments deposited for each of the equal time intervals assuming a constant rate of sediment supply and that the depositional profile is gently sloping. The sediments deposited over the time interval between the minimum relative sea-level and the more pronounced increase in sea-level form the LST. (c) Geometry and features of the LST along a margin with a shelf break. Initially, submarine slope fans may be deposited (shown as $t_{16}$–$t_{18}$) until the gradient is low enough and sea-level is high enough that the shoreline can prograde out into the basin ($t_{18}$–$t_{21}$) and coastline sediments can be deposited. (d) Geometry and features of the LST along a ramp margin. (c) and (d) are not to scale.

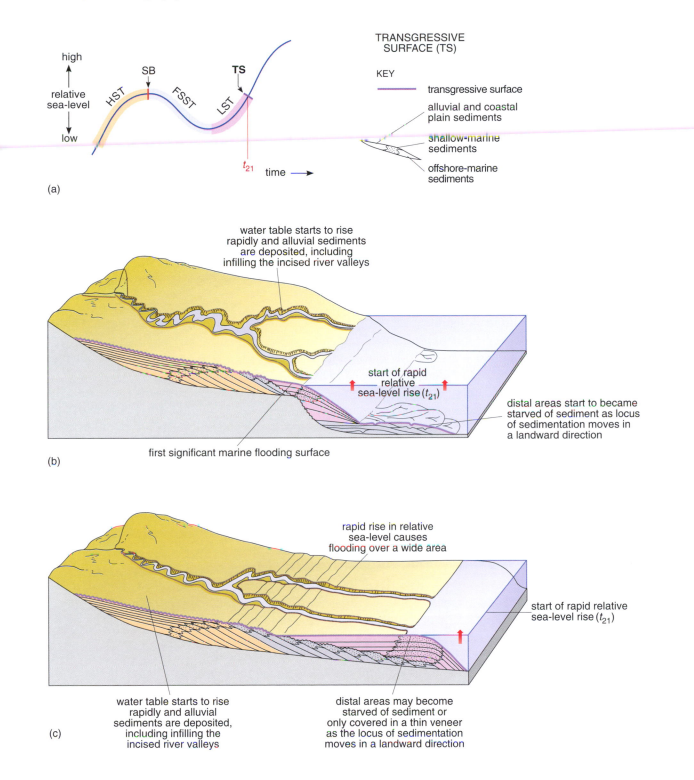

**Figure 4.16** The geometry and features of the transgressive surface (TS). (a) Position on the theoretical relative sea-level curve where the transgressive surface starts to form. (b) Geometry and features of the transgressive surface along a shelf break margin. (c) Geometry and features of the transgressive surface along a ramp margin. (b) and (c) not to scale.

### The transgressive systems tract (TST)

The sediments immediately overlying the transgressive surface form the *transgressive systems tract* (TST) and are all deposited during the interval when the rate of increase in accommodation space is greater than the rate of sediment supply (in this case, $t_{21}$–$t_{23}$, Figure 4.17). Note that more sediment is usually deposited in proximal areas than distal areas during this period because the locus of sedimentation has been moved in a proximal direction (Figure 4.17b–d). The facies belts and parasequences of the TST will retrograde; retrograding parasequences are diagnostic of the TST in most siliciclastic and some carbonate environments. In the case of carbonate depositional environments, transgression often results in large shallow-marine areas being flooded and made available for colonization by carbonate-producing communities such as coral reefs. This can lead to an *increase* in the rate of carbonate production and deposition which will result in aggradational or even progradational parasequence stacking patterns (see Chapters 11 and 12). Transgression of the sea will lead to infilling of the incised valleys created during falling relative sea-level.

Variations in sediment supply, the rate of relative sea-level rise and the exact nature of the depositional profile determine the type of deposits in the TST and its timing. During transgression, siliciclastic sediment supply tends to be lower than at other times because sediment is trapped in proximal areas and there is no incision. If sediment supply is low, the TST may be thin or even absent, or comprise reworked sediments rich in fossils. Over continental shelves or in shallow seas, such as those that existed for much of the Mesozoic, deposition of organic carbon-rich mudrocks was common and these deposits are often interpreted to represent TSTs. This is because relative sea-level rise is thought to have led to increased organic productivity as more nutrients were available from the newly flooded area. The earliest time the TST can stop forming is at the maximum rate of relative sea-level rise ($t_{22}$) but because sediment supply is here assumed to be constant and we have a sinusoidal curve, the top and bottom of the TST will be symmetrical about $t_{22}$.

### The maximum flooding surface (MFS)

As the rate of relative sea-level rise increases, distal parts of the depositional profile may be completely, or almost completely, starved of siliciclastic sediment, because the locus of sedimentation has moved so far landward that no, or very little, sediment reaches the deeper parts of the basin (Figure 4.18 overleaf). This starvation reaches its most landward position between the maximum rate of relative sea-level rise and the maximum sea-level ($t_{22}$ and $t_{29}$ respectively on our theoretical relative sea-level curve), depending on the particular conditions. In this case, it occurs at $t_{23}$. Sediment starvation in the distal area will continue longer than in the proximal area (pink area on Figure 4.18c and e). The low sediment supply results in formation of a condensed bed; such beds are often highly fossiliferous as there is less sediment to 'dilute' the fossils which continue to be deposited on the sea-floor. The fossils' preservation potential is increased by the likelihood of cementation and precipitation of authigenic minerals like phosphate, as sedimentation rate decreases. The top of the condensed section and/or submarine unconformity is termed the *maximum flooding surface* (MFS) (Figure 4.18). In proximal areas, the MFS is associated with the most landward position of the shoreline and the most extensive marine sediments (e.g. thin brackish or marine sediments in deltaic successions). Marine sediments are often deposited in proximal areas which have previously been entirely non-marine. In addition, flooding of more proximal areas will cause the most pronounced rise in the water table, affecting deserts and causing further realignment in the depositional profile of alluvial systems. If deltas are present, the

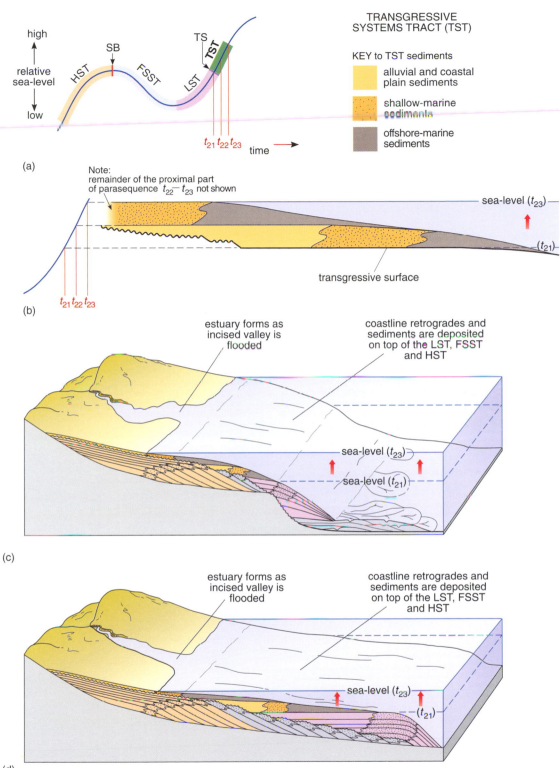

**Figure 4.17** Features of the transgressive systems tract (TST). (a) Interval on the theoretical relative sea-level curve shown in Figure 4.8 during which the TST is deposited. (b) Detail of the relative sea-level curve (blue) and TST sediments deposited. The curve spans a phase where increase in the rate of relative sea-level rise is greater than the rate of sediment supply. The relative sea-level curve is divided into equal time units (red lines $t_{20}$, $t_{21}$ etc.). Dashed horizontal black lines in the middle indicate the amount of accommodation space created during each time step in the relative sea-level rise. The right-hand part shows sediments deposited for each of the equal time intervals assuming a constant rate of sediment supply. These parasequences show a retrogradational pattern. Sediments deposited over the time interval between pronounced increase in the rate of creation of accommodation space and maximum rate of relative sea-level rise form the TST. (c) Geometry and features of the TST along a margin with a shelf break. (d) Geometry and features of the TST along a ramp margin. (c) and (d) not to scale.

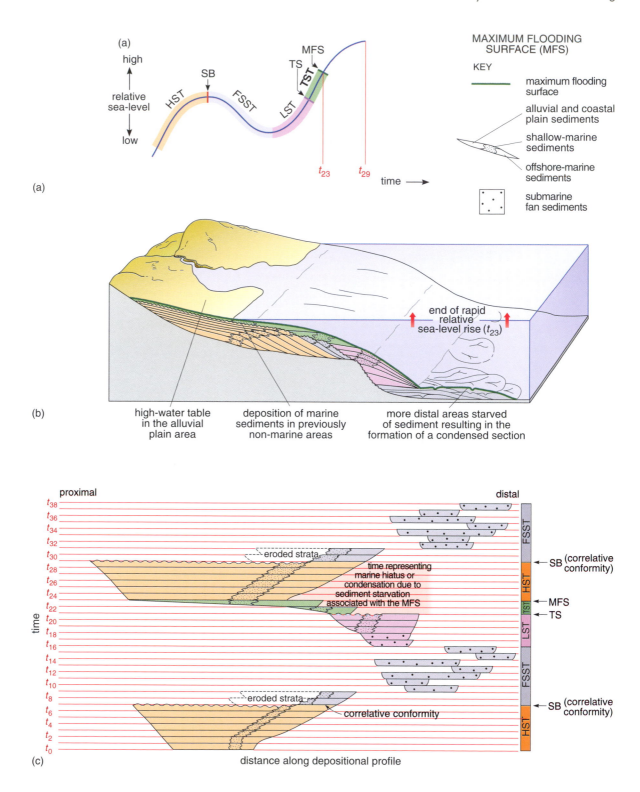

(a)

MAXIMUM FLOODING
SURFACE (MFS)

KEY

— maximum flooding
surface

alluvial and coastal
plain sediments

shallow-marine
sediments

offshore-marine
sediments

submarine
fan sediments

(b)

high-water table
in the alluvial
plain area

deposition of marine
sediments in previously
non-marine areas

more distal areas starved
of sediment resulting in the
formation of a condensed section

end of rapid
relative
sea-level rise ($t_{23}$)

(c)

proximal                                           distal

time representing
marine hiatus or
condensation due to
sediment starvation
associated with the MFS

eroded strata

eroded strata

correlative conformity

SB (correlative
conformity)

MFS
TS

SB (correlative
conformity)

distance along depositional profile

rise in the water table, due to maximum flooding, will increase the occurrence of swamps, flooding of floodplains and avulsion* of distributary channels. This often leads to the formation and preservation of peat in proximal areas because the waterlogged conditions promote anoxia, thus reducing the chance of the organic matter being oxidized. The next river avulsion or relative sea-level fall will deposit siliciclastic sediments on the peat and following burial this will become preserved as coal.

* Lateral displacement of stream/river from its main channel into a new course on the floodplain.

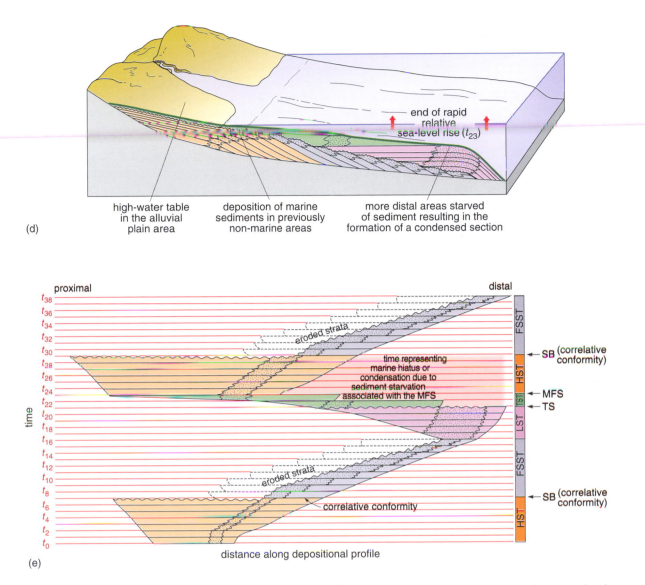

(d)

high-water table in the alluvial plain area

deposition of marine sediments in previously non-marine areas

more distal areas starved of sediment resulting in the formation of a condensed section

end of rapid relative sea-level rise ($t_{23}$)

(e)

**Figure 4.18** Features of the maximum flooding surface (MFS). (a) Position ($t_{23}$) on the theoretical relative sea-level curve shown in Figure 4.8 where the maximum flooding surface forms in this case. The exact position of the MFS depends on the balance between the rate of relative sea-level rise and rate of sediment supply. (b) Geometry and features of the maximum flooding surface along a margin with a shelf break. (b and d) Sediments that may be deposited simultaneously with formation of the maximum flooding surface are not shown. (c) Chronostratigraphical diagram from $t_0$–$t_{38}$ to show the time represented by the maximum flooding surface and condensed section (shaded pink). (d) Geometry and features of the maximum flooding surface along a ramp margin. (e) Chronostratigraphical diagram from $t_0$–$t_{38}$ showing similar features to (c). Note that for simplicity the hemipelagic and pelagic sediments that will be deposited in the deeper part of the basin are not shown. (b) and (d) not to scale. See text for further explanation.

When the long-term rate of increase in accommodation space is again balanced or exceeded by the rate of sediment supply, in this case at $t_{23}$, then the next highstand system tract will be deposited. The earliest this can start to happen is just after the maximum rate of relative sea-level rise and the latest is at maximum relative sea-level ($t_{22}$ and $t_{29}$ respectively in this case).

The cycle will then repeat itself again. The package of sediment between two successive sequence boundaries forms a *depositional sequence* (or commonly just a '*sequence*'). The sequence stratigraphy model presented here is the most current at the time of writing. A brief summary of how the model has been updated and variations on it are discussed in Box 4.1 (p.86).

### 4.3.2 Field examples of key surfaces and systems tracts

Figure 4.19 shows photographs of several features of depositional sequences in the field from a variety of depositional settings in southern England.

○ Assuming that Figure 4.19a and b are equivalent in age, and that they were deposited in areas with the same subsidence history, which is the more proximal section and why?

● Figure 4.19b is likely to be the more proximal section because at this locality the sequence boundary is marked by a sharp surface (possibly an unconformity) whereas in Figure 4.19a it is a correlative conformity which is always more basinward than the unconformity.

(a) Cliff *c.* 40 m high.

(b) Cliff *c.* 3 m high.

(c) Cliff *c.* 16 m high.

(d) Cliff *c.* 2.5 m high.

**Figure 4.19** Features of depositional sequences. (a) Succession showing upward gradational increase in amount of silt and fine-grained sand interpreted to represent a correlative conformity, Kimmeridgian, Dorset. (b) Sharp surface interpreted as a sequence boundary between open marine mudstones and shallow-marine/estuarine sandstones, Oxfordian, Dorset. (c) Micrites and biomicrites with chert nodules (black) overlain by cross-stratified oosparites (white bed at clifftop) from a carbonate ramp succession. The base of the oosparites is interpreted as a sequence boundary as the ooids were deposited in much shallower water than the biomicrites, Portlandian, Dorset. (d) Intensely bioturbated biomicrites containing a very rich, diverse fauna; interpreted as a condensed TST from a carbonate ramp succession, Portlandian, Dorset.

(e) Ammonite = 40 cm diameter.

(f) Nodule = 8 cm across.

(g) View c. 25 cm across.

(h) Ruler = 25 cm length.

**Figure 4.19 (continued)**   (e) An oyster and serpulid encrusted ammonite indicating condensation from the maximum flooding surface at the top of the TST shown in (d), Portlandian, Dorset. (f) A carbonate nodule from within a mudrock succession. The formation of nodules is often related to a pause or decrease in the rate of deposition. These nodules are interpreted to mark a maximum flooding surface, Kimmeridgian, Dorset. (g) Cross-section through a hardground within a chalk succession showing nodules and possibly pebbles coated with the authigenic minerals glauconite (green) and phosphate (yellow). There is intense bioturbation and some of the nodules have also been bored. The top of this bed is interpreted to represent a maximum flooding surface, Cretaceous, Isle of Wight. (h) A shallow-marine shell bed containing a mixture of grains including mudstone pebbles, ooids and quartz sand grains representing the transgressive lag formed during shoreface erosion at the base of a TST, Oxfordian, Dorset. ((a)–(f), (h) Angela Coe, Open University; (g) Simon Grant, British Petroleum.)

○  What does the evidence in Figure 4.19g indicate about (a) the sedimentation rate and (b) the time represented by the bed?

●  (a) The precipitation of authigenic minerals indicates a very low to zero sedimentation rate. (b) The bed probably represents quite a long period of time because of the boring and the intense authigenic mineralization. Also some of the nodules appear to have been rolled around to form pebbles.

○  Figure 4.19d contains a rich and diverse fauna, some of which is broken and reworked and some of which is not. How does this evidence fit in with the interpretation that it is a TST?

●  The rich and diverse fauna, both broken and whole, fits with reworking of sediment during transgression when the shells would become broken and there would be a relatively low sediment input as most of the sediment was probably trapped in more proximal areas during relative sea-level rise.

## Box 4.1   Controversies in the sequence stratigraphy model

### Three systems tracts versus four, Type 1 and Type 2 sequence boundaries

In the original sequence stratigraphical model devised in the late 1970s, and for over a decade after, only three systems tracts were recognized: lowstand, transgressive and highstand systems tracts. The highstand systems tract was interpreted to form between the maximum rate of relative sea-level rise and maximum rate of relative sea-level fall. Thus, the HST included the lower part of the falling stage systems tract where it was preserved. The lowstand systems tract was interpreted to form between the maximum rate of relative sea-level fall and start of the rapid rate of rise (marked by the transgressive surface).

In the earlier versions of the model, two types of sequence boundary were recognized (Types 1 and 2). Type 1 boundaries were interpreted to form when there was a relative sea-level fall (now termed a forced regression) at the depositional-shoreline break[*]. In the specific case of Type 1 sequence boundaries that formed on shelf break margins, the 'lowstand systems tract' was subdivided into the basin-floor fan (equivalent to the upper part of the falling stage systems tract, Section 4.3.1), slope fan and lowstand wedge (equivalent to the lowstand systems tract, Section 4.3.1). Type 2 boundaries were interpreted to form when there was what is now termed a regression at the depositional-shoreline break. In this case, the lowstand systems tract was referred to as the shelf-margin wedge. Lowermost systems tract was yet another term used all-inclusively to describe the different 'lowstand' deposits.

In the early 1990s, it was noted that the model failed to account properly for sediment packages deposited during relative sea-level fall on a ramp type margin, like those shown in Figure 4.13, and particularly for features of carbonate successions. Thus, the distinction between forced regressions and regressions was noted and received prominence; several researchers then proposed a fourth systems tract to account for sediments interpreted to have been deposited during falling relative sea-level rather than just low sea-level. This systems tract has had various names but is now commonly referred to as the falling stage systems tract. Many examples of falling stage systems tracts, in addition to basin-floor submarine fans, are now recognized; for examples see Sections 6.4, 9.2 and 13.1. Research on other successions together with computer modelling has led to wider acceptance of the falling stage systems tract. However, some authors still divide their sequences deposited during forced regressions into three systems tracts.

[*] The depositional-shoreline break is the position on the shelf proximal from which the depositional surface is above base level and distal from which the depositional surface is below base level.

### Position of the sequence boundary

In the 1970s' to 1980s' model, much emphasis was placed on the fact that the sequence boundary formed at the maximum rate of relative sea-level fall. Though interesting, some of the chronostratigraphical figures from these papers do not show this but do fit well with the model presented in Section 4.3.1. Whilst most researchers now accept the four systems tract model for forced regressions, there is still some controversy as to where the sequence boundary should be placed. The three options are:

(i)   At the base of the falling stage systems tract (Section 4.3.1). In this case, the correlative conformity forms at the start of the relative sea-level fall and equates to the start of erosion in the proximal areas and therefore the onset of formation of the unconformity.

(ii)   At the top of the falling stage systems tract. In this position, the correlative conformity forms at the absolute low of relative sea-level and is equivalent to the lowest surface of incision in the proximal sections. In this case, the unconformity starts to form before the formation of the correlative conformity. The problem with putting the sequence boundary in this position is that its timing is dependent on the position along the depositional profile, correct distinction between the falling stage and lowstand systems tract, and the sedimentation rate.

(iii)   At the maximum rate of relative sea-level fall. In this case, the timing is hard to define in proximal sections because it lies in the middle of the unconformity; however, the maximum rate of relative sea-level fall in some cases equates to the time when there was the most pronounced major basinward shift of facies.

### Genetic stratigraphic sequences

An alternative sequence stratigraphical model was proposed by Galloway (1989) in which the maximum flooding surfaces were chosen as the boundaries to the packages of sediment rather than the sequence boundaries. He termed the sediment packages between maximum flooding surfaces 'genetic stratigraphic sequences'. Though this model has raised much discussion because, often, maximum flooding surfaces are prominent and easily correlated over wide areas, the model has not been widely accepted because it results in the unconformity being in the middle of the package. Nor does it give prominence to the only key surface that is independent of sediment supply (i.e. the sequence boundary). For further discussion of this model, see references in the further reading list.

Miall (1997) and Posamentier and Morris (1999) both contain a description of the history of the sequence stratigraphy model and describe most of the controversies.

### 4.3.3 Natural variability and summary of the features of sequences

The four systems tracts — FSST, LST, TST and HST — are all composed of several co-existing and linked depositional systems or environments ranging from the deep marine to coastal and fluvial systems. Therefore, sequences do not divide the sedimentary record on the basis of different sedimentary systems or facies belts but rather on key surfaces (the sequence boundary, transgressive surface and maximum flooding surface) which each represent a particular time at which *all* the depositional systems change in response to a particular change in the relative sea-level cycle. The exact nature of the systems tracts and key surfaces within the depositional sequence depend upon: (i) the shape of the equilibrium profile of the sedimentary systems; (ii) the rate and amplitude of relative sea-level change; (iii) the sedimentary system present (e.g. delta, strandplain or carbonate ramp); the rate of sediment supply.

Consequently, systems tracts and sequences will vary in thickness, internal character and in the time taken to deposit them, depending on the combined effects of these different factors. Systems tracts may often only be partially preserved or in fact be entirely absent. In these cases, key surfaces become superimposed. Some of the variations in the sequences resulting from the first three points listed above have been introduced earlier in this Chapter and will be covered in the case studies later in the book. Sediment supply is considered in a bit more detail below.

#### Sediment supply and its control on sequence architecture

If we consider three extreme situations in which sediment supply is either zero, very high, or highly variable, it is not hard to see that the morphology of the systems tracts and sequences in our model will alter dramatically. Quite simply, if there is no sediment supply, there can be no deposition of new sediment, regardless of what relative sea-level is doing. So what will happen during one cycle of relative sea-level rise and fall in a case where sediment supply is zero? As relative sea-level falls, the previous depositional surface will be eroded as the wave-bases are lowered, in addition to the subaerial erosion of newly exposed rock. The eroded products will be deposited on the sequence boundary as the FSST and LST. As relative sea-level rises, marine erosion of the FSST and LST may occur in which case all that will be left is a thin coarse-grained 'lag' deposit representing the TST. The HST will be limited to thin, condensed intervals, formed from whatever debris settled out of the water column. Such conditions will lead to a condensed sequence or simply an unconformity surface.

If sediment supply is very high, accommodation space will quickly be filled, even during the highest rates of relative sea-level rise. This could mean that there is too much sediment entering the basin for aggradation and retrogradation to occur. Thus, rather than the sequence being identifiable on the basis of its progradational, aggradational and retrogradational parasequence stacking patterns, systems tracts would instead only exhibit varying degrees of progradation (Section 13.1 describes an example with no retrogradation).

In general, the rate of siliciclastic sediment supply will drop during a relative sea-level rise, because flooding of the land reduces the potential for subaerial erosion close to the shoreline. Conversely, a relative sea-level fall elevates the land with respect to the sea, exposing areas formerly submerged. The potential for subaerial erosion is therefore enhanced and so is siliciclastic sediment supply. For carbonate environments, the situation can be very different. The production of carbonate in shallow-marine areas can keep pace with relative sea-level rise, and because of the increase in flooded area the supply of carbonate sediment increases. During relative sea-level fall, the carbonate production area often decreases, areas that are subaerially exposed become cemented

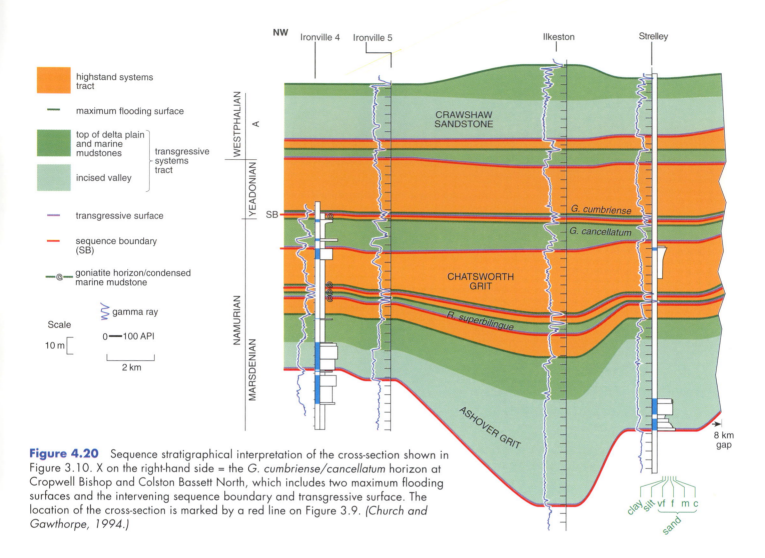

**Figure 4.20** Sequence stratigraphical interpretation of the cross-section shown in Figure 3.10. X on the right-hand side = the *G. cumbriense/cancellatum* horizon at Cropwell Bishop and Colston Bassett North, which includes two maximum flooding surfaces and the intervening sequence boundary and transgressive surface. The location of the cross-section is marked by a red line on Figure 3.9. *(Church and Gawthorpe, 1994.)*

and the supply of carbonate sediment decreases. These differences between carbonates and siliciclastics are discussed further in Chapters 11 and 12.

Relative sea-level fluctuations can control the deposition of sediment at two scales; (1) high-frequency fluctuations responsible for the deposition of parasequences, which are superimposed on (2) lower-frequency fluctuations responsible for the deposition of sequences. This may seem complex enough, but in fact even lower-frequency (longer-term) sea-level changes act to modify the way in which sequences stack together. We will consider this in Chapter 5. In Section 4.4, we will reconsider the Carboniferous succession described in Section 3.5 to illustrate some of the many new terms and concepts that have been introduced in this Chapter.

## 4.4 The Carboniferous example revisited

In Section 3.5, we described sedimentary cycles from the Namurian, formed in a deltaic environment, and concluded that deposition was probably controlled by fluctuations in relative sea-level. You will remember that each cycle is composed of a progradational phase (delta), a transgressive phase (the transition back to delta front and prodelta sediments and the capping by a goniatite-bearing marine mudstone) and some have an erosive phase (incised valleys), most likely caused by a drop in relative sea-level.

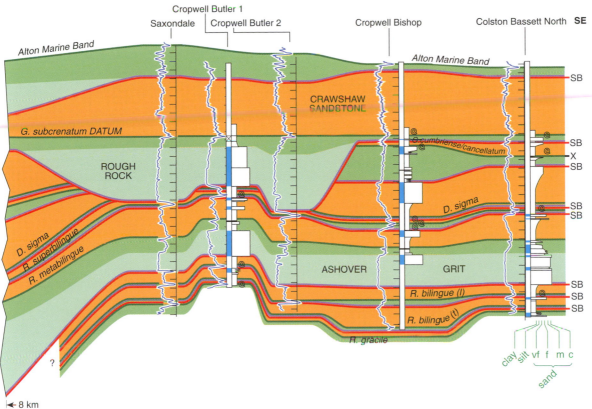

○ From what you have learnt in this Chapter, how did the prominent surface that defines the base of the incised valleys form and what surface does it represent?

● The incised valleys were created during a relative sea-level fall, so the base of each valley is interpreted as a sequence boundary.

○ During what part of the relative sea-level change were the incised valleys filled with sediment, and to what systems tract would you assign fill?

● The incised valleys were infilled during relative sea-level rise and so most likely belong to the TST.

However, depending on the position of the incised valley and the magnitude of the relative sea-level rise, it is possible that the incised valley fills represent the LST, or even the lower HST. But most incised valley fills found in the geological record are now interpreted to be TSTs.

Figure 4.20 shows one possible sequence stratigraphical interpretation of the data presented in Figure 3.10. Rather unsurprisingly, the goniatite-bearing marine mudstones are also assigned to a TST. However, it is likely that the transgression would not have been an instantaneous event. Periodic flooding of the delta as relative sea-level started to rise, forced rivers to burst their banks (or 'levées'), prior to the final marine inundation and abandonment of the delta. Consequently, the underlying delta plain deposits, at the point where mouth bar/delta front/delta top sandstones show a transition back to prodelta mudstone and delta front sandstone, are also assigned to the TST (Figure 4.20). This is because the increased flooding of the delta would lead to the eventual dominance of muddier facies (seen as the change from yellow below to grey above in Figure 3.10). The gamma-ray log reflects this with progressively higher API values (i.e. the log deflects to the right), so its response can be useful in defining the base of the TST.

The TST terminates with the goniatite-bearing marine mudstones. As explained in Section 4.2.1, these goniatite-bearing mudstones are condensed due to their great distance from any sediment supply. This reflects movement of the point of sediment input into the basin (i.e. the mouth bars) further and further landward. The goniatite-bearing mudstones are therefore interpreted to represent a condensed section associated with the maximum flooding surface.

Overlying the maximum flooding surface, the main phase of delta progradation is interpreted as the HST. This would have continued until the next relative sea-level fall subaerially exposed the delta plain and fluvial incision commenced once more (creating the next sequence boundary).

○ Why do some cycles have incised valleys, whilst others do not?

● There are two possible reasons for this. It is possible that they all do, but the cross-section in Figure 4.20 fails to intersect all of the valleys, i.e. they lie beyond the line of our intersection. Alternatively, the deposition of some sequences may have been controlled by a regression rather than a forced regression, in which case there would be no fluvial incision.

The identification of sequences with and without incision will help us to determine the magnitude of the relative sea-level curve that controlled deposition during this time. Sediment transported through these eroded valleys and out into the basin would form part of the FSST and LST of a sequence (Figures 4.12, 4.15). Such deposits are not shown in Figure 4.20 because the section is too proximal (i.e. too far to the left of Figures 4.12 and 4.15).

## 4.5 Lithostratigraphy versus chronostratigraphy, seismic stratigraphy, and the geometry of sequence stratigraphical surfaces

○ Consider the flooding surfaces between the parasequences marked in Figure 4.18. Are these chronostratigraphical or lithostratigraphical boundaries?

● They are chronostratigraphical boundaries as they mark the time of flooding and cut across different lithologies.

In a progradational parasequence set, the chronostratigraphical flooding surfaces that gently dip in an offshore direction are often referred to as *clinoforms* (Figure 4.21). Clinoforms form as sedimentary systems such as deltas prograde into the basin. During each time interval, sediment is deposited on the delta plain, delta front and prodelta. On a smaller scale, the lateral accretion surfaces in point bars of meandering river systems are a type of clinoform; similarly, the front surface of a subaqueous or subaerial dune as it builds forward to form cross-stratification is also a type of clinoform. Geometrically, clinoforms are inclined surfaces whose dip lessens at their top and base until they run tangential to the horizontal (Figure 4.21).

Clinoforms are widely recognized in seismic reflection studies (Box 4.2) which are used to study sedimentary deposits on a basin-wide scale. These clinoforms often represent the progradation of a coastal succession. The recognition of clinoforms on seismic sections brings us to another important point, in fact one of

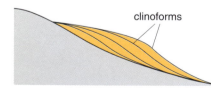
clinoforms

**Figure 4.21** Cartoon to show clinoforms.

the major factors that led to the development of sequence stratigraphy in the 1970s and 1980s. This is that seismic reflectors are nearly always *chronostratigraphical* surfaces rather than lithostratigraphical surfaces.

○ Why are seismic reflectors more likely to be chronostratigraphical than lithostratigraphical surfaces?

● Because the acoustic impedance contrast (Box 4.2) is usually greater across chronostratigraphical compared to lithostratigraphical surfaces, the latter tending to be gradational as one lithology is progressively replaced by another.

For instance, the change from offshore transition zone mudstones to shoreface sandstones is gradational, which leads to no marked impedance contrast, whereas if foreshore sandstones overlie offshore transition zone mudstones due to a relative sea-level fall, the acoustic impedance contrast will be greater and therefore the boundary is likely to be expressed as a seismic reflector.

You should now be able to see how we can take this further; unconformities, such as those that mark sequence boundaries, will show up as seismic discordances where the underlying strata have been partly removed by erosion or no sediment has been deposited. Reflectors within the underlying packages will terminate against the unconformity, and this type of reflector termination is called *erosional truncation* (Figure 4.22).

**Figure 4.22** (a) Different types of seismic reflector termination. Some of these geometric relationships can also be seen in large exposures and cross-sections. (b) An example of a seismic section with some of the different types of reflector termination marked. The two sequence boundaries delineate a complete sequence. The horizontal scale shows the position of the equally spaced shot points of the seismic source. In this case, each of the shot points is 251 m apart. ((a) Bally, 1987; (b) data courtesy of WesternGeco.)

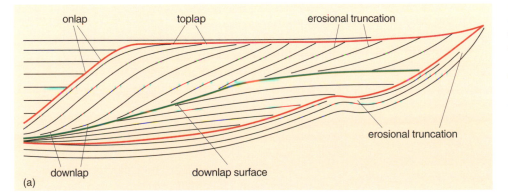

(a)

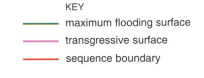

KEY

— maximum flooding surface
— transgressive surface
— sequence boundary

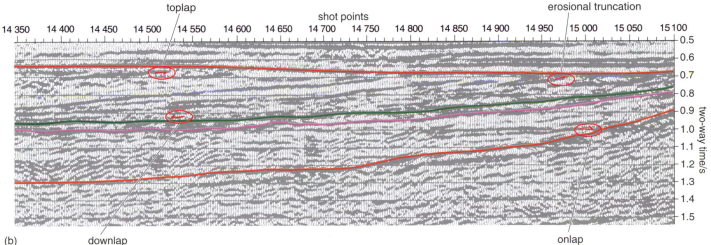

(b)

○ What will happen on the seismic section as a sequence boundary is traced out laterally in a distal direction?

● The seismic reflector will become less well marked because the erosional truncation and change in lithology will gradually become less until the unconformity becomes a correlative conformity.

As a sequence boundary is traced from an unconformity in the proximal sections to a correlative conformity in distal sections, the type of reflector termination underlying the sequence boundary will change from erosional truncation to *toplap* (Figure 4.22). An analogue to this is a perfect set of cross-stratified beds dipping towards the right with the top and bottom of the beds curving to a shallower angle so that they run tangentially to the horizontal. If we now erode the top of the cross-stratified beds along an inclined surface dipping towards the left (Figure 4.23), we are left with erosional truncation on the left-hand side and toplap on the right-hand side. Sediments deposited on top of the erosional truncation surface will show an *onlap* geometry (Figure 4.22).

Maximum flooding surfaces can often be picked out on seismic sections if the overlying HST is composed of clinoforms, the toes of which run tangentially into the maximum flooding surface; this kind of termination is called *downlap* (Figures 4.22 and 4.23).

Thus, the reflector termination characteristics of chronostratigraphical surfaces can be used to delineate key sequence stratigraphical surfaces and hence systems tracts and individual sequences. Most commonly, sequence boundaries are characterized by erosional truncation and toplap below the surface and onlap on top of the surface (Figure 4.22); transgressive surfaces are characterized by onlap; and maximum flooding surfaces are characterized by downlap of the overlying sediments onto the flooding surface (Figure 4.22). These features may also be recognizable within individual exposures, especially large-scale cliffs, and can also sometimes be deduced by compiling many sections.

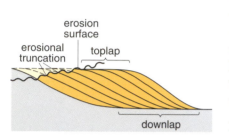

**Figure 4.23** Cartoon showing the difference between erosional truncation and toplap.

---

### Box 4.2  Seismic reflection surveying

Seismic reflection surveying is a geophysical technique used by the oil industry and research groups to derive subsurface geological information. It has been used particularly in the study of sedimentary basins, and can be done on land or at sea. Seismic waves are sound waves generated by a seismic source; they then travel through the subsurface, are reflected at geological boundaries within the subsurface and return to the surface where they are recorded. This is shown as wave 1 reflected from reflector 1 (Figure 4.24a). Wave 2 has been both reflected by reflector 2 and refracted (or bent) by reflector 1. Thus, a picture of the subsurface geology can be compiled. The record of ground motion (or pressure variation at sea) with time when plotted on paper is called a seismic trace (Figure 4.24b). The amplitude of a reflection is a measure of the strength of the ground motion, shown by the deviation of the seismic trace. Individual seismic traces are the result of a complex stacking of many seismic traces (the details of which we will not consider here), which act to reduce background noise (e.g., from nearby roads in land surveys

or ships in marine surveys). A number of seismic traces displayed side by side form a seismic section (Figures 4.22b, 4.24c,d). This is a representation of a slice of the Earth and is produced by moving the source and detectors along the line of survey. Shots are fired at a regular horizontal distance apart (usually around 25 m). In order to make the reflections more easily visible, the right-hand half of the amplitude of the seismic wave trace is usually coloured black (compare Figure 4.24c with Figure 4.24d).

The horizontal scale of a seismic section is a measure of horizontal distance along the line of survey. The vertical scale is the two-way time of the seismic wave, i.e. the time taken for the wave to travel down into the ground and back up again after reflection.

When interpreting seismic sections, the section is examined for reflection continuity. Continuity is where a reflection on a trace can also be recognized on neighbouring traces, with only small changes in the arrival time. Because half of the seismic trace is coloured black, such continuities appear as black or white lines running across the section (Figure 4.22b: note that in this Figure, the black has been converted to grey so the interpretation can be seen). These continuities are termed reflectors and are generated by interfaces where the density and/or acoustic velocity (speed of sound in the material) of the rock and/or its fluid content changes. The product of the density and sonic velocity of a material is termed its *acoustic impedance.* The interfaces may be bedding planes but may also be fault planes or any other extensive boundary between rock types. In some areas, there may be few or no seismic reflectors because there are no reflecting interfaces; too deep for sufficient seismic energy to reach (seismic energy is attenuated as it travels through the Earth); or a very complex structure.

The fact that the vertical scale is measured in two-way time and not in depth is important because it means that a seismic section is *not* a true geological cross-section through the Earth.

○   What extra information would we need to convert a seismic section into a geological cross-section?

●   The velocity of each of the rock layers because $v = d/t$ where $v$ = velocity, $d$ = distance and $t$ = time.

As this velocity will vary for each different lithology, this is no easy task. Oil companies need such information because they have to calculate how deep to drill in order to intersect an oil reservoir within a geological structure. Whilst it can be obtained by computer modelling, the only precise way to calculate velocities is to drill a borehole somewhere along the line of the seismic section and then to measure directly the time taken for a sound wave to pass through each of the rock units within the borehole. This gives the time it takes a seismic wave to reach any specified depth in the borehole. Where this depth corresponds to a major change in rock type capable of generating a seismic reflector, we can use this time to locate the corresponding reflector on the seismic section. This is called 'tying the well to the seismic section' and allows information obtained on the various rocks in the borehole to be extrapolated along the seismic section. The configuration, continuity, amplitude, frequency and interval velocity of seismic reflection patterns can be used to predict the lithological content of the subsurface packages (seismic facies analysis).

It should be realized that this is a very simplified explanation of how seismic sections can be used to derive subsurface geological information. In reality, there are other subtle differences that distinguish seismic sections from a geological cross-section and much computer time is required to remove 'artefacts' which are inherent in the acquisition and processing of seismic data. However, there is insufficient space to detail these here.

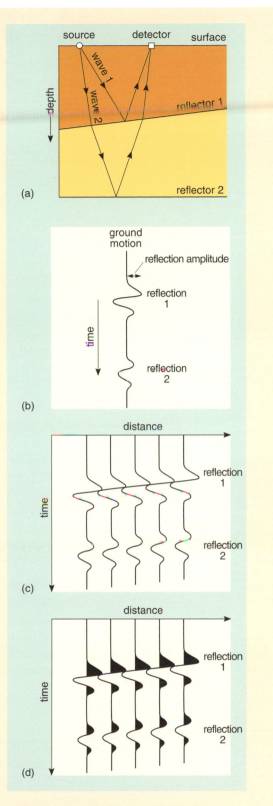

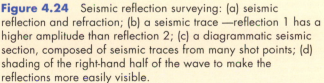

**Figure 4.24**   Seismic reflection surveying: (a) seismic reflection and refraction; (b) a seismic trace —reflection 1 has a higher amplitude than reflection 2; (c) a diagrammatic seismic section, composed of seismic traces from many shot points; (d) shading of the right-hand half of the wave to make the reflections more easily visible.

## 4.6 Global sea-level change and eustatic sea-level charts

The previous Sections in this Chapter should give a clear indication that subtle changes in relative sea-level can be detected through analysis of the sedimentary record. However, analysis of single sedimentary successions or even one particular basin does not indicate that the changes in sea-level detected are global in extent. This can only be established by examining the same age successions from a number of sedimentary basins in different tectonic regimes all over the world.

This then raises the problem of stratigraphical correlation of the different sections at the time-resolution of at least the sequence boundaries. Various stratigraphical techniques, but particularly biostratigraphy (Section 2.1.2), have been used extensively to try to test the global correlation of various parts of the stratigraphical record. This has given mixed results; whilst some depositional sequences appear to correlate globally, others do not. However, it is often not clear that the biostratigraphy (or other dating technique) is of sufficient resolution or precision. The breakthroughs in proving the global correlation of the cycles (or not) will probably only come from using an integration of stratigraphical dating techniques. This integration would potentially allow both a higher resolution correlation and the robustness of the correlation to be demonstrated by independent techniques. The other problem with testing the global correlation of sequences is separating out the local effects from the possible global signal. For instance, if a particular area or a whole basin has a very high sediment supply or a complex local tectonic history, then the timing of the individual sequences will be different from a succession with a moderate sediment supply and simple subsidence history.

At the time of the development of the sequence stratigraphy model in the late 1970s, Peter Vail and colleagues from Exxon Production Research, USA and elsewhere, compiled a global sea-level chart for the Mesozoic and Cainozoic, showing (at the resolution of about 1 Ma) how they interpreted global sea-level to have changed through time. Since its first publication (see Payton, 1977), the chart has been revised and republished several times, each version incorporating more data, some of which is from classical exposures around the world. One of the most extensively referenced versions of this chart is the 'Mesozoic–Cenozoic Cycle Chart' * published in 1988 by Haq *et al.* in Wilgus *et al.* These charts have produced much discussion partly because the data that support the interpretations have never been made fully available as they were based largely on confidential seismic sections and because researchers around the world have spent much time comparing particular sections with these 'global' sea-level charts.

The chart's most recent reincarnation (Hardenbol *et al.*, 1998) is slightly more modest because it only claims to be a European sea-level chart. Still, though, a full data set supporting the interpretations has never really been made widely available. It is interesting to note that publication of these sea-level charts was always, at least in part, intended to be provocative, and it certainly has got stratigraphers and sedimentologists discussing whether or not there is a global signal in the stratigraphical record.

However, the global nature of sequence stratigraphical cycles for much of geological time has yet to be proven. Global sea-level changes have really only been proven for the Quaternary and much of the Neogene where the oxygen-isotope record is good and can be directly linked to global sea-level change.

* Cenozoic is the American spelling of Cainozoic.

# 4.7   Summary

- Parasequences are the smallest bed-scale cycles within sedimentary successions. They comprise shallowing-upward progradational successions and are bounded by flooding surfaces. A flooding surface separates younger from older strata across which there is evidence of an increase in water depth.

- Parasequences stack together in patterns according to the longer-term changes in the balance between the rate of sediment supply and the rate of change of accommodation space (relative sea-level). The stacking pattern may be either progradational (parasequences build in a basinward direction), or aggradational (parasequences remain in the same lateral position), or retrogradational (parasequences build in a landward direction). Groups of parasequences showing the same stacking pattern are termed parasequence sets.

- Depositional sequences or sequences comprise relatively conformable packages of sediment bounded by sequence boundaries.

- The sequence boundary, transgressive surface and maximum flooding surface are all key surfaces associated with depositional sequences.

- Sequences are divided into systems tracts, each of which comprise a linkage of contemporaneous depositional systems or facies belts.

- The features of key surfaces and systems tracts are summarized in Table 4.1 (overleaf).

- The Namurian succession of northern England can be interpreted using the sequence stratigraphy model. Sequence boundaries formed during forced regressions are represented by the base of the incised palaeovalleys, and where fluvial incision has not occurred by subaerial exposure of the delta plain. During the transgression, the incised valleys are infilled and delta front, prodelta and marine mudstones are deposited across the top of the delta plain. Maximum flooding is interpreted to be represented by the goniatite-bearing marine mudstones. The HST is marked by the main phase of delta progradation.

- Seismic reflectors are chronostratigraphical surfaces.

- Sequence stratigraphical analysis of seismic sections is greatly aided by recognition of different types of seismic reflector termination. These include onlap, erosional truncation, downlap and toplap. Similar types of stratal termination and geometries may be recognized in large-scale cross-sections.

**Table 4.1** Summary of the features of systems tracts and key surfaces. Note that for simplicity the timing of the systems tracts and key surfaces are given with reference to a relative sea-level cycle assuming that the sedimentation rate is constant but, the same features, except for the FSST, could form due to changes in the sedimentation rate with relative sea-level staying constant. However, in most geological situations these sequence stratigraphical features are a combination of the balance between relative sea-level change and sediment supply.

| Key surface or systems tract | Timing within the relative sea-level cycle | | | Bounding surface | | Parasequence stacking pattern and geometry of the key surface | Typical sedimentary features that may develop |
|---|---|---|---|---|---|---|---|
| | Period | Earliest starting point | Latest end point | Lower | Upper | | |
| Highstand systems tract (HST) | Stable to decreasing rate of relative sea-level rise | Maximum rate of relative sea-level rise | Maximum relative sea-level | MFS | SB | HST comprises aggradational to progradational parasequence sets. It usually downlaps onto the MFS and may onlap onto the SB. | Fairly stable depositional environments; widespread facies belts. |
| Maximum flooding surface (MFS) | Between maximum rate of relative sea-level rise and maximum relative sea-level | Maximum rate of relative sea-level rise | Maximum relative sea-level | – | – | MFS represents period when distal areas are starved of sediment. The period of starvation will increase in duration in a distal direction. Strata overlying the MFS downlap onto it. | Condensed section; precipitation of authigenic minerals (e.g. phosphate and glauconite); abundant and diverse fauna; deepest-water facies; widest landward extent of marine facies; last significant flooding surface in the sequence. |
| Transgressive systems tract (TST) | Increasing rate of relative sea-level rise | Pronounced rise in relative sea-level | Maximum rate of relative sea-level rise | TS | MFS | Provided the rate of relative sea-level rise is greater than the supply of sediment, which is often the case, the TST comprises a retrogradational parasequence set. | In many successions, the TST comprises fairly condensed, reworked sediments, with an overall deepening-upward trend. However, for some carbonate successions, it may represent the period of highest sediment productivity. Typically comprises very widespread facies belts. |
| Transgressive surface (TS) | Start of increased rate of relative sea-level rise | Pronounced rise in relative sea-level | Maximum rate of relative sea-level rise | – | – | TS marks onset of pronounced relative sea-level rise. It is the first significant marine flooding surface. In most siliciclastic and some carbonate successions, it is the base of the first retrogradational parasequence. Usually overlain by onlapping strata, and may be detected by increase in the amount of onlap. | Sharp erosion surface in shoreface zone; winnowed lag of fossils and/or clasts; initiation of deepening-upward trend. |

| | | | | | | | |
|---|---|---|---|---|---|---|---|
| Lowstand systems tract (LST) | Initial very slow rate of relative sea-level rise or stillstand | Minimum relative sea-level | Pronounced rise in relative sea-level | – | TS | LST comprises progradational to aggradational parasequence sets. The LST will onlap the FSST. | Deposition of sediments on top of the underlying FSST (if present). Similar facies to underlying FSST. |
| Falling stage systems tract (FSST) * | Falling relative sea-level (forced regression) | Start of relative sea-level fall | Minimum relative sea-level | SB | – | FSST comprises a progradational parasequence set or redeposited sediments. | Deposition of sediments lower down the depositional profile than the underlying HST. FSST either comprises downstepping parasequences (unattached to attached) or redeposited material below the shelf break. |
| Sequence boundary (SB) | Falling relative sea-level (forced regression) or inflexion point in the temporary decrease in the rate of relative sea-level rise | Start of relative sea-level fall (forced regressions) or inflexion point in the temporary decrease in the rate of relative sea-level rise | Minimum relative sea-level | – | – | Depositional sequences are bounded by sequence boundaries. SB is an unconformity formed by subaerial exposure and erosion in proximal areas (for forced regressions), and/or marine erosion in coastal areas. The unconformity passes into a correlative conformity as it is traced laterally into stratigraphically complete distal sections or areas with a higher subsidence rate. Underneath the sequence boundary there may be erosional truncation of strata due to erosion and strata may onlap onto the sequence boundary. | Unconformity (often marked by a biostratigraphical gap) or its correlative conformity; evidence for extensive subaerial exposure and/or erosion; incised valley formation; basinward shift of facies. |

* FSSTs are a consequence of forced regressions; they do not form during regressions.

# 4.8 References

## Further reading

EMERY, D. AND MYERS, K. J. (eds) (1996) *Sequence Stratigraphy*, Blackwell Science, 297pp. [An advanced textbook on sequence stratigraphy and its application to different sedimentary environments, but becoming outdated.]

MIALL, A. D. (1997) *The Geology of Stratigraphic Sequences*, Springer-Verlag, 433pp. [An advanced textbook on the sequence stratigraphy model, the history of its development, the driving mechanisms for sequence development and the status of global cycle correlation.]

POSAMENTIER, H. W. AND ALLEN, G. P. (1999) *Siliciclastic Sequence Stratigraphy–Concepts and Applications*, SEPM (Society for Sedimentary Geology), Concepts in Sedimentology and Palaeontology No. 7, 204pp., and indexes. [An excellent advanced-level book covering up-to-date sequence stratigraphy concepts and their application to siliciclastic successions.]

## Other references

CHURCH, K. D. AND GAWTHORPE, R. L. (1994) 'High resolution sequence stratigraphy of the late Namurian in the Widmerpool Gulf (East Midlands, UK)', *Marine and Petroleum Geology*, **11**, 528–544.

CHURCH, K. D. AND GAWTHORPE, R. L. (1997) 'Sediment supply as a control on the variability of sequences: an example from the late Namurian of northern England', *Journal of the Geological Society, London*, **154**, 55–60.

GALLOWAY, W. E. (1989) 'Genetic stratigraphic sequences in basin analysis I', *American Association of Petroleum Geologists Bulletin*, **73**, 125–142.

HARDENBOL, J., THIERRY, J., FARLEY, M. B., JACQUIN, T., DE GRACIANSKY, P. C. AND VAIL, P. R. (1998) 'Mesozoic and Cenozoic sequence stratigraphical framework of European basins', in DE GRACIANSKY, P. C. *et al.* (eds) *Mesozoic and Cenozoic Sequence Stratigraphy of European Basins*, SEPM (Society for Sedimentary Geology) Special Publication No. 60, 3–13 and enclosures.

HELLAND-HANSEN, W. AND MARTINSEN, O. J. (1996) 'Shoreline trajectories and sequences: description of variable depositional-dip scenarios', *Journal of Sedimentary Research*, **66**, 670–688.

HUNT, D. E. AND TUCKER, M. E. (1992) 'Stranded parasequences and the forced regressive wedge systems tract: deposition during base-level fall', *Sedimentary Geology*, **81**, 1–9.

LEEDER, M. R. (1999) *Sedimentology and Sedimentary Basins*, Blackwell Science, 592pp.

PAYTON, C. E. (ed.) (1977) *Seismic stratigraphy — applications to hydrocarbon exploration*, American Association of Petroleum Geologists Memoir, No. 26, 516pp.

PLINT, A. G. AND NUMMEDAL, D. (2000) 'The falling stage systems tract: recognition and importance in sequence stratigraphic analysis', in HUNT, D. AND GAWTHORPE, R. L. (eds), *Sedimentary Responses to Forced Regressions*, Geological Society Special Publication No. 172, 1–17.

POSAMENTIER, H. W., ALLEN, G. P., JAMES, D. J. AND TESSON, M. (1992) 'Forced regressions in a sequence stratigraphic framework: concepts, examples, and exploration significance', *Bulletin of the American Association of Petroleum Geologists*, **76**, 1687–1709.

POSAMENTIER, H. W. AND MORRIS, W. R. (2000) 'Aspects of stratal architecture of forced regressive deposits', in HUNT, D. AND GAWTHORPE, R. L. (eds) *Sedimentary Responses to Forced Regressions*, Geological Society Special Publication No. 172, 19–46.

VAN WAGONER, J. C., MITCHUM, R. M., CAMPION, K. M. AND RAHMANIAN, V. D. (1990) *Siliciclastic sequence stratigraphy in well logs, cores and outcrops*, American Association of Petroleum Geologists Methods in Exploration Series No. 7, 55pp.

WHEELER, H. E. (1958) 'Time stratigraphy', *American Association of Petroleum Geologists' Bulletin*, **42**, 1047–1063.

WILGUS, C. K., HASTINGS, B. S., KENDALL, C. G. ST. C., POSAMENTIER, C. A., ROSS, C. A. AND VAN WAGONER, J. C. (eds) (1988) *Sea-level changes: an integrated approach*, Special Publication of the Society of Economic Palaeontologists and Mineralogists No. 42, 407pp.

# 5  Processes controlling relative sea-level change and sediment supply

*Kevin D. Church and Angela L. Coe*

Sequence stratigraphy is a holistic model for viewing and dividing the sedimentary record, because it considers both the packages of sediment and the surfaces between them at a range of scales from groups of beds to sedimentary basin fills. It is also clear that many of the major features of the sedimentary record are linked to relative changes in sea-level. Before we go on to present a range of case studies illustrating how the sequence stratigraphy model can be applied, we will consider further what controls relative sea-level change, sediment type and supply.

Sediment supply and relative sea-level change are dominantly controlled by climatic and tectonic processes: their influences are shown in Figure 5.1. Tectonic and climatic processes affect relative sea-level change and sediment generation on a variety of different time-scales and at different magnitudes. Most of these processes are cyclic and, in addition, some are regular.

○  What do we mean by a regular process?

●  Regular means that the process or phenomenon recurs over a *fixed* period of time, e.g. a new day every 24 hours.

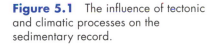

**Figure 5.1**  The influence of tectonic and climatic processes on the sedimentary record.

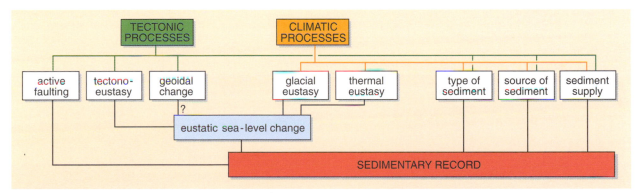

The regularity or irregularity of the processes, together with an understanding of what magnitude and frequency of change each of the processes are capable of, helps us to distinguish which process is responsible for the change in relative sea-level and development of the sedimentary record.

There are about five orders of cycle commonly recognized in the sedimentary record: first order: *c.* 50 to *c.* 200+ Ma; second order: *c.* 5 to *c.* 50 Ma; third order: *c.* 0.2 to *c.* 5 Ma; fourth order: *c.* 100 to *c.* 200 ka; fifth order: *c.* 10 to *c.* 100 ka.

Whilst the orders are not rigidly defined because of the wide range of conditions, parasequences are generally the high-frequency fourth and fifth order cycles, and depositional sequences are, by definition, one order higher than the parasequences of which they are composed. The majority of sequences are third order cycles. First and second order cycles are described in Section 5.5. Different processes have been interpreted to control the various orders of cyclicity and it is the interaction of these processes that produces the complex sedimentary record. Figure 5.2 summarizes the key features of these different orders of cyclicity.

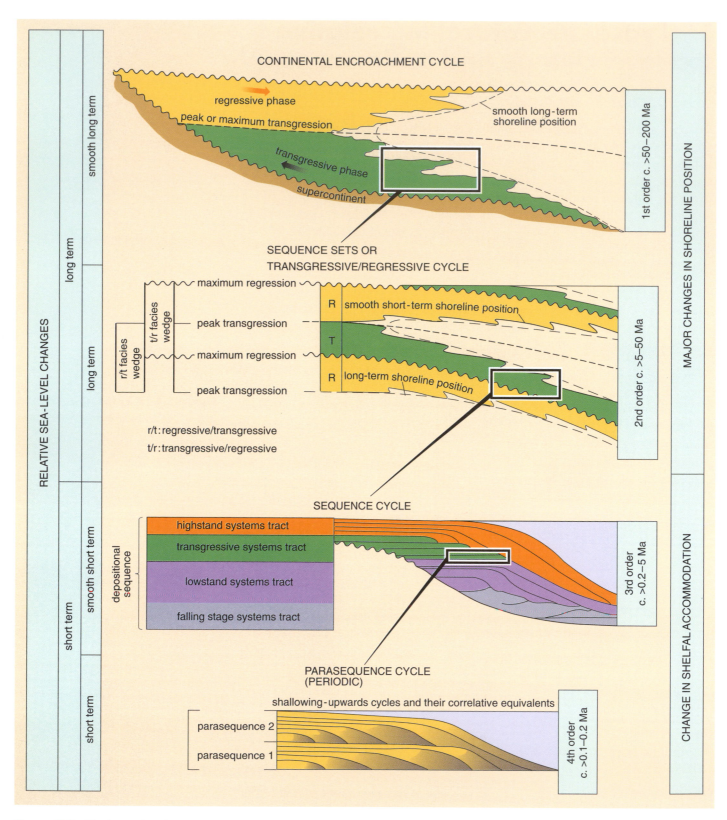

**Figure 5.2** The hierarchy of stratigraphical cycles. *(Duval, 1992.)*

# 5.1    Climatic processes

Figure 5.1 shows that climatic processes influence the sedimentary record by controlling glacio-eustasy, thermal expansion and contraction of the ocean, sediment supply, type of sediment, and the source of the sediment.

○  What are the six main factors controlling climate change on the Earth?

●  (i) Changes in the Earth's orbital parameters (i.e. Milankovich cycles, Box 2.2); as the Earth's orbital parameters change, they affect the amount of solar energy that reaches the Earth which thereby controls the climate. (ii) Distribution of the continents; this affects the distribution of the solar energy and the position of ocean currents. (iii) Changes in the composition of the atmosphere. (iv) Ice sheet growth and melting. (v) Changes in the oceans such as circulation and upwelling. (vi) Changes in the biosphere, including forest growth and peat bog formation.

## 5.1.1    Glacio-eustasy

Calculations have shown that the total melting of the present-day Antarctic ice cap would result in a sea-level rise of about 60–80 m and that the growth of the Quaternary ice sheets may have forced a glacio-eustatic sea-level fall of $c$. 120 m. Concomitant rates of sea-level change of $c$. 1 cm yr$^{-1}$ have been proposed. This is up to a thousand times more rapid than the average rate for tectono-eustatic mechanisms, so glacio-eustasy can therefore much more readily control the (short-term) third- and probably fourth- and fifth-order cycles. The repeated slow growth and rapid melting of ice for about the last 800 ka resulted in the sea-level change being highly variable in amplitude and highly asymmetrical. This is illustrated by the 'saw-tooth' pattern of slow rates of glacio-eustatic fall (up to 5 m ka$^{-1}$) followed by rapid marine transgression (up to 4 m per 100 years, Figure 5.3). At other times in the past, the growth and melting of ice appears to have been more symmetrical.

○  How would we test if these Quaternary changes in eustatic sea-level were Milankovich-driven?

●  By testing if the cycles are regular or not and ascertaining their periodicity.

Section 3.2.1 described how the oxygen-isotope record can be used as an indicator of the growth and melting of ice sheets and described one example of how the sedimentological record could be tied to these data. Similar studies have been completed on other Quaternary and Neogene successions.

## 5.1.2    Eustatic sea-level changes during greenhouse conditions

The link between changes in the volume of ice sheets and sea-level is obvious. However, during many periods of Earth history the mean global temperature was warmer and thus polar ice caps either did not exist or were very small. These periods of Earth history are referred to as being characterized by greenhouse conditions (as opposed to icehouse when extensive ice caps could form). The dominance of greenhouse or icehouse conditions through geological time is shown in Figure 5.4 overleaf.

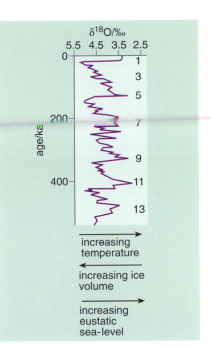

**Figure 5.3**   Oxygen-isotope variations over the past 500 ka from Deep Sea Drilling Program Site 607. This curve can be interpreted directly in terms of eustatic sea-level change. Note that 1‰ change in δ$^{18}$O approximates to $c$.100 m change in eustatic sea-level. Odd numbers refer to globally recognized oxygen-isotope stages interpreted as warm periods (see Figure 3.3 and Section 3.2.1 for further explanation). *(Ruddiman et al., 1989.)*

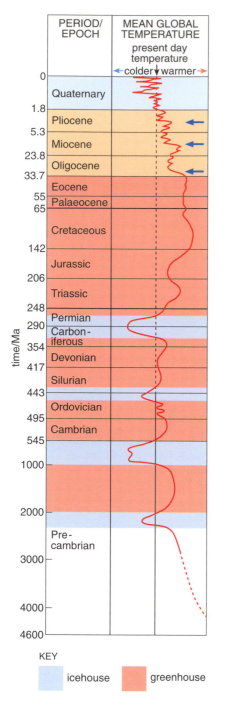

It is interesting to note that sedimentary cycles are just as common in the parts of the sedimentary record deposited under greenhouse conditions as those deposited under icehouse conditions. In addition, many of the sedimentary cycles deposited under greenhouse conditions have been found to be regular and can be linked to Milankovich frequencies. So, what are the possible mechanisms for changing eustatic sea-level during these periods? It has been demonstrated that the volume of ocean water, and hence eustatic sea-level, can fluctuate with temperature; this results in sea-level rises of 1–2 m for each 1 °C change in temperature.

Other potential mechanisms for changing eustatic sea-level are the storage and release of water as mountain ice or in inland lakes or aquifers. The volume of ocean water can also change if juvenile water is added to the system. However, this is likely to be balanced by the effect of ocean plate subduction which removes from the system pore waters contained in the oceanic crust. Calculations have shown that the charging and emptying of groundwater stored in sedimentary aquifers could cause sea-level to fluctuate by as much as 50 m. The 1–2 Ma time-scale on which this is thought to occur would make it a possible mechanism for second- or third-order cyclicity. All of the mechanisms discussed in this Section await further modelling and direct linkage to the sedimentary record.

It has been demonstrated that, in many cases, the short-term eustatic sea-level changes recorded in sedimentary rocks deposited under icehouse conditions are much greater in amplitude than those under greenhouse conditions. Figure 5.5 shows a summary of some of the differences.

### 5.1.3 The role of climate in sediment supply

#### Siliciclastic sediments

For siliciclastic sediments, the role of climate in sediment supply is relatively simple. According to the climatic conditions, the amount of precipitation and the range in temperature vary and these changes, in turn, control the supply of sediment. The degree of precipitation affects the type and abundance of vegetation and, therefore, both the weathering and erosion rate of the hinterland and transportation of resultant products. Whilst water is essential for chemical weathering, the effect of increased rainfall is partially counteracted by the fact that heavier precipitation tends to lead to more plants which bind the sediment together, inhibiting erosion. Plants, however, increase biological weathering. The greater the range in temperature either side of 0 °C, the greater the degree of physical weathering because the extremes of temperature lead to frost shattering. Several studies have found that during periods of climate change, more sediment tends to be produced than during stable climatic periods. This is interpreted to be because more sediment is produced as a new equilibrium profile is formed.

**Figure 5.4**  Changes in mean global temperature relative to that of the present day (15 °C) since the Precambrian. The curve is based on an array of geological evidence that allows the broad division into periods when icehouse conditions prevailed and extensive ice caps could form because of the low global temperature and intervals with a higher global temperature termed greenhouse periods. Note that the subdivision between icehouse and greenhouse conditions is not well defined. Temperature changes pre-Oligocene are less well known than those post-Oligocene. The current icehouse conditions represent the culmination of a long period of cooling from the early Oligocene. Pronounced cooling steps in this trend are indicated with arrows: each of these correspond to the growth of ice sheets in the Northern and/or Southern Hemisphere. Icehouse conditions could be described as initiating at any one of these times because of the vague definition of icehouse versus greenhouse conditions. The time-scale is non-linear. (Frakes, 1979.)

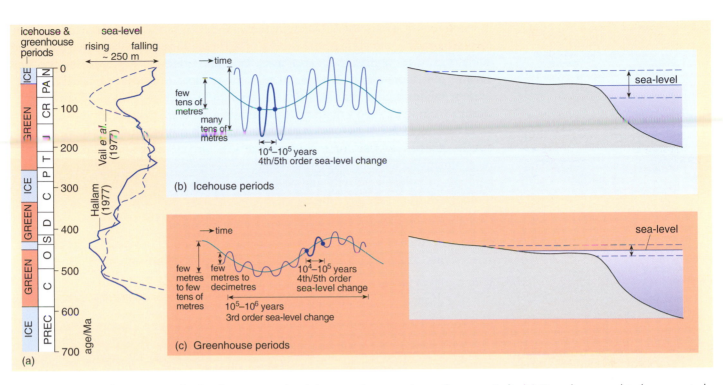

**Figure 5.5** The change in amplitude of eustatic sea-level during icehouse and greenhouse periods. (a) Greenhouse and icehouse periods throughout geological time together with the long-term sea-level curves of Hallam and Vail *et al.* (b) and (c) Curves and cross-sections (not to scale) showing differences in mean elevation and amplitude of sea-level change during icehouse and greenhouse periods. (b) Lower mean sea-level under icehouse conditions is accompanied by high-amplitude short-term changes in eustatic sea-level so that the upper parts of continental slopes, continental shelves and coastal plains are subjected to frequent flooding and subaerial exposure. (c) In contrast, high mean sea-level during greenhouse periods is accompanied by low-amplitude short-term changes in sea-level that result in extensive epeiric seas covering platforms and low-angle ramps. The short-term changes in sea-level cause less extensive flooding or subaerial exposure of large land areas. Thus, potentially during icehouse periods, short-term fourth/fifth order cycles can produce amplitudes of relative sea-level change that are equivalent to the amplitudes produced during longer-term third-order cycles during greenhouse periods. PREC = Precambrian; C = Cambrian; O = Ordovician; S = Silurian; D = Devonian; C = Carboniferous; P = Permian; T = Triassic; J = Jurassic; CR = Cretaceous; PA = Palaeogene; N = Neogene and Quaternary. *(Wilson, 1988, left part of (b) and (c) based on Tucker, 1993.)*

## Carbonate sediments

For carbonates, climate is even more significant for sediment supply as it determines whether or not the conditions are appropriate for carbonate production. The majority of carbonate is produced biologically *in situ* and the organisms responsible require specific climatic conditions to maintain and enhance productivity. Important factors include: a high enough level of light penetration into the water; water temperature and salinity; circulation to supply nutrients to carbonate-secreting organisms present; and well-oxygenated waters. In general, there also needs to be no, or a very low, supply of siliciclastic sediments to prevent filter-feeding organisms choking on sediments. Chemically produced carbonate grains such as ooids, and the direct precipitation of calcium carbonate mud, also require particular conditions. The importance of climate for carbonate production is discussed more fully in Chapter 11.

## 5.2 Tectonic processes

### 5.2.1 Tectono-eustasy

Tectono-eustatic mechanisms which affect the volume of basins containing seawater have mainly been used to explain the longer-term, first- and second-order global sea-level cycles. In particular, the volume of ocean basins is reduced by continent fragmentation, which increases the length of ocean ridges, and displaces water onto continental shelves. High sea-floor spreading rates buoy up oceanic crust for a greater distance away from the ridge, further reducing the volume of ocean basins. Continental collision reduces the total continental area, thereby increasing the volume of basins containing seawater. Rates of sea-level change from these mechanisms do not usually exceed 1 cm in 1000 years. Sea-level falls of 11–13 m have been estimated for each million square kilometres of continent area removed by stacking thrust belts during collisions. Other mechanisms include the changing volume of oceanic subduction trenches, large amounts of volcanic activity and ocean floor bulges caused by the rise of mantle plumes. All of these mechanisms are thought to have had a negligible effect on short-term sea-level variations but they affect long-term variations.

Figure 5.6 illustrates estimates of the long-term variation in sea-level since the start of the Jurassic. The curve illustrates two important points:

1   Sea-level change in the geological past is hard to interpret, with the result that different interpretations exist.

2   The long-term rise in sea-level through the Jurassic and Cretaceous occurred at the same time as the break-up of the supercontinent Pangea and the associated increase in sea-floor spreading.

**Figure 5.6**   Estimates of long-term variations in sea-level since the Jurassic from three different sources. For the long-term (first order) curve of Haq *et al.* (1988) that is shown, the water locked up in late Cainozoic ice-caps has not been allowed for, resulting in present-day sea-level being shown at *c.* 60 m.

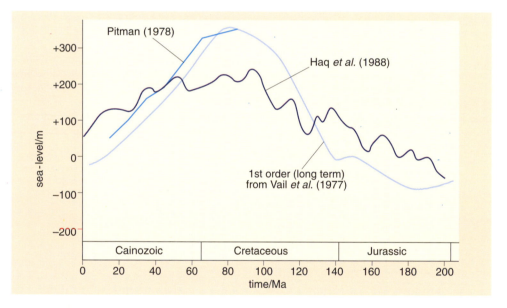

### 5.2.2   Intraplate stresses

Some researchers have suggested that the interaction between horizontal intraplate stresses and the deflection of the lithosphere by sediment loading can cause regionally extensive, short-term variations in sea-level. They suggest that the stresses can be relayed between the continental margin and intracratonic basins through the lithosphere, therefore resulting in a regional effect. The researchers have suggested that the intraplate stresses are caused by subduction 'pull' and mid-ocean 'push'. Whilst intraplate stresses potentially provide a shorter-term mechanism than mid-ocean ridge spreading or continental collision, modelling and observation suggest that the frequency is greater than 2 Ma. Thus, this mechanism may explain some, but not all, third-order sequence cycles.

## 5.2.3 Inhomogeneities in the mantle and geoidal sea-level change

Over the past few decades, it has been discovered that inhomogeneities within the mantle can cause vertical movements of the Earth's surface and hence sea-level changes. This is in addition to the sea-level changes caused by the movement of tectonic plates on the Earth's surface, which have ultimately been driven by the convection cells in the mantle. One of the most dramatic examples is centred on southern Africa. This area of continent and the surrounding sea-floor has been rising over the past 100 Ma and it has been calculated that it has risen by about 300 m over the past 20 Ma. This is despite the fact that southern Africa has not been involved in a tectonic collision for over 400 Ma. The rock record also shows that other parts of the Earth's surface have bowed down by hundreds of metres during periods when they were also not subjected to large plate tectonic movements: these include Australia and the east coast of North America.

The driving mechanism behind these vertical movements of the Earth's crust has been formulated and investigated using the vibrations, or seismic waves, from earthquakes. By measuring the time taken for the seismic waves to travel from thousands of earthquake epicentres (points of origin) to recording stations at the Earth's surface, scientists have been able to infer the temperature and density of the Earth's mantle and thus build up a 3-D picture. This is because the velocity of the seismic waves is governed by the chemical composition, temperature and pressure of the rocks that the waves travel through. There are basically two scenarios. (i) Continental and oceanic crust can rise where it is underlain by less dense rock because it is hot or of a less dense composition (referred to as a hot spot). This uplift will lead to a relative sea-level fall. (ii) Continental and oceanic crust can sink where it is underlain by dense rock because it is cold or of a more dense composition, for instance, where a subducted tectonic plate sinks through the mantle but remains too cold and therefore too dense to mix with the surrounding rock. As the tectonic plate sinks, it creates a downward movement of material pulling the continental crust with it. This can happen quite a while after the subduction zone has been active at the surface such that ghosts of previous subduction zones can cause the continental crust to be pulled down. The depression of continental crust will cause a rise in relative sea-level.

The other line of evidence that has helped to resolve these vertical movements in the continental and oceanic crust is the influence of mantle density on the Earth's gravitational field. The Earth's gravitational field is represented by the geoid; this is a map of the Earth's gravitational field on which departures from the theoretical gravity norm at sea-level (the spheroid, as calculated using the International Gravity Formula), caused by inhomogeneities in the mantle, are plotted and contoured. Surprisingly, the geoid shows that the gravitational field is stronger over hot spots (rather than weaker as might be expected over hotter and therefore less dense rocks), and weaker over cold areas. This phenomenon has been explained by the fact that as a low-density (hot) fluid rises up in the mantle, the force of the flow pushes higher density fluid above it which creates an excess of mass and hence a stronger gravitational field. Similarly, the gravitational field is lower over cold dense material as it drags mass down at the surface, causing a lower gravity.

○ Will the vertical movements of the continental crust and oceanic crust always give rise to eustatic sea-level changes?

● No, because the changes are not globally uniform. Relative sea-level fall could occur adjacent to one continent due to uplift of that continent whilst relative sea-level rise due to a sinking of a continent was occurring elsewhere.

Thus, this mechanism is of sufficient magnitude to control the lower-order longer-term cycles. However, the signal is unlikely to be global in extent.

### 5.2.4  Active faulting (neotectonics)

Active faulting is distinct from tectono-eustasy because it covers the very rapid creation of sediment accommodation space during fault movement in tectonically active areas (e.g. in Greece and Turkey today). In active (fault-controlled) rift basins, rates of subsidence and uplift may approach those of glacio-eustasy.

○ Is active faulting likely to have a regular or irregular periodicity?

● Irregular: there is no reason why faults should move at a set time period.

○ Is active faulting likely to have a local or global effect?

● A local effect.

It can therefore be concluded that although neotectonics may cause local changes in relative sea-level, the cycles would not be correlatable across a wider or global area, and that the cycles would not have a regular periodicity.

### 5.2.5  Tectonics and sediment supply

Besides climate, tectonic movement is the other factor controlling sediment supply. Actively uplifting regional areas or even local fault systems will provide increased sediment supply to the sea via the high gradient alluvial systems. As we explored in Section 4.1 and Figure 4.2, alluvial systems constantly try to maintain an equilibrium profile. High sediment supply associated with uplift can be seen in the active mountain belts of today, e.g. in the Himalayas (where sediment is being supplied to the sea from the Ganges, Brahmaputra and Indus rivers), and along active faults in California. Evidence such as the proximity and distribution of debris flows and turbidites to fault scarps can indicate an area of active tectonics.

## 5.3  Sediment compaction and its control on relative sea-level

The effects of sediment compaction can be difficult to distinguish from tectonic subsidence. As mentioned in Section 4.1, mudstones compact by up to 80% and sandstones up to about 30%, and much of this compaction occurs within the first few hundred metres of burial, as shown in Figure 5.7. Sediment compaction is thus a dynamic process that can potentially play an important role in creating accommodation space even though relative sea-level may be constant.

This mechanism has mainly been considered in deltaic environments. Maps of the Mississippi delta in the USA (Figure 5.8) clearly show how, over time, each

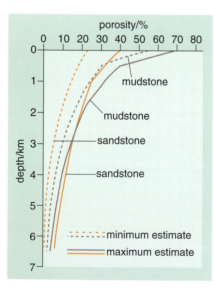

**Figure 5.7** Porosity/depth curves for mudstone and sandstone, showing the maximum and minimum rate of porosity reduction with depth as the sediments become compacted. *(Bond and Kominz, 1984.)*

delta lobe has prograded and then been abandoned because the nearshore accommodation space is filled. This has caused the river to switch its course (avulse) and a new lobe to start forming, a process called lobe switching (Figure 5.8). If a borehole was drilled in any of the lobes, it would show a succession of coarsening-upward parasequences of deltaic sediments (Figure 4.5b). As the new delta lobe progrades, the sediments in the previous lobe will become compacted due to the weight of the water. This, together with any regional tectonic subsidence, will create new accommodation space which might be filled at a later time when the river avulses again. Other possible mechanisms for lobe switching are local subsidence due to the sediment load, and eustatic sea-level change altering the position of the available accommodation space.

○   How would we distinguish between local compaction/tectonic loading
     mechanisms and eustatic sea-level changes?

●   By examining whether or not the cycles are local or global in extent.

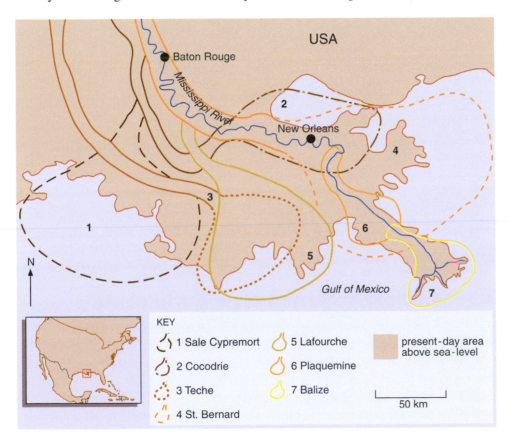

**Figure 5.8**   The position and relative age of the different sediment lobes in the Mississippi delta, USA. The seven lobes shown all formed during the Holocene. (Leeder, 1999 based on Kolb and Van Lopik, 1958, and Coleman, 1976.)

## 5.4   How do we unravel these controls?

The evidence is that most sedimentary successions plot within a field defined by eustatic, tectonic and sediment supply end-members. The sequence architecture depends on the relative influence exerted by each of these factors (Figure 5.9). We can make the following general comments:

1   Thick sedimentary packages and the dominance of progradational stacking patterns suggest sediment supply as an important control. Sediment flux in turn may be controlled either by climate (e.g. conditions favourable to high carbonate production rates or weathering and erosion of siliciclastics) or tectonic movements (e.g. sedimentation adjacent to an active fault).

2   Angular unconformities, syndepositional faults, evidence for abrupt deepening/uplift and locally developed sequences suggest that tectonic processes are an important control.

3   Widespread interplate correlation and demonstrable synchroneity of sequences and sequence boundaries suggest eustasy as an important control.

4   Regular cyclicity suggests Milankovich-controlled climate changes that influence eustatic sea-level changes or sediment supply.

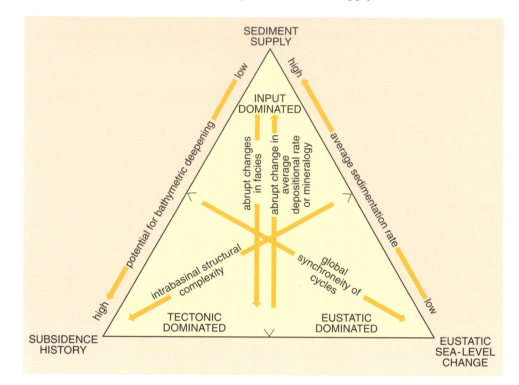

**Figure 5.9**   Principal factors influencing sequence development. *(Galloway, 1989.)*

## 5.5   Multiple order sea-level fluctuations

In a similar manner to the stacking pattern of parasequences within sequences (Section 4.3), sequences are, in turn, influenced by longer-term lower orders of cyclicity. Figure 5.10a–c shows three eustatic sea-level cycles of different amplitude and wavelength. We can take the curve in Figure 5.10a as the one that controls the deposition of parasequences and that in Figure 5.10b as the one controlling sequences. A further cycle of sea-level change of still greater amplitude and longer wavelength is shown in Figure 5.10c. This longer-wavelength cycle controls how the sequences stack together to form *sequence sets*. Figure 5.10d represents the combination of the curves shown in Figure 5.10a–c.

Depending on the rate of long-term eustatic sea-level change, it is possible to form falling stage, lowstand, transgressive and highstand sequence sets (analogous to the production of FSSTs, LSTs, TSTs and HSTs that make up sequences). In this book, the falling stage and lowstand sequence sets are grouped together for simplicity. In the following three Sections, we will examine three specific portions of Figure 5.10d and describe how each type of sequence set is produced.

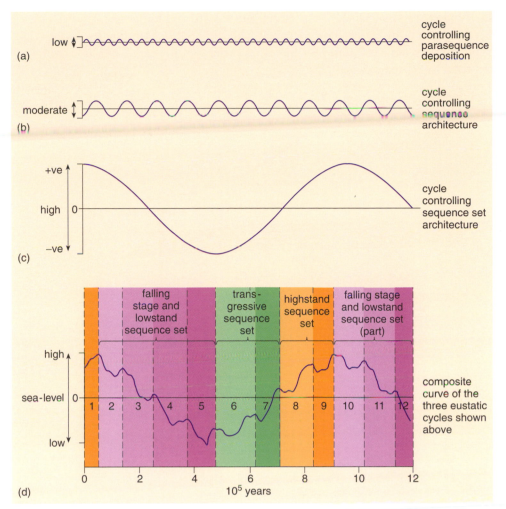

**Figure 5.10** Three orders of eustatic sea-level cyclicity responsible for the deposition of (a) parasequences, (b) sequences and (c) stacked sequences and sequence sets. (d) A wave train produced by the combination of curves (a)–(c), showing 10 complete eustatic sea-level cycles and two partially complete cycles. The sequences deposited during each of the 12 cycles are shown in Figure 5.12. See text for further explanation of (d). Note that so we can focus on the relationship between different amplitudes and frequencies of eustatic sea-level change, these curves are shown as eustatic curves and not relative sea-level curves. *(Curves based on Van Wagoner et al., 1990.)*

## 5.5.1 Deposition of successive sequences during a longer-term eustatic sea-level fall — the falling stage and lowstand sequence set

The major eustatic sea-level falls in Figure 5.10d (cycles 2 to 5 and 10 to 12) are punctuated by relatively short-lived, higher-order sea-level rises. Often, the cycles which control the deposition of sequences are so distorted by the longer-term lower-order sea-level fall that they fail to show any rise at all; instead, either the rate of rise is reduced to zero (i.e. the curve is horizontal) and a 'stillstand' is developed, or the fall continues at a slightly reduced rate (e.g. cycle 11).

○ Under conditions dominated by long-term eustatic sea-level fall, the development of which systems tract(s) will be favoured in the sequences that make up this sequence set?

● The FSST and LST(s).

It is reasonable to assume therefore that the FSST and LST will predominate as shown in Figure 5.11a. This falling stage and lowstand sequence set will be characterized by well-developed fluvial channels cutting down into older sequences to create incised valleys which will be filled with sediment during the subsequent transgression. Sediment transported along these valleys will be deposited either as submarine fans within the basin if sea-level falls below the shelf break (Figures 5.11a, 5.12a, Sequences 2 to 5, and Figure 4.12c) or along the shoreline as downstepping sequences (Figure 4.12d). Consecutive sequences dominated by the FSST and LST therefore form a falling stage and lowstand sequence set (Figure 5.12b).

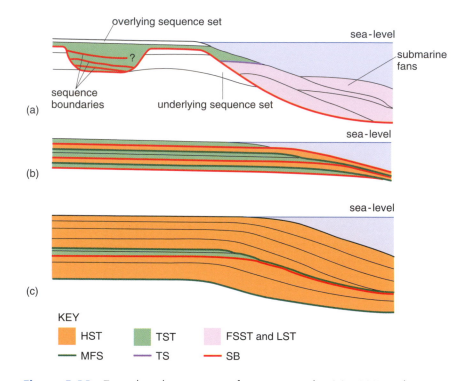

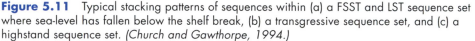

**Figure 5.11** Typical stacking patterns of sequences within (a) a FSST and LST sequence set where sea-level has fallen below the shelf break, (b) a transgressive sequence set, and (c) a highstand sequence set. *(Church and Gawthorpe, 1994.)*

The deposition of individual HSTs and TSTs will be limited to rarer shorter-term (higher-order) rises in sea-level on the overall falling low-order trend. As these rises are of short duration, these systems tracts will be correspondingly thin.

○ What is the likely preservation potential of these systems tracts? Give a reason for your answer.

● Their preservation potential is low because a high amplitude sea-level fall invariably follows from such small rises. This will cause erosion of any HST and TST deposits as a new phase of fluvial downcutting is initiated (Figure 5.12a, Sequences 2 to 5 and 10 to 12).

In this way, as successive sequences are deposited within this falling stage and lowstand sequence set, so the FSST and LST of each sequence is preserved at the expense of HSTs and TSTs. Thus, many FSSTs and LSTs may be stacked on top of each other (Figures 5.11a, 5.12a, Sequences 2 to 5, 10 to 12). Another consequence of the bias for erosion is that FSSTs and LSTs from earlier sequences may be completely removed during erosion associated with the deposition of new sequences.

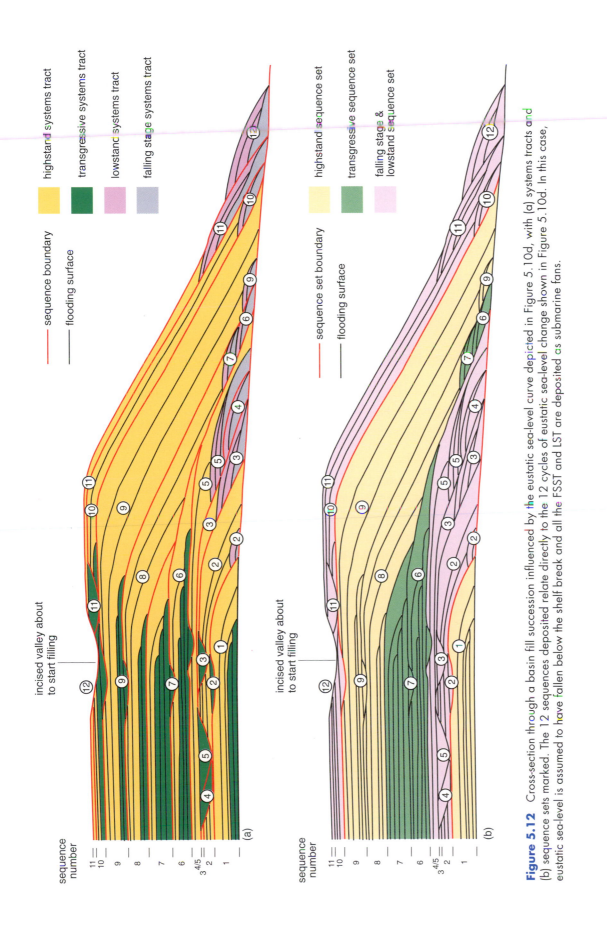

**Figure 5.12** Cross-section through a basin fill succession influenced by the eustatic sea-level curve depicted in Figure 5.10d, with (a) systems tracts and (b) sequence sets marked. The 12 sequences deposited relate directly to the 12 cycles of eustatic sea-level change shown in Figure 5.10d. In this case, eustatic sea-level is assumed to have fallen below the shelf break and all the FSST and LST are deposited as submarine fans.

### 5.5.2 Deposition of successive sequences during longer-term eustatic sea-level rise — the transgressive sequence set

Rather unsurprisingly, when considering the major rises in long-term eustatic sea-level (Figure 5.10d, cycles 6 and 7), the opposite logic applies to that detailed for falling stage and lowstand sequence sets. Here, for the majority of the time, accommodation space is being created faster than sediment supply can fill it due to eustatic sea-level rise. Thus, each sequence will preferentially develop a TST, so that overall a transgressive sequence set will form (Figure 5.12b, Sequences 6 and 7). It is quite possible that each sequence in the TST may also show a HST; however, these will be thin because the rate of sea-level rise to its maximum will be rapid (Figure 5.11b). The eustatic sea-level rise is punctuated by relatively minor eustatic sea-level falls (Figure 5.10d). There is thus little potential for the development of FSSTs or LSTs, though where these do occur their preservation potential will be very high because of the overall rise.

### 5.5.3 Deposition of successive sequences during a longer-term eustatic sea-level high — the highstand sequence set

The theory behind the development of the highstand sequence set follows logically from that of the transgressive sequence set described above. As the long-term (low-order) sea-level curve in Figure 5.10d climbs away from the inflection point, the rate of eustatic sea-level rise starts to decrease (cycle 8). As this rate of rise slows down (eventually to zero at the peak of the sea-level curve, cycle 9), sediment supply will increasingly outpace eustatic sea-level rise. As a consequence of this, sediment will be transported to sites of deposition further and further into the basin and exhibit the aggradational to progradational parasequence stacking patterns which typify HSTs (Figure 5.12a, Sequences 8 to 9). The HST therefore becomes increasingly significant as the sea-level curve in Figure 5.10d reaches its maximum. Sequences 8 and 9 form a highstand sequence set (Figure 5.12b).

By considering FSST, LST, TST and HST sequence sets, we are referring to the prevalence of one particular type of systems tract within sequences that comprise the sequence set. However, as Figure 5.12 illustrates, the precise composition of each sequence will depend upon its exact position on the sea-level curve. So, for example, sequences formed early in a falling stage and lowstand sequence set will be dominated by FSSTs, whereas those formed later in the same sequence set will have better-developed LSTs and TSTs (compare Sequences 3 to 4 with Sequence 5 in Figure 5.12).

### 5.5.4 So what about subsidence?

So far in this Section, we have been talking solely in terms of eustatic sea-level change. This we have done by assuming that no subsidence occurred in the hypothetical sedimentary basin that we have been filling. Whilst useful for the purposes of discussing sequence sets, this assumption is unrealistic because without subsidence all the sequences in Figure 5.12 would not be preserved. Basins subside either in a 'pulsed' fashion during active rifting, or more gradually as a result of cooling in the crust/upper mantle underneath (the so-called 'sag' or 'thermal relaxation' phase). Once basins begin to fill, the ever-increasing weight of sediment becomes an additional contributory factor, resulting in isostatic subsidence.

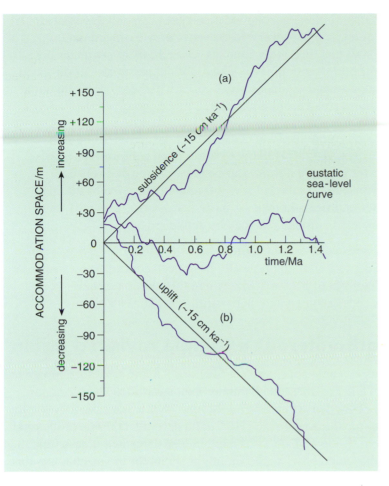

**Figure 5.13** The eustatic sea-level curve of Figure 5.10d modified to show the effects of (a) subsidence and (b) uplift. Both (a) and (b) are therefore relative sea-level curves. *(Van Wagoner et al., 1990.)*

As discussed in Chapter 4, a consistent subsidence rate within a sedimentary basin will have the effect of adding a uniform amount of accommodation space, so we can quite easily modify our eustatic sea-level curve in Figure 5.10d to make it a relative sea-level curve. This has been done in Figure 5.13a, which shows the same curve as Figure 5.10d, but the curve has additionally been sheared upwards to simulate the effect of subsidence within the basin of 15 cm ka[-1], which represents a fairly typical value.

○   Which two of the three sequence sets that you considered in Sections 5.5.1–5.5.3 do you think will dominate the sedimentary fill in such a basin?

●   The overall gradient of the curve depicted in Figure 5.13a has been increased to such an extent that in the middle there is no potential for any fall in relative sea-level. This suggests that accommodation space is always being created (though by varying amounts). Any relative sea-level falls are limited to the start and end of the curve, but even here the overall trend is still upward. It is therefore clear that the sedimentary fill resulting from such a relative sea-level curve will be dominated by transgressive and highstand sequence sets at the expense of the falling stage and lowstand sequence sets.

○   If the basin was being uplifted by 15 cm ka[-1] (Figure 5.13b), rather than subsiding by that amount, which sequence set do you think would dominate the basin fill?

- If the basin is being uplifted, accommodation space can only be created where the rate of relative sea-level rise exceeds the rate of tectonic uplift. Figure 5.13b shows that this rarely happens. Overall, there is greater potential for relative sea-level fall and this will lead to loss of accommodation space. falling stage and lowstand sequence set deposits will therefore be favoured at the expense of transgressive and highstand sequence sets.

In general, subsidence is more likely to occur during the active filling of a sedimentary basin, whereas uplift is likely to result from compressive forces acting on the basin, for example during continental collision and related mountain-building episodes. As the latter will frequently turn the basin into an area of positive relief (a process known as 'basin inversion'), it usually halts sediment deposition, and is instead replaced by active erosion.

The reason why we see cycle stacked upon cycle, with often little evidence of erosion, is because deposition might typically have been controlled by a relative sea-level curve in the form of Figure 5.13a, where the addition of new accommodation space is favoured and there is little potential for erosion.

## 5.6   The Carboniferous example revisited again

In the Carboniferous example considered in Sections 3.5 and 4.4 and summarized in Figure 4.20, the TSTs are often much thinner than the HSTs. The former are partially composed of condensed marine mudstones, so it is not easy to relate the thickness of these to the thickness of the HST delta deposits in terms of the time taken to deposit them. If the TST deposits were indeed deposited in less time than the HST deposits, this could have important implications for the mechanism that caused them.

○ In Section 3.2.1, we discussed the likelihood of a glacio-eustatic influence on these cycles. How might these observations relate to such a mechanism?

- Glaciers tend to expand gradually over many thousands of years, yet there is growing evidence to suggest that during some periods of Earth history they have melted far more rapidly, over time-spans of hundreds of years. Such behaviour will be reflected in a 'saw-tooth'-shaped, asymmetric eustatic sea-level curve of rapid sea-level rise followed by a gradual fall, which could conceivably produce the thin TSTs and thick HSTs observed.

A more detailed examination of Figure 4.20 shows that the sequences are of three different forms:

1   Sequences dominated by incised valleys (e.g. Rough Rock cycle) which formed during deposition of a falling stage and lowstand sequence set (cf. Section 5.5.1). Although these sequence sets are not actually preserved on this part of the cross-section, the pronounced incision observed is probably the result of several relative sea-level falls and many systems tracts are likely to be present in more distal sections.

2   Sequences dominated by both thin TSTs and thin HSTs (e.g. *R. gracile* and *R. bilingue* (t) cycles). Consecutive sequences dominated by such systems tracts would constitute a transgressive sequence set (cf. Section 5.5.2).

3   Sequences dominated by thick HSTs (e.g. Chatsworth Grit cycle).
    Consecutive sequences dominated by such systems tracts would constitute a
    highstand sequence set (cf. Section 5.5.3).

If we assign each sequence in Figure 4.20 to one of the above sequence sets, then
the resultant relative sea-level curve might resemble that in Figure 5.14. This
curve shows the two orders of relative sea-level change responsible for the
development of sequences and sequence sets. The position of the maximum
flooding surface for each sequence is marked with a dot. Note that although there
is the potential for incised valleys to develop during the relative sea-level fall of
any sequence, it is during the falls associated with the longer-term relative sea-
level falls that hold the greatest potential. Such long-term falls are represented by
the Ashover Grit, the Rough Rock and the Crawshaw Sandstone (Figures 4.20,
5.14). Whilst each incised valley is interpreted as belonging to a single sequence
on Figure 4.20, the interpretation in Figure 5.14 suggests that each incised valley
could alternatively result from the stacking of several sequences (a sequence set;
Section 5.5.1) in which incised valley formation is dominant.

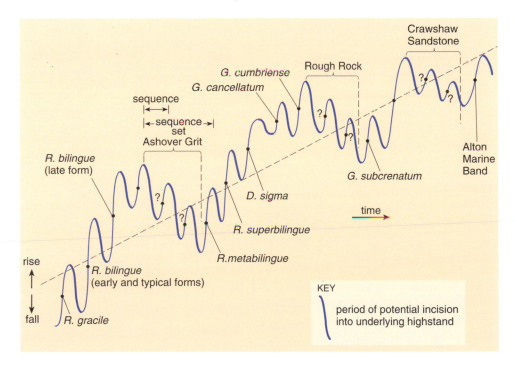

**Figure 5.14**   Relative sea-level
interpretation for the Namurian, based
partly on Figure 4.20. The relative sea-
level curve is schematic and so neither
axis has values assigned. However, 50 m
eustatic sea-level rise and 200 000 years
for each sequence may serve as
approximate values. *(Church and
Gawthorpe, 1994.)*

○   If this is true, why is it unlikely that deposits of each sequence will be
    detectable within each incised valley fill?

●   Figure 5.14 suggests that during each long-term (lower-order) relative sea-
    level fall, sea-level reaches its lowest point during deposition of the last
    sequence within that fall. Consequently, fluvial incision during this sequence
    will remove most of the deposits of earlier sequences within the same valley.
    At the time of its final abandonment, the valley fill will therefore consist
    predominantly of deposits from the last sequence of the sequence set.

The interpretation presented here is just one possibility. However, it demonstrates
how sequence stratigraphy can be used to interpret potentially complex sea-level
fluctuations. So far, we have used only relative terms; there are no values
assigned to the axes on Figure 5.14. It is interpreted that each sequence was

deposited over a period of very approximately 200 000 years (based on average figures for goniatite biozone lengths). Calculating the magnitude of the eustatic sea-level change between successive sequences (i.e. from one maximum flooding surface to the next) is more problematical. Thickness measurements have to be made between a delta plain (known to have been at sea-level), and the first occurrence of the next delta plain in the overlying delta. A minimum estimate of the magnitude of eustatic sea-level rise $SL_r$ (Figure 5.15) is given by:

$$SL_r = (T_m + T_{mc}) - S \qquad (5.1)$$

where:

$T_m$ is the thickness of deposits between the two successive delta plains;

$T_{mc}$ is the additional increment that accounts for the compaction of sediment after deposition; and

$S$ is the amount of the tectonic subsidence that occurred during deposition.

$T_m$ can be measured at outcrop and $T_{mc}$ may be calculated from sediment compaction curves. $S$ may be estimated if the subsidence history of the coastal plain site is known. The estimate is only a minimum because it does not account for (1) the amount of relative sea-level rise before the top of the delta plain is flooded ($SL_b$) (Figure 5.15) or (2) erosion of the top of the upper delta plain during a succeeding relative sea-level fall. Using this estimate, values for $SL_r$ of 45 m have been obtained. Looking at the problem from a different angle, the volumes of Gondwanan and Quaternary ice sheets have been calculated and compared as an alternative way of calculating $SL_r$, and a figure of around $60 \pm 15$ m has been obtained. This compares quite favourably with the value of 45 m mentioned above.

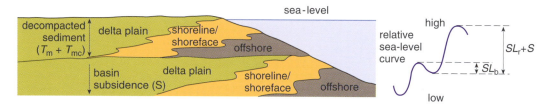

**Figure 5.15**   Illustration of the terms defined in Equation 5.1. Figure 4.8a shows the difference between a relative and eustatic sea-level curve.

This illustrates the difficulties involved in making quantitative determinations of sea-level change from the sedimentary record, though detecting qualitative trends, as we have seen, can be done.

## 5.7   Summary

- Five orders of cyclicity are commonly recognized in the sedimentary record: first order: $c$. 50 to $c$. 200+ Ma; second order: $c$. 5 to $c$. 50 Ma; third order: $c$. 0.2 to $c$. 5 Ma; fourth order: $c$. 100 to $c$. 200 ka; fifth order: $c$. 10 to $c$. 100 ka.

- Sequences stack together to form sequence sets; these are the result of changes in sea-level that are longer term (lower order) than those that produce sequences.

- Eustatic sea-level cycles (operating at different frequencies and magnitude) and basin subsidence are capable of controlling the deposition of parasequences, sequences and sequence sets. These ideas embody the concept of sequence stratigraphy.

- Climatic and tectonic processes are the most important controls on eustatic sea-level and sediment supply.

- Sediment compaction can cause relative sea-level change.

- Eustatic sea-level changes are generally larger in amplitude during icehouse conditions compared to during greenhouse conditions.

- During icehouse conditions, changes in eustatic sea-level are caused mainly by changes in volume of the ice caps which, in turn, are driven by Milankovich climate forcing. During greenhouse conditions, the mechanism controlling eustatic sea-level change is less clear. Possibilities include Milankovich-driven changes in the volume of ocean water due to thermal expansion and contraction, salinity changes, the storage and release of water from mountain ice, inland lakes and aquifers.

- Changes in mid-ocean ridge spreading rates and continental break-up are important in controlling second- and first-order changes in tectono-eustasy. Intraplate stresses may also be important in controlling long-term changes. Active local faulting may cause rapid relative sea-level changes but the changes will be restricted in their lateral extent and non-regular.

- Milankovich-driven climatic processes have a regular and predictable periodicity.

- The Namurian of the Widmerpool Gulf and East Midlands Shelf can be divided into sequences and sequence sets. The cycles were most likely controlled by glacio-eustatic sea-level changes, and appear to be of a similar magnitude to Quaternary glacio-eustatic cycles.

# 5.8   References

## Further reading

MIALL, A. D. (1997) (see further reading list for Chapter 4).

NICHOLS, G. (1999) (Ch. 21) (see further reading list for Chapter 2).

PLINT, A. G., EYLES, N. AND EYLES, C. H. (1992) 'Control of sea level change', (Chapter 2) in WALKER, R. G. AND JAMES, N. P. (eds) *Facies Models: Response to Sea Level Change*, Geological Association of Canada, 15–26. [This provides a concise summary of the controls on sea-level change.]

## Other references

CHURCH, K. D. AND GAWTHORPE, R. L. (1997) (see other references for Chapter 4).

COLLINSON, J. D. (1988) 'Controls on Namurian sedimentation in the Central Province Basins of northern England', in BESLEY, B. M. AND KELLING, G. (eds) *Sedimentation in a Synorogenic Basin Complex: The Upper Carboniferous of NW Europe*, Blackie. Glasgow and London, 85–101.

FRASER, A. J. AND GAWTHORPE, R. L. (1990) 'Tectono-stratigraphic development and hydrocarbon habitat of the Carboniferous in northern England', in HARDMAN, R. F. P. AND BROOKS, J. (eds) *Tectonic events responsible for Britain's oil and gas reserves*, Geological Society Special Publication No. 55, 49–86.

GURNIS, M. (2001) 'Sculpting the Earth from inside out', *Scientific American*, 34–40, W. H. Freeman.

LEEDER, M. R. AND STRUDWICK, A. E. (1987) (see other references for Chapter 3).

LEEDER, M. R. (1999) (see other references for Chapter 4).

# 6 Case study: Quaternary of the Gulf of Mexico

## Angela L. Coe and Kevin D. Church

This Chapter draws together many of the concepts that we have covered in this book so far in order to examine a deltaic sedimentary succession from the north-east Gulf of Mexico, some 60 km east of the modern Mississippi delta, USA (Figure 6.1). The succession is in a distal location, on the shelf edge of the Gulf of Mexico, and was deposited during the glacial and interglacial conditions of the Quaternary. This is an important area in which extensive work has been carried out to improve our understanding of when and why different facies are deposited in response to changing sea-level. It is no coincidence that this is an area of intense oil exploration, and a research consortium consisting of most major oil companies joined forces to produce the data and interpretation. In addition to understanding when and where facies were deposited in response to changing sea-level, the consortium also constructed a depositional model for the shelf edge sedimentary deposits. This model can be used to predict the location of sandstones that act as potential reservoirs for hydrocarbons. The data and interpretation were published in a scientific paper in 1998 which is referenced at the end of this Chapter.

**Figure 6.1** Location map for the north-east Gulf of Mexico, showing two boreholes (Main Pass 288 and Viosca Knoll 774) and the seismic section line shown in Figures 6.6 and 6.8. Contours (m) indicate the thickness of the delta ('isopachs') deposited during its last major period of progradation which occurred towards the end of the last glaciation. The basinward limit of progradation is also shown. The 80 m water depth contour marks the location of the present-day shelf edge. The 'bull's-eye' patterns shown by the contours were caused by the more recent upward movement of the sedimentary deposits due to salt diapirs. (Winn et al., 1998.)

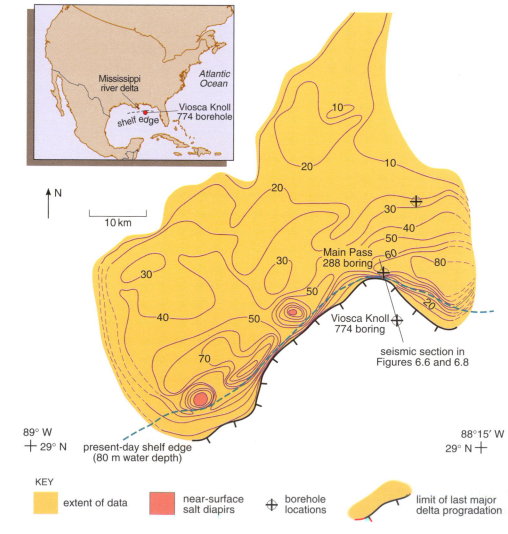

KEY

extent of data

near-surface salt diapirs

borehole locations

limit of last major delta progradation

This study utilizes many different techniques to date the sedimentary succession and to measure eustatic sea-level fluctuations. It considers gamma-ray logs together with core samples taken from a borehole; the latter were analysed for sedimentology, microfossil content and oxygen-isotope data. In addition to the borehole data, seismic reflection data allowed consideration of the regional-scale geology, and the integrated data set has enabled a sequence stratigraphical model to be developed for the succession.

## 6.1   Geological setting

Historical records show that the Mississippi river delta has avulsed (changed course) many times during the past few centuries. However, of greater significance are changes that have occurred in the recent geological past over periods of tens to hundreds of thousands of years (Figure 5.8). These longer-term changes are associated with glacio-eustatic sea-level fluctuations. During falling relative sea-level (i.e. during glaciation), the reduction in accommodation space forced the delta to prograde rapidly towards the edge of the continental shelf, down onto the shelf slope and into much deeper water, thereby increasing the extent of the shelf.

○  What effect would relative sea-level fall have had on older deltaic deposits further landward?

●  Older deltaic deposits would be incised by rivers on the delta plain as relative sea-level fell. (This is exactly the same as the Carboniferous example described in Sections 3.5, 4.4 and 5.6.)

During deglaciation and rising relative sea-level, the incised valleys were filled with sediment, and fine-grained deposition occurred across most of the shelf, slope and deep basin as the deltas were forced in a landward direction (i.e. to the north-west). As the glacial conditions returned and relative sea-level fell, the delta started to prograde again.

The data gathered from the borehole, called Viosca Knoll 774 boring (Figure 6.1), tied to seismic reflection profiles, has been used to calibrate these proximal–distal translocations of the deltaic shoreline to known periods of glaciation.

## 6.2   The Viosca Knoll borehole

The Viosca Knoll borehole was drilled in the Gulf of Mexico, on the shelf slope in 182 m of water about 10 km beyond the present-day shelf edge (Figure 6.1). The borehole was positioned here because it enabled the consortium to examine the effect of sea-level change on the progradation of deltas close to the shelf edge. It was drilled to a depth of 245 m and was continuously cored, though only 85% of the sedimentary succession was actually retrieved for laboratory analysis.

### 6.2.1   Sedimentology

Figure 6.2 shows a summary lithological log of the core, a natural gamma-ray log (Box 3.2) and a graph of the composition of the sand-sized fraction of the sediment. The uncoloured part of the right-hand graph represents the percentage of grains >63 μm composed of diagenetic minerals (e.g. carbonate, glauconite (c. 0–30%) and shell fragments (c. 0–10%)).

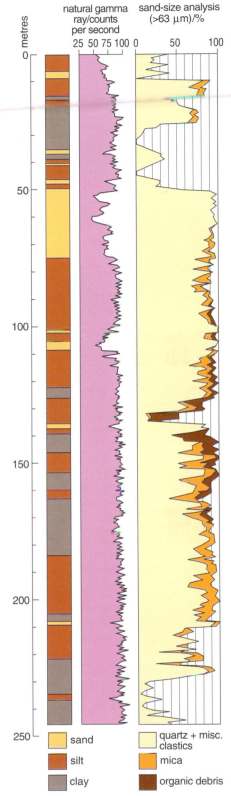

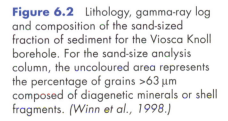

**Figure 6.2** Lithology, gamma-ray log and composition of the sand-sized fraction of sediment for the Viosca Knoll borehole. For the sand-size analysis column, the uncoloured area represents the percentage of grains >63 μm composed of diagenetic minerals or shell fragments. *(Winn et al., 1998.)*

The graphs in Figure 6.2 show both sudden and gradational transitions from mudstone to sandstone and *vice versa*. In Section 3.5, we considered a vertical succession through a delta and saw how a gradational increase in the sand-to-mud ratio over tens of metres can result from the progradation of a delta. The reverse, over a thinner interval, might suggest the gradual abandonment of a delta, initiated perhaps by slow rates of relative sea-level rise. Additionally, a rapid transition from sand to mud might suggest the sudden abandonment of a delta, initiated perhaps by rapid relative sea-level rise.

It is postulated that fluctuations in relative sea-level have caused the sudden changes in the sedimentary succession in the borehole. Such observations will prove to be crucial when we consider the sequence stratigraphy of this area in Section 6.4.

### 6.2.2 Age and palaeoclimates

Though analysis of both the sedimentology and gamma-ray log helps us to identify the changes in depositional environment, it gives only a limited indication of the climate prevailing at the time of deposition and no indication of the age of deposition. Such information is crucial if the sedimentary succession within the Viosca Knoll borehole is to be correlated with those nearby to give a regional picture of how patterns of sedimentation changed with time. To resolve these issues, we must use evidence provided by microscopic calcareous organisms called nanofossils (also sometimes spelt 'nanno-' in the literature) and foraminifers, which are present in their thousands in cores recovered from the borehole.

Four biostratigraphical datums based on nanofossils are recognized in the Viosca Knoll borehole (see right-hand side of nanofossil abundance column on Figure 6.3). These datums (oldest first) and how they correlate to the Pleistocene time-scale (Figure 6.4 overleaf) can be explained as follows:

1   *Pseudoemiliania lacunosa* LAD; this indicates the last appearance datum (LAD) of this particular species and represents either its extinction in time and/or migration out of the area. Correlation of this datum with the Pleistocene time-scale (Figure 6.4) indicates a numeric age of about 460 ka (middle of the Middle Pleistocene).

2   *Emiliania huxleyi* FAD; this indicates the first appearance datum (FAD) of this species and represents either its evolution in time and/or migration into this area. Correlation of this datum to the Pleistocene time-scale (Figure 6.4) together with the datum described in point 3 below indicates a numeric age of 260 ka (late Middle Pleistocene).

3   *Gephyrocapsa caribbeanica* dominance 50% uphole reduction; this indicates that this species shows a 50% reduction in its dominance at this point in the Viosca Knoll borehole. It is correlated with *Gephyrocapsa caribbeanica* reduction (50%) on the Pleistocene time-scale (Figure 6.4). This is not a particularly good datum for correlation but nevertheless provides further biostratigraphical information.

4   *Emiliania huxleyi* Acme; this represents a rapid increase (and subsequent decrease) in the quantity of this species relative to others. Correlation of this datum to the time-scale indicates a numeric age of 90 ka (Late Pleistocene).

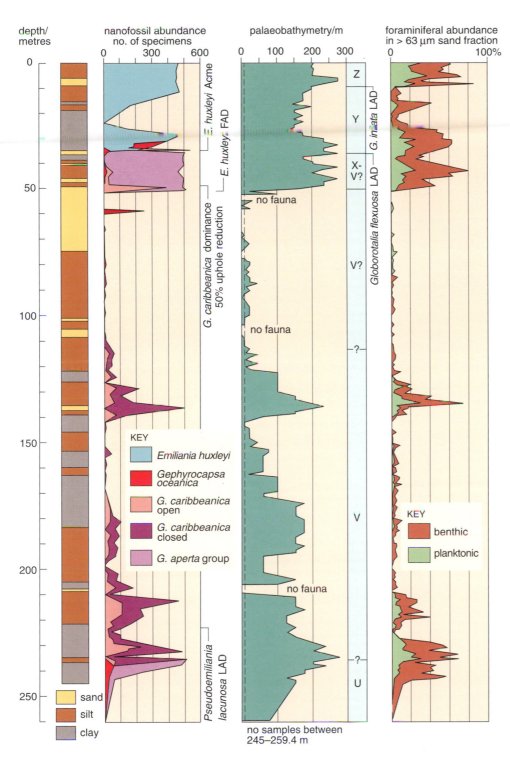

**Figure 6.3** Biostratigraphical data from the Viosca Knoll borehole. This includes the nanofossil abundance, the datums of index nanofossil species, palaeobathymetric estimates, the foraminiferal Ericson biostratigraphical zones (U–Z), two foraminiferal species datums, and a graph of the abundance of foraminifers. Palaeobathymetric estimates were derived from analysis of benthic foraminifers found in the core. (Winn et al., 1998.)

Additional biostratigraphical zonation of the Viosca Knoll borehole is provided by the presence and abundance of planktonic foraminifers of the *Globorotalia menardii/tumida* group and the *Globorotalia inflata* group. These have been used to identify the foraminiferal Ericson biostratigraphical zones (U–Z) shown on the Pleistocene time-scale (Figure 6.4). The position of these foraminiferal zone boundaries in the Viosca Knoll borehole is shown in Figure 6.3.

Though some controversy exists over the exact positioning of the biostratigraphical datums and the numeric ages assigned to them (Figure 6.4), the use of nanofossils and foraminifers would suggest that the strata penetrated near

**Figure 6.4** The mid-Pleistocene to Holocene oxygen-isotope stages, planktonic foraminiferal zones (Ericson zones), nanoplankton datums, oxygen-isotope curve and interpreted eustatic sea-level curves. Even-numbered isotope stages = cool periods; odd-numbered isotope stages = warm periods. In the majority of cases, even and odd stage numbers refer to conventional glacial and interglacial episodes respectively inferred from the terrestrial Quaternary stratigraphical record, although it should be noted that isotope stage 3 is anomalous in that, although recognized as a 'warm' stage, it is not regarded as a full interglacial. In addition, two of the 'interglacial' stages have been subdivided into separate warmer and colder episodes. Stage 5 contains five substages, with warmer/colder intervals. Likewise for Stage 7. As explained in the caption to Figure 3.3, isotope stages are defined by the maximum and minimum δ18O values rather than the boundaries between each stage. However, the boundaries are reproduced here as they appeared in the original research paper. *(Winn et al., 1998.)*

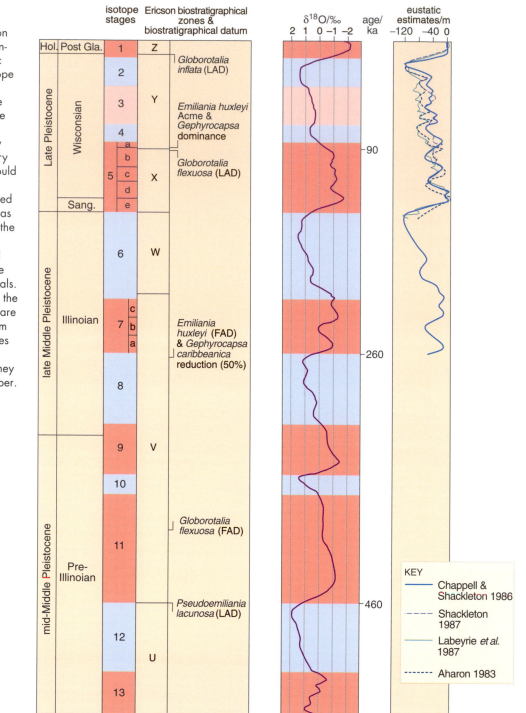

the bottom of the borehole were deposited within the mid-Pleistocene and that at the top was deposited in the Late Pleistocene.

The planktonic foraminifers can also yield direct evidence of temperature change, with different species dominating during warmer (interglacial) and cooler (glacial) periods. *Globorotalia inflata* is a cool-water foraminifer, whilst *Globorotalia menardii* and *Globorotalia tumida* represent tropical to subtropical surface water temperatures. The relative abundance of these foraminifers at different levels in the borehole (not shown on any of the Figures presented here) has been used to support palaeoclimatic evidence gained from other sources (Section 6.2.4).

## 6.2.3   Palaeobathymetry

In the Carboniferous example (Sections 4.4 and 5.6), it was not easy for us to estimate the water depths into which the deltas were prograding. However, in this case study, water depths for the sedimentary deposits in the core were inferred by comparing benthic foraminifers found at certain water depths in the Gulf of Mexico today with the same benthic foraminifers found in the core. The percentage of planktonic foraminifers was also used in palaeodepth estimates: a greater percentage of planktonic foraminifers is generally associated with deeper water.

○   Why must we be wary of interpretations made from foraminifer species recovered from sediments deposited on slopes?

●   Foraminifers inhabiting a certain depth range on a slope may subsequently become displaced downslope as they are incorporated into the sediments. On that evidence alone, the sediments from which they were recovered might be assigned to a lesser water depth (based on the depth at which the foraminifers live today) than was actually the case.

○   How can we overcome this problem?

●   By using the species indicating the maximum water depth.

Samples were taken at closely spaced intervals in order to create a log of palaeobathymetry versus depth in the borehole (Figure 6.3). This clearly shows that sedimentary rocks down to 50 m in the borehole were deposited in depths around 200 m, sedimentary rocks from 50 to 120 m down the borehole were deposited in very shallow water depths at around 20 m, whilst below 120 m down the borehole, water depth fluctuated between 0 and 280 m when the sediments were deposited. This information does not directly indicate the rise and fall of relative sea-level because the water depth information is also a function of the sediment being deposited in that part of the basin at that time. The data therefore indicate those times when accommodation space is being created (e.g. time represented by rocks at 50 m depth, Figure 6.3) or destroyed, due either to relative sea-level fall or to delta progradation (e.g. time represented by the rocks at 135–110 m depth).

## 6.2.4   Oxygen isotopes

During alternations between glacial and interglacial conditions, the oxygen-isotope composition of seawater is dominated by the amount of evaporation and precipitation as the ice-caps wax and wane (Section 3.2.1). During glacial conditions, seawater has higher, less negative $\delta^{18}O$ values than during interglacial conditions (Figure 6.4). Figure 6.5 (overleaf) shows a plot of $\delta^{18}O$ values taken from samples of the planktonic foraminifer *Globigerinoides ruber* in the Viosca Knoll core, versus depth. The pattern of oxygen-isotope variation within *G. ruber* is considered to reflect global changes in the isotopic composition of seawater caused by changes in the volume of glacial ice (this is based on other studies and comparison with benthic foraminiferal records which are usually used, Section 3.2.1). In general, for the late Quaternary, values greater than +0.2‰ $\delta^{18}O$ for *G. ruber* represent full glacial conditions, –0.45 to –0.9‰ $\delta^{18}O$ represent a transition between glacial and interglacial conditions, and values less than –1.5‰ $\delta^{18}O$ (i.e. more negative) suggest full interglacial conditions in the absence of glacial meltwater, high-latitude rainwater brought down by rivers, or diagenetic alterations. Local variations in temperature and salinity of the seawater in the Gulf of Mexico are not thought to have significantly affected the Viosca Knoll borehole isotopic record.

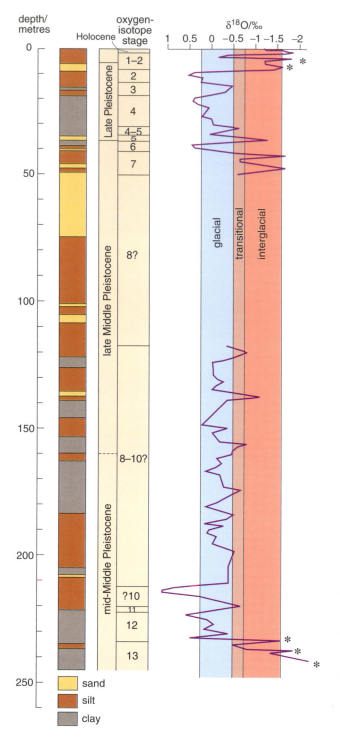

**Figure 6.5**  Oxygen-isotope stratigraphy and interpretation of the oxygen-isotope stages, glacial, transitional and interglacial conditions from Winn *et al.* (1998) for the Viosca Knoll borehole data. The isotopic measurements were all made on the planktonic foraminifer *Globigerinoides ruber* relative to the PDB standard (Section 3.2.1). Note that the range of values used to define the glacial, transitional and interglacial conditions here are different from those generally used for the late Quaternary (see p. 123). The relevance of the asterisks on the $\delta^{18}O$ curve is discussed in the text. *(Winn et al., 1998.)*

The oxygen-isotope data in Figure 6.5 show a progressive shift from interglacial to glacial conditions from the bottom of the borehole at *c.* 240 m up to *c.* 215 m. From 215 to 120 m, glacial conditions alter in magnitude but interglacials never became established. No $\delta^{18}O$ values were obtained between 120 and 50 m.

○  Consider the composition of the sand-sized fraction of the sediments (Figure 6.2) and palaeobathymetry (Figure 6.3) for the interval 120 to 50 m and suggest why no oxygen-isotope data are available for this part of the borehole.

●  Figure 6.2 shows that the composition of this interval was dominated by quartz sand, especially towards the top. Figure 6.3 shows that the palaeobathymetry during this interval was often less than 20 m, and towards the top it was virtually zero. Both these facts suggest the development of a high-energy environment (possibly a shoreline or delta mouth bar) in which these microfossils would have had a very low preservation potential. So, it is likely that no $\delta^{18}O$ values were obtained between 120 to 50 m because there were no foraminifers to sample.

Above 50 m, the $\delta^{18}O$ values suggest that glaciation became progressively established, with a few intervening periods of interglacial conditions. The top 10 m of the borehole contains strongly negative values of $\delta^{18}O$, representing the interglacial period that we are currently experiencing.

In general, the $\delta^{18}O$ curve in Figure 6.5 is quite jagged, suggesting rapid alternation of conditions between glacial and transitional or interglacial and transitional conditions. It is likely that some of these $\delta^{18}O$ shifts are not due entirely to the changes in ice volume, instead being caused by the introduction of meltwater and/or high-latitude precipitation run-off with highly negative values of $\delta^{18}O$ into the area.

○  What is the likely origin of these waters in the Gulf of Mexico?

●  The Mississippi river.

Figure 6.1 shows that our area of interest lies some 60 km east of the mouth bar of the modern Mississippi river. In the past, this mouth bar would have been even more proximal, especially during times of low sea-level, and it is likely that the sandier intervals in Figure 6.2 represent times when the mouth bar reached the Viosca Knoll borehole site. This river would have introduced large quantities of freshwater into the Gulf of Mexico with very low $\delta^{18}O$ values compared to those of the surrounding seawater. The northern headwaters of the Mississippi would have drained glaciers during the early stages of deglaciation, to produce meltwaters with $\delta^{18}O$ values below $-20.0$‰, and high-latitude rainwater would have introduced water into the river with $\delta^{18}O$ values of $-7.0$‰. Combined with the seawater $\delta^{18}O$ value, this would produce the sharp negative spikes of $-1.0$ to $-2.0$‰ in Figure 6.5, giving the illusion of short interglacial periods where none in fact existed. Suspect spikes are marked with an asterisk (*) in Figure 6.5.

## 6.2.5   Correlating Viosca Knoll with known periods of glaciation

We now have a reasonably good idea of the sedimentary succession of the Viosca Knoll borehole, and have used microfossils to interpret its age and variations in bathymetry during deposition of the sediments. In addition, both oxygen isotopes and microfossil identifications have revealed changes in the prevailing climate.

For the Viosca Knoll data to be relevant on a global scale, we need to tie the borehole isotopic variation to the oxygen-isotope stages as shown on Figure 6.4. This correlation is shown in Figure 6.5 and discussed further in Section 6.4. As we know approximately how long each isotope stage lasted, it is possible (knowing the thickness of sedimentary rocks at Viosca Knoll) to estimate the rate at which sediment was deposited. Also, in theory, by comparing the thickness of any particular isotope stage in the Viosca Knoll borehole with the thickness of the same stage in other boreholes, it should be possible to produce contour maps to identify areas of the Gulf of Mexico that were centres of sediment deposition (i.e. the position of delta lobes). By comparing such maps for each isotope stage, it would be possible to see how these centres changed with time. Unfortunately, in order to generate a sufficient quantity of data to be statistically meaningful, we would have to drill an enormous number of wells and analyse their cores in detail. As a cheaper and less time-consuming alternative, it is possible to use the seismic reflection technique to gather similar thickness data.

## 6.3   Seismic reflection data

Figure 6.6 (overleaf) is a north-west–south-east oriented seismic section produced from a seismic reflection survey (Box 4.2) in the Gulf of Mexico. The positions of the Viosca Knoll (VK 774) and another borehole MP (Main Pass) 288 are marked on the seismic section. In general, the seismic reflectors dip seaward to the south-east. MP 288 lies just on the break of slope between the present-day shelf (to the north-west) and the continental slope (to the south-east). As also indicated on Figure 6.1, VK 774 lies some 9 km from MP 288 on the shelf slope. Before we interpret this seismic section, two other features that it shows are worthy of note. First, the reflectors marked 'multiples', crossing the MP 288 borehole track at 0.2 s two-way time, are not 'real' reflectors. We are interested only in those seismic ray paths that travel downwards, are reflected once by the rock layers and travel back up to the detector. However, there are other ray paths that take more tortuous routes; these produce 'multiples'. In Figure 6.6, the multiples are produced by a ray path being reflected upwards off the sea-bed, then reflected downwards again off the sea-surface (which acts as a very efficient reflector), then repeating its first journey over again. This doubles the overall travel time, producing a false reflector at a depth twice that of the time depth of the sea-bed reflection. Secondly, three faults are interpreted towards the middle of Figure 6.6. Because of their steep angle of dip, faults are rarely imageable on seismic sections. Instead, their presence coincides with

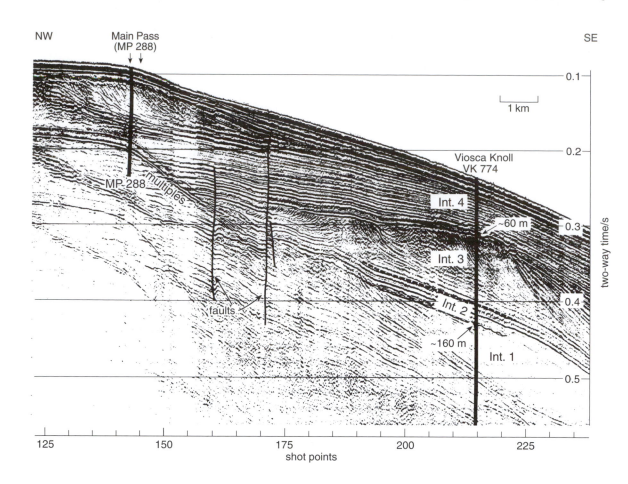

**Figure 6.6** North-west–south-east oriented seismic section through the Viosca Knoll borehole (VK 774). The positions of this seismic section and another borehole, MP (Main Pass) 288, are shown on Figure 6.1. The depths of 60 m and 160 m in VK 774 were derived from analysis of the acoustic properties of the rocks within the borehole (see Box 4.2). *(Winn et al., 1998.)*

displacements in the reflectors on either side of the fault. By matching up these reflectors from one side to the other, it is possible to measure the throw of the fault (though of course this is measured in time, not depth).

At VK 774, the seismic section shows four distinctive intervals (labelled 1–4) based on variations in the angle of dip of the reflectors. The boundary between Interval 3 and Interval 4 at 0.315 s two-way time is especially clear, being marked by the truncation of steeply dipping reflectors (clinoforms) at the top of Interval 3 by more gently dipping reflectors representing Interval 4. Because the seismic section is measured and displayed in time and not depth, the dip of the reflectors is *greater* than would be shown on a geological cross-section. Interval 4 is characterized by gently dipping reflections that continue up to the present day sea-bed. A similar relationship exists between Intervals 1 and 2 at about 0.430 s two-way time. Both boundaries (between Intervals 1 and 2, and that between Intervals 3 and 4) are likely to represent unconformities in the core taken from VK 774. The clinoforms of Intervals 1 and 3 are interpreted as prograding deltas, whereas the more gently dipping reflectors of Intervals 2 and 4 represent the deposition of near-horizontal beds on the shelf/slope during less active periods of sedimentation.

○ Why is the boundary between Intervals 2 and 3 harder to define?

● Because the change in dip angle is not as great. This is because it represents the change from the gently dipping shelf/slope beds of Interval 2 and the conformable downlapping of the clinoforms of Unit 3, compared to the unconformable top of Intervals 1 and 3.

The gently dipping reflectors of Intervals 2 and 4 may appear to represent a fairly simple depositional history. However, the lateral equivalents of these reflectors in MP 288 suggest that this is an oversimplification.

○   Why should this be so?

●   Below 0.11 s in MP 288, there lies another area of steeply dipping clinoforms, which are younger than the Interval 3 clinoforms at Viosca Knoll.

The position of the prograding clinoforms in this delta defines the boundary between the present-day shelf and slope. As with the other two deltas, this delta is overlain by an unconformity and the offshore equivalent of this must intersect VK 774 at about 0.26 s, within Interval 4. This portion in the Viosca Knoll borehole is, therefore, more complex than the reflectors close to the borehole would have us believe. This shows the value of using seismic sections to correlate and interpolate between boreholes. The complexity of Interval 4 should come as no surprise if you glance back at the different facies shown in the graphic log in Figure 6.5. Analysis of the acoustic properties of the rocks within VK 774 (see Box 4.2) suggests that the base of Interval 4 (at 0.315 s) lies at a depth of about 60 m. On Figure 6.5, the interval down to 60 m in the core represents eight oxygen-isotope stages (1–8), implying a significant variation in temperature and climatic conditions which would lead to large changes in eustatic sea-level and hence the sediments deposited. Combining the borehole and seismic data enables a sequence stratigraphical model of the area to be constructed.

## 6.4   Sedimentology and sequence stratigraphy of the Viosca Knoll borehole

In the Carboniferous case study (Sections 3.5, 4.4 and 5.6), the sequence stratigraphical interpretation was based solely on the sedimentology of the deltas. In contrast, this study benefits greatly from data which allow us to interpret directly palaeowater depths (from benthic foraminifers) and periods of glaciation (from oxygen-isotope values). It is therefore possible to identify specific systems tracts relating to high, intermediate and low relative sea-level and then to examine the sedimentology of the deposits that make up each systems tract.

### 6.4.1   Depositional history

An idealized depositional sequence for this succession showing the characteristics of each systems tract is shown in Figure 6.7a (overleaf). The sequence stratigraphical interpretation of the entire Viosca Knoll borehole is shown on Figure 6.7b. The interpretation presented here differs slightly from that in the published paper on this data because it takes into account recent advances in sequence stratigraphy and is presented as an interpretation consistent with the theory presented in this book. Figure 6.8 (overleaf) is the seismic section you have already studied (Figure 6.6), showing a possible sequence stratigraphical interpretation.

In total, there are seven sequences present in Viosca Knoll, each composed of up to four systems tracts of greatly varying thickness: LST, TST, HST and FSST (Figure 6.7b). It is sometimes difficult to distinguish between the FSSTs and LSTs, so these have been grouped together; however, the graduation in colour on Figure 6.7b is intended to give some idea where the boundary between these two systems tracts might lie. The succession has been divided into intervals A to O. These are shown in the middle column of Figure 6.7b, which is also coloured to show the individual key surfaces and systems tracts.

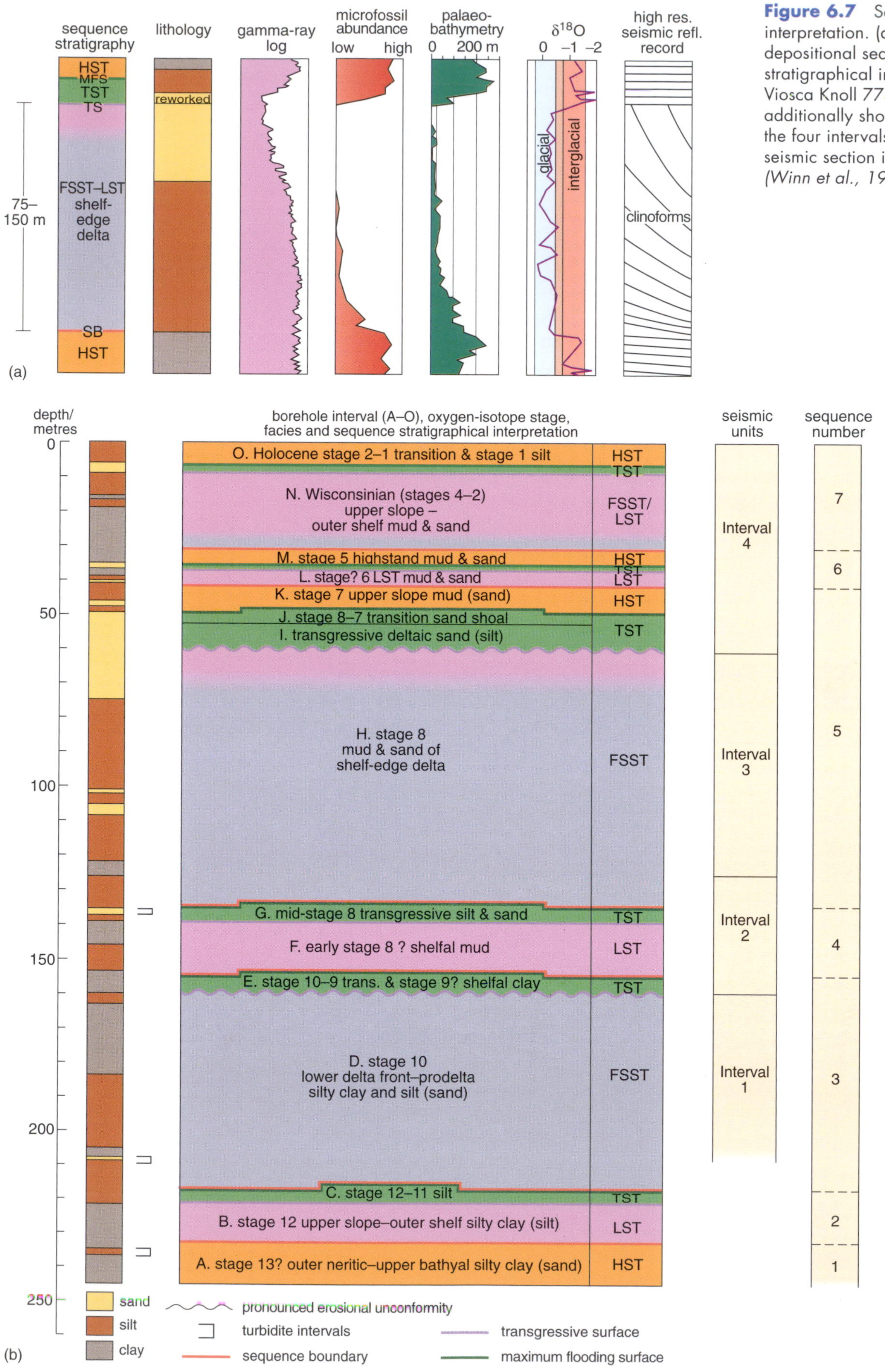

**Figure 6.7** Sequence stratigraphical interpretation. (a) An idealized depositional sequence. (b) Sequence stratigraphical interpretation of the Viosca Knoll 774 borehole, additionally showing the position of the four intervals defined on the seismic section in Figures 6.6 and 6.8. (Winn et al., 1998.)

Interval A (245–232 m core depth, Figure 6.7b) is interpreted as a HST: its low $\delta^{18}O$ values make it most likely that it corresponds to isotope stage 13 (Figure 6.5). This correlation is not entirely clear because of the possible overprint of the oxygen-isotope glacio-eustatic signal by the Mississippi salinity signal. Increasing microfossil abundance and palaeobathymetry within this interval (Figure 6.3) are all consistent with rising sea-level.

There is a sudden positive shift in the $\delta^{18}O$ values at the contact between intervals A and B (232 m core depth; Figure 6.5) suggesting that there is a gap in the section at Viosca Knoll and that the gradual increase in $\delta^{18}O$ between isotope stages 13 and 12 is absent (Figure 6.4). Interval B (232–221 m core depth, Figure 6.7b) is interpreted as a LST: this fits with the decreasing abundance of microfossils and decreasing palaeobathymetry (Figure 6.3).

Interval C (221–218 m depth, Figure 6.7b) is interpreted as a TST because it has intermediate $\delta^{18}O$ values that most likely correspond to part of isotope stage 11. It is thought that this isotope stage may not be well expressed due to bioturbation mixing the signal or insufficient sample density. This transgression is marked by a moderate increase in the abundance of microfossils which is consistent with the transitional conditions suggested by the isotope data (Figures 6.3, 6.5). The overlying HST is either thin or absent, and this is interpreted to be because the delta did not prograde into this area at this time.

Interval D (218–160 m core depth, Figure 6.7b) corresponds to high $\delta^{18}O$ values indicating glacial conditions and low eustatic sea-level (Figure 6.5). This is the lowest level shown on the seismic section (Interval 1; Figure 6.8) and it is characterized by clinoforms consistent with rapid progradation of a shelf margin delta during falling relative sea-level; it is thus interpreted as a FSST. Interpretation of this interval as a FSST also fits with the low microfossil abundance (Figure 6.3).

**Figure 6.8** North-west–south-east oriented seismic section as shown in Figure 6.6, showing a possible sequence stratigraphical interpretation. For key to colours, see Figure 6.7. (Winn et al., 1998.)

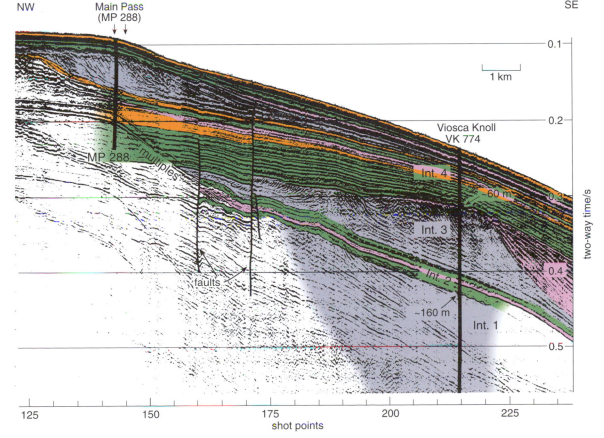

The seismic section (Figure 6.8) indicates pronounced onlap between seismic intervals 1 and 2 which correlates with the top of the prograding clinoforms of Interval D and its contact with the overlying Interval E (160–156 m core depth, Figure 6.7b). This surface is interpreted as a transgressive surface and the base of a TST. Interval E shows a decrease in $\delta^{18}O$ values consistent with rising relative sea-level (Figure 6.5).

The siltstones of interval F (156–139 m core depth, Figure 6.7b) are again characterized by intermediate $\delta^{18}O$ values indicating glacial conditions, but there is no recorded change in palaeobathymetry or microfossil abundance. This interval is interpreted as a LST.

Interval G (139–139 m core depth, Figure 6.7b) shows a pronounced negative shift in $\delta^{18}O$ values (probably isotope stage 9) indicating interglacial conditions and hence higher relative sea-level. This is consistent with the pronounced increase in microfossil abundance, increase in palaeobathymetry (Figure 6.3) and reworking of the sediments to give coarser-grained lithologies. This interval is interpreted as a TST. Similar to sequences 2 and 3, the HST for this sequence does not seem to be preserved in this area.

Interval H (136–60 m core depth, Figure 6.7b) corresponds to interval 3 on the seismic section which is an area of prograding clinoforms. In this case, it is clear that the delta has moved down the depositional profile and then prograded into the study area. This indicates a lowering of relative sea-level as do the coarse-grained sediments (Figure 6.2), very low palaeobathymetry estimates, and low microfossil abundance (Figure 6.3). The angular contact seen on the seismic section between the underlying HST and this FSST suggest that it is a detached FSST (Figure 6.9). The uppermost part of interval H probably represents a LST.

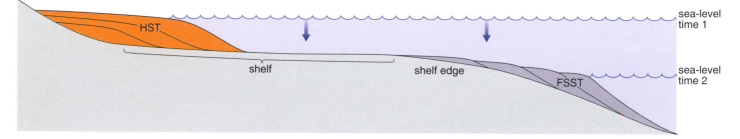

**Figure 6.9** Model for the deposition of a delta during a substantial relative sea-level fall on a marine shelf. If relative sea-level falls fast enough, the site of sediment deposition will shift rapidly from the shelf to the shelf edge and cause the HST and FSST components to become detached. As relative sea-level falls to its lowest point, erosion of the HST will commence.

Intervals I and J (60–50 m core depth, Figure 6.7b) are interpreted as a TST because of the increase in microfossil abundance, the palaeobathymetry estimates (Figure 6.3), the onlap shown on the seismic section (Figure 6.8), and the correspondence with low $\delta^{18}O$ values indicating the interglacial conditions of isotope stage 7 (Figure 6.5).

Interval K (50–42 m core depth, Figure 6.7b) is interpreted to represent a HST deposited during the last part of isotope stage 7. The mixture of silt and sand fits with less reworking than in the coarser-grained underlying TST.

Interval L (42–37 m core depth, Figure 6.7b) is interpreted as a thin LST deposited during isotope stage 6 which infers glacial conditions and low sea-level.

Interval M (37–31 m core depth, Figure 6.7b) is interpreted as a HST with possibly a thin TST at the base where there is sand with reworked fossils. The $\delta^{18}O$ values are high and indicate part of isotope stage 5 (Figure 6.5).

Interval N (31–9 m core depth, Figure 6.7b) contains high $\delta^{18}O$ values indicating glacial conditions (Figure 6.5). This fits with the reduced microfossil abundance and palaeobathymetry (Figure 6.3). However, the changes are not as well marked as in the FSST/LST of intervals H and D. Examination of the seismic section also shows no prograding clinoforms at the Viosca Knoll site, although the seismic section does show prograding clinoforms about 10 km to the north-west in equivalent age strata. This indicates that the delta did not prograde as far as Viosca Knoll at this time and provides an explanation for the higher palaeobathymetry values than in the underlying FSSTs.

Interval O (9–0 m core depth, Figure 6.7b) has low $\delta^{18}O$ values (Figure 6.5) but, like interval A, they are probably enhanced by the signal from Mississippi meltwater during this last interglacial phase. The microfossil abundance and palaeobathymetry estimates (Figure 6.3) all fit with rising relative sea-level and deposition of this TST and HST. This is consistent with the belief that we are currently in an interglacial period. The succeeding FSST/LST of this sequence will be deposited in the forthcoming ice age.

### 6.4.2  Summary of the general features of the systems tracts

#### General features of the highstand systems tracts

Planktonic foraminifers within the HSTs indicate warm or warming climatic conditions and warm ocean temperatures. Benthic foraminifers indicate deposition generally in water depths greater than 150 m (palaeobathymetry; Figure 6.3). Oxygen-isotope compositions are low, consistent with full interglacial conditions (Figure 6.4). Sharp negative $\delta^{18}O$ spikes in sequences 1 and 7 (Figure 6.7b) suggest meltwater events, bringing glacial meltwaters into an otherwise unglaciated area (see Section 6.2.4). All of these factors point to periods of reduced ice volume, which resulted in high eustatic sea-level.

The sediments deposited during the HSTs are generally thin, highly bioturbated and very fossiliferous clays and silts that settled out of suspension distant from the deltas. These are in fact sometimes so thin that they are not distinguished on the core log for all sequences. HST deposits correspond to low angle, seaward-dipping reflectors on the seismic section (Figures 6.6 and 6.8). In contrast to the proximal highstand deposits of the Carboniferous delta (Section 4.4), the Viosca Knoll deposits are much more distal and therefore much thinner. The Viosca Knoll HST deposits are the toes of prograding HST clinoforms.

#### General features of falling stage and lowstand systems tracts

The oxygen-isotope composition is at its maximum in the FSST and LST deposits. Thus, the isotope values are more positive than in the previous HSTs indicating mainly glacial periods and low sea-levels. Benthic foraminifers generally indicate deposition in water depths less than 150 m (Figure 6.3). The cool water planktonic foraminifer *Globorotalia inflata* is rare to abundant in these systems tracts.

The sediment thickness for the FSST and LST tract between different sequences is highly variable, ranging from 5 m up to 75 m. However, in general, the FSST and LST are the thickest systems tracts in the Viosca Knoll borehole. In the thickest examples, bioturbated mudstones grade upwards to partially contorted, parallel-laminated and structureless coarse-grained sandstone, interpreted as prodelta and delta front deposits.

The tops of some FSST deltas appear to have been eroded. This can be clearly seen from the way in which the steeply dipping reflectors in Figures 6.6 and 6.8 appear to be erosionally truncated by the more gently dipping reflectors which overlie them. (This erosion marks the boundary between seismic Intervals 1–2 and 3–4.) These erosion surfaces were created by: (i) subaerial erosion as relative sea-level fell further below the shelf edge to expose the delta top whilst the delta was still prograding into the more distal area; and (ii) submarine erosion which would have occurred as relative sea-level rose and transgressed over the area. This interpretation has far-reaching implications, as it suggests this relative sea-level fall was capable of exposing the entire shelf margin. This erosion marks the time of lowest sea-level, which corresponds to the period of maximum glaciation.

○ Why are the FSSTs and LSTs of such variable thickness?

● Their thickness is related to the degree of delta progradation into the area. As sea-level fell, delta lobes prograded out towards the shelf edge. However, some may have moved in other directions, never reaching the Viosca Knoll borehole site, whilst others may have been heading in the right direction, but failed to prograde far enough.

An example of the latter is the steeply dipping reflectors near the top of MP 288 borehole (two-way time 0.11 s–0.18 s, Figures 6.6, 6.8). The contours given in Figure 6.1 represent the thickness of this delta ('isopachs') deposited during the last major period of progradation (i.e. the last ice age). Figure 6.1 shows that the FSST delta in MP 288 in fact lies at the western edge of a large lobe which exceeds 80 m in thickness.

### General features of the transgressive systems tracts

The TSTs are characterized by benthic foraminifers which indicate water depths were increasing (Figure 6.3). In some cases, the change in palaeobathymetry is very rapid, e.g. at the base of the TST in sequence 6. Decreasing $\delta^{18}O$ values suggest a change to interglacial conditions, which is consistent with a rise in sea-level as the glaciers melted.

Like the deposits of the HSTs, the TSTs are lithologically variable, thin and bioturbated. Their composition is related to the lithology over which transgression occurred. This suggests that as relative sea-level rose and the shoreline transgressed over the previously deposited FSSTs/LSTs (themselves eroded during subaerial exposure), there was much reworking of the underlying sediment by marine waves and currents. Thus, most of the TSTs are condensed. This is generally supported by the high abundance of nanofossils (Figure 6.3). Like the HST, the TST deposits correspond to low-angle, seaward-dipping reflectors on the seismic section (Figures 6.6, 6.8).

## 6.5   Conclusions and summary

- The integration of different techniques has improved our understanding of the history of deposition of the Viosca Knoll area of the Mississippi delta during the past 500 ka. These techniques include gamma-ray logs and cores taken from a borehole. From the latter, the microfossil content, oxygen-isotope composition and the sedimentary facies have been analysed. Foraminifers have helped to constrain the water depths at the time of deposition and changes in the prevailing climate. Nanofossils and foraminifers have allowed biostratigraphical dating of the sedimentary succession. The oxygen-isotope composition of the foraminifers has been used to infer directly eustatic sea-level fluctuations. By correlating this information to the sedimentology of the core, we have been able to assess how changing sea-level has controlled the style of sediment deposition. The boreholes have been extended to the regional scale by studying seismic reflection data.

- Sequence stratigraphical analysis of the data shows that the succession can be divided into seven sequences representing major cycles of glaciation and deglaciation.

- For each glacial/interglacial cycle, the onset of an interglacial period shifted the shoreline landward out of the study area. This is interpreted as the transgressive surface and TST. The landward shift in facies has resulted in the TST in the study area comprising reworked sediments. As the rate of relative sea-level rise slowed down, HST deltas prograded out over the shelf. However, unlike the Carboniferous case study (Sections 3.5, 4.4, 5.6), the Viosca Knoll study area is too distal and the deltas never reached this far out during deposition of the HSTs. Instead, the HSTs are here represented by thin, bioturbated shelf mudstones and siltstones. In common with the thinner FSST/LST deposits, they are depicted by gently seaward-dipping reflectors. Such reflectors are present in seismic intervals 2 and 4 where they represent groupings of these three systems tracts. As relative sea-levels start to fall at the onset of the next glacial period, the reduction in accommodation space on the shelf increases the rate of delta progradation into the basin and into the study area; these deltaic sediments are interpreted as the FSST/LST systems tracts. This progradation corresponds to the high-angle seaward-dipping reflectors of seismic intervals 1 and 3.

- Whilst the sequence stratigraphical model presented appears to be consistent with all the data available, there are many gaps in the data which mean that it will almost certainly require modification. Still, to date, few studies have been as comprehensive in aiding our understanding of the response of sediment deposition to the waxing and waning of the Pleistocene/Holocene ice sheets.

## 6.6   Reference

WINN, R. D., JR., ROBERTS, H. H. AND KOHL, B. (1998) 'Upper Quaternary strata of the Upper Continental Slope, northeast Gulf of Mexico: Sequence stratigraphical model for a terrigenous shelf edge', *Journal of Sedimentary Research*, **68**, 579–595. [This is the research paper on which this Chapter is based.]

# PART 3 SILICICLASTICS CASE STUDY: THE BOOK CLIFFS

# 7 Tectonic setting, stratigraphy and sedimentology of the Book Cliffs

*John A. Howell and Stephen S. Flint*

Chapters 7 to 10 present a case study of Cretaceous siliciclastic sediments deposited in a foreland basin and now exposed in the Book Cliffs, Utah, USA. This case study will consider the applicability of the sequence stratigraphical concepts, introduced in Chapters 4 and 5, to sediments deposited in a convergent plate margin setting during greenhouse climate conditions. Chapters 7 to 10 are also designed to give a virtual 'hands-on' experience of assembling the practical 'toolbox' of methodologies with which to undertake sequence stratigraphical analysis anywhere in the world. To this end, Section 7.4 is a refresher of shallow-marine and non-marine sedimentology applicable to the Book Cliffs because an understanding of the sedimentary facies always forms the starting point for outcrop-based sequence stratigraphy.

## 7.1 Introduction to the geology of the Book Cliffs

The Book Cliffs of eastern Utah contain some of the best-exposed shallow-marine and coal-bearing coastal deposits in the world. A cliff up to 300 m high runs almost continuously for 300 km from the town of Helper in Utah (Figure 7.1 overleaf) to Grand Junction in Colorado. The cliff is cut by numerous canyons, which add an important three-dimensional quality to the exposures. The strata within the cliffs dip gently at between 3° and 7° towards the east and there are no major folds or faults (Figure 7.2 overleaf). Much of the area is semi-desert and consequently there is very little vegetation covering the rocks. This is especially true in the south-east of the area whereas the area to the north-west is higher in elevation and slightly more vegetated.

There is a long history of geological study in the Book Cliffs area, initially fuelled by interest in the extensive coal measures in the north-west of the region. Subsequent attention turned to the shallow-marine deposits that interfinger with the coals.

As facies models became the accepted method for studying sedimentary rocks during the 1970s, the Book Cliffs were frequently cited as the classic example of shoreface and wave-dominated delta systems. Since the late 1980s, with the advent of high-resolution sequence stratigraphy, attention has returned to the area, which has provided an ideal field laboratory for furthering, testing and teaching sequence stratigraphical concepts. Staff of Exxon Production Research Company carried out much of the pioneering work. More recently, a plethora of universities and oil company research groups have continued these investigations and the work done by these groups forms the basis for this case study.

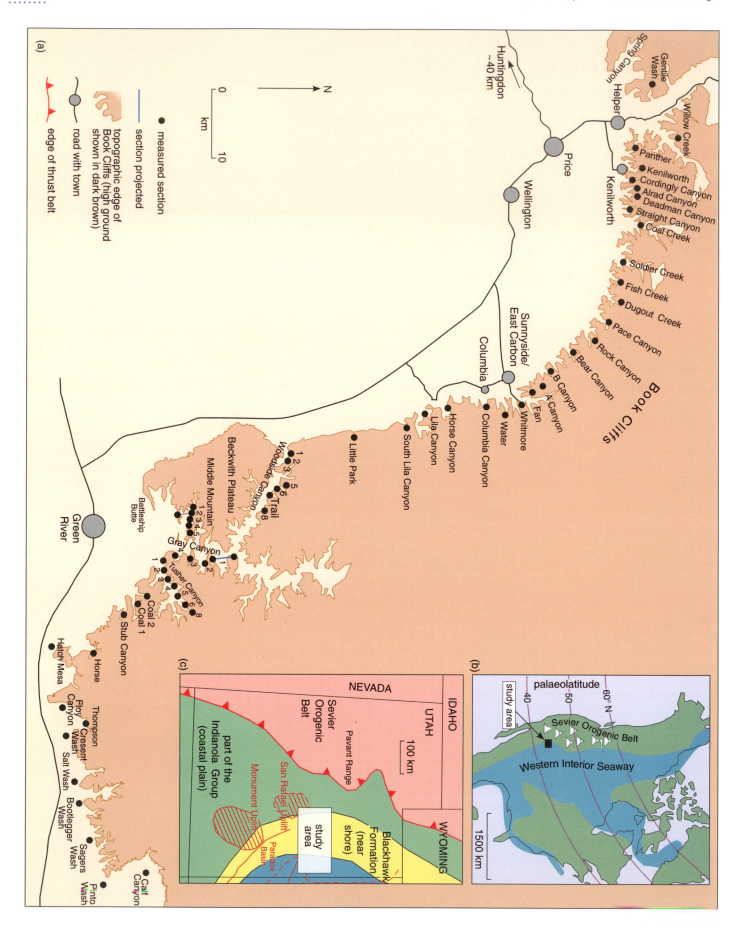

(a)

**Legend:**
- ● measured section
- | section projected
- ◠ topographic edge of Book Cliffs (high ground shown in dark brown)
- ⌐ road with town
- ◣ edge of thrust belt

N →

km
0
10

**Map labels (a):**
Huntingdon ~40 km

Book Cliffs:
Gentile Wash, Spring Canyon, Willow Creek, Panther, Helper, Kenilworth, Cordingly Canyon, Alrad Canyon, Deadman Canyon, Straight Canyon, Coal Creek, Soldier Creek, Fish Creek, Dugout Creek, Pace Canyon, Rock Canyon, Bear Canyon, B Canyon, A Canyon, Fan, Whitmore, Water

Price, Wellington, Kenilworth, Columbia, Sunnyside/East Carbon, Columbia Canyon, Horse Canyon, Lila Canyon, South Lila Canyon, Little Park

Green River, Battleship Butte, Middle Mountain, Beckwith Plateau, Woodside Canyon, Trail, Gray Canyon, Tusher Canyon, Coal 2, Coal 1, Stub Canyon, Horse, Hatch Mesa, Floy Canyon, Thompson, Cresent Wash, Salt Wash, Bootlegger Wash, Sagers Wash, Pinto Wash, Calf Canyon

(c)
NEVADA
IDAHO
UTAH
WYOMING
Sevier Orogenic Belt
Pavant Range
part of the Indianola Group (coastal plain)
San Rafael Uplift
Monument Uplift
Paradox Basin
Blackhawk Formation (near shore)
study area
100 km

(b)
palaeolatitude
40  50  60° N
study area
Sevier Orogenic Belt
Western Interior Seaway
1500 km

(a)

**Figure 7.2** (a) Beckwith Plateau, south of Woodside Canyon. This general view shows the exposure quality in the central and southern part of the Book Cliffs. The continuous pale brown bands are shallow-marine shoreface deposits and fluvial sandstones; the greyish-brown intervals are the offshore deposits of the Mancos Shale. The main cliff is *c.* 300 m high.
(b) Woodside Canyon and the modern-day Price River. This aerial view shows the superb three-dimensional exposure in the canyons which dissect the main Book Cliffs. The cliffs are *c.* 170 m high. *((a–b) John Howell, University of Bergen.)*

(b)

○ Why would an oil company be interested in areas such as the Book Cliffs: is there likely to be any oil there?

● Actually, there are minor amounts of oil in the Book Cliffs region but that is not why the oil companies are interested. They study there because these superb exposures are the best way to understand the anatomy of geologically similar sedimentary systems elsewhere.

**Figure 7.1 (opposite)** Maps to show the location and generalized Cretaceous palaeogeography of the Book Cliffs. (a) Topographic edge of the Book Cliffs showing location of canyons named in the text. (b) Cretaceous palaeogeography of the USA showing the Western Interior Seaway and the Sevier Orogenic Belt. Position of the study area within the State of Utah is highlighted. (c) Campanian (Late Cretaceous) palaeogeography of Utah. Sediment was eroded from the uplifted mountain belt to the west and deposited as a broad coastal plain and shoreface system within the Western Interior Seaway. *((c) O'Byrne and Flint, 1995.)*

The Book Cliffs are used as analogues to subsurface hydrocarbon reserves in parts of the Gulf of Mexico, the Niger Delta, the Brent reservoirs of the North Sea and elsewhere. Concepts such as sequence stratigraphy are directly exportable from one set of rocks to another.

## 7.2 Summary geological history and regional tectonic setting — foreland basin, Sevier and Laramide tectonics

During the Cretaceous, opening of the North Atlantic Ocean was associated with a westward drift of the North American Plate. A large subduction zone lay in the region that is now California and a thrust and fold belt formed on the leading edge of the North American Plate as the Pacific Plate was forced beneath it. This period of mountain building was termed the Sevier Orogeny and generated a narrow range of mountains that ran north–south from Alaska to New Mexico. Associated with the growth of this mountain range was the formation of the Western Interior foreland basin to the east of the mountains. Foreland basins form when compressional plate movements create mountain chains that load the lithosphere (Box 3.1). As the lithosphere is elastic, it flexes downward under the weight of the mountain belt and a basin is formed via a process termed flexural loading. As a loose analogue to this process, if we support a thin 2 m long wooden plank at each end and load several house bricks onto it, the plank flexes. If the wooden plank were substituted for a thick railway 'sleeper', the same load would produce much less flexure. This difference in flexural response is referred to as the elastic thickness of the crust. Modern-day analogues to this type of basin include the Sub-Andean basins of Colombia and Argentina. Foreland basins are a prime site for sedimentation because the basin provides the accommodation space and the new mountain chain provides a ready sediment supply.

The oldest deposits (Jurassic) of the Western Interior Basin are predominately non-marine. Continued subsidence, coupled with unusually high global sea-levels during the Cretaceous, led to marine flooding of the basin. This flooding started in the north during the late Jurassic and progressed south through Canada and eventually into Utah. Palaeogeographical reconstruction for the period shows a large embayment in the area around the Book Cliffs. After the flooding, there was a period of offshore, shelf sedimentation that was interspersed with periods of deltaic shoreline progradation from the west, south and north. The shallow-marine and alluvial deposits of the Book Cliffs represent some of these phases of delta progradation.

During the Late Cretaceous, Central Utah occupied a subtropical palaeolatitude (42 °N). This, coupled with the warm, humid global climate of the Cretaceous (greenhouse conditions, Figure 5.4), resulted in delta plains that were covered in luxuriant vegetation. This included peat swamps, which were preserved as extensive coal seams in the northern part of the Book Cliffs, the Wasatch Plateau to the west and the Kaiparowits Plateau to the south. The delta plains were probably very similar to the modern-day Niger or Paraibo deltas (Figure 7.3a), the only key difference being that dinosaurs (the footprints of which are preserved in the coal seams) occupied the Utah coal swamps. Studies of spores, pollen and leaf fossils within the coals show no change in the assemblage through the Book Cliffs succession.

**Figure 7.3** Modern analogues for the Book Cliffs. (a) Satellite image of the modern Paraibo delta (Brazil), a classic wave-dominated delta system comparable to the Blackhawk shoreline system. The coastline shown is c. 20 km in length. Note the ancient beach ridges, luxuriant vegetation on the coastal plain and relatively straight shoreline geometry compared to a river-dominated delta (Figure 5.8). (b) Modern-day coast of Brunei: a wave-dominated shoreline (c. 20 m wide) with a narrow, linear beach and luxuriant coastal plain. This is what the Book Cliffs probably looked like 80 Ma ago. ((a) NASA; (b) John Howell, University of Bergen.)

(a)

(b)

○ The plants changed little through the Late Cretaceous. What does that suggest about the climate?

● These observations indicate a stable climate through the Late Cretaceous.

During the Late Cretaceous, high sediment supply, related to the wet climate and the uplifting mountains, caused the depositional system to prograde into the basin and the shoreline moved eastward over time. Consequently, on a large scale, the basin fill becomes increasingly non-marine upward. This transition to non-marine conditions was accelerated by a change in the tectonic style along the mountain front. A reduction in the angle of subduction to the west resulted in very broad, slower and much more areally extensive subsidence (Late Cretaceous), followed by regional uplift (latest Cretaceous/early Palaeogene). This is because a shallower angle of subduction results in a broader zone of interaction between the oceanic plate and overlying continental plate, resulting in increased deformation in the overriding plate. In Utah, the initial phase of this uplift was associated with the partitioning of the basin into a number of smaller sub-basins. These were sites of continued deposition. However, by this time (early Palaeogene) the depositional systems were fully non-marine and the fluvial drainage systems fed a series of very large lakes within the mountain range (intermontane lakes). The sediment deposited in and around the margins of these lakes makes up the Roan Cliffs that lie behind and are stratigraphically younger than the Book Cliffs. Continued uplift (Eocene–Oligocene) terminated this deposition although the subsequent collapse of the mountain chain has initiated a number of extensional basins such as the late Cainozoic Great Salt Lake Basin.

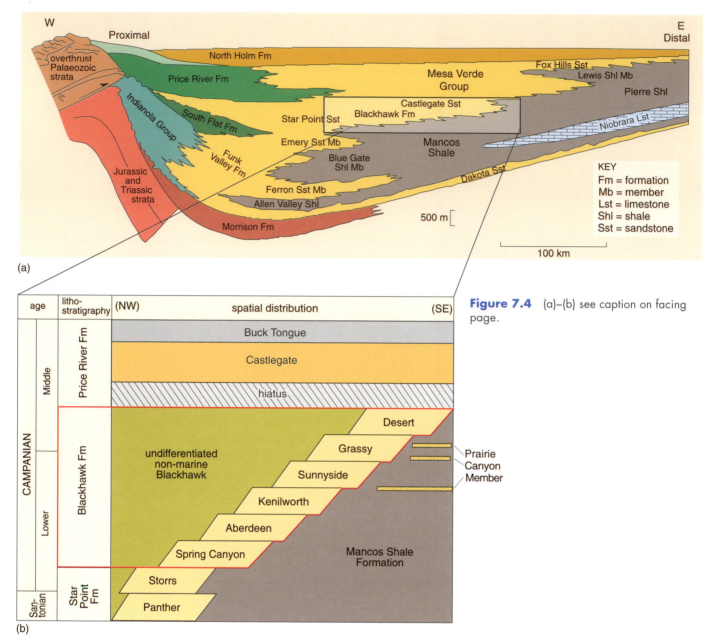

(a)

(b)

**Figure 7.4** (a)–(b) see caption on facing page.

One of the interesting features of the Book Cliffs region is that it has been uplifted by at least 6000 m and has undergone virtually no tectonic deformation. With the exception of minor tilting associated with movements on the underlying Permian salt, the area has undergone none of the contractional deformation that is typical of foreland basins. This fortuitous situation allows stratal surfaces and rock bodies to be easily traced for many kilometres, without encountering structural discontinuities. This has enabled detailed facies analysis and understanding of stratal relationships in a way unparalleled almost anywhere else on Earth.

## 7.3 Summary of the Book Cliffs stratigraphy

The formal lithostratigraphy of the Book Cliffs region is not especially important for your understanding of the sequence stratigraphical interpretation. However, it is useful to review the existing lithostratigraphy as some of the terminology is used later in this part of the book and is prevalent throughout the published literature. The lithostratigraphy is summarized in Figure 7.4.

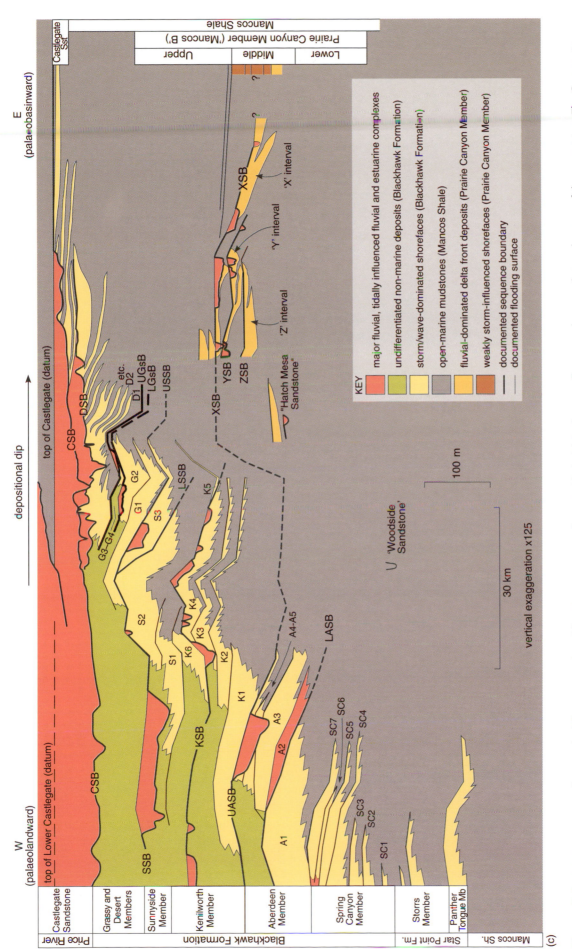

**Figure 7.4** (a) Cross-section showing the lithostratigraphy of the Western Interior Basin in Utah. Note the overall eastward progradational nature of the succession. The Campanian Blackhawk Formation and Castlegate Sandstone are the focus of this case study. (b) Lithostratigraphy of the Book Cliffs. The strata are divided into four formations, and the Blackhawk Formation is outlined in red. The formation is subdivided into six members (Spring Canyon, Aberdeen etc.), each of which comprises a shallow-marine sandstone. These members are the traditional way of correlating the strata. Broadly age-equivalent to the shallow-marine sandstone members are the eastward-thinning package of non-marine strata that lies to the west (undifferentiated non-marine Blackhawk) and the offshore deposits of the Mancos Shale which lies to the east. (c) Depositional cross-section through the Book Cliffs. Note that the shallow-marine members are in turn divided into smaller lenses of shallow-marine sandstone. Note also the presence of the major fluvial/estuarine complexes. The data in this cross-section form the basis for this case study. The letters SC1, A2, S3 etc., refer to the parasequences within each member, while LASB, SSB etc., all refer to sequence boundaries; these labels are referred to in Chapters 7–10. ((a) and (b) Young, 1955; (c) Hampson, 2001.)

The shallow-marine sediments that were deposited along the western edge of the Western Interior Seaway are collectively known as the Mesa Verde Group (Figure 7.4a). Towards the east, these deposits overlie and interfinger with offshore-marine deposits of the Mancos Shale. The Mesa Verde Group in the Book Cliffs is further divided into the Star Point and Blackhawk formations. The Star Point Formation comprises river-dominated deltas of the Panther Tongue and distal shoreface deposits of the Storrs Member. The overlying Blackhawk Formation is comprised of six formally defined members, each comprising a wedge of shallow-marine sandstone (Figure 7.4b). These members are separated from one another by tongues of Mancos Shale. There is a 300 m-thick wedge of undifferentiated non-marine coastal plain and fluvial deposits to the west (landward) of the shallow-marine members (Figure 7.4b,c). Individual members pass eastward into the marine shale and successive members sit further basinward. The wedge of non-marine Blackhawk Formation deposits thins in an easterly (basinward) direction and is unconformably overlain by braided fluvial deposits of the Castlegate Sandstone. The Mancos Shale which lies beneath and east of the Blackhawk and Star Point shoreline deposits also includes a number of sandstone-rich horizons termed the Prairie Canyon Member. The significance of these will become apparent later.

○ Using Figure 7.4, determine the overall stacking pattern of the shallow-marine members of the Blackhawk Formation: are they progradational, aggradational or retrogradational?

● The stacking pattern is progradational, because successively younger units lie further to the east (basinward).

Ammonite-based correlation with the geological time-scale indicates that the Book Cliffs succession represents about 9 million years of the Late Cretaceous spanning from the Late Santonian to the Campanian (c. 84–75 Ma). We shall pay most attention to the Star Point and Blackhawk Formations (c. 84–79.5 Ma). The entire succession is up to 800 m thick and the calculated subsidence rate of (150 m/Ma) was relatively low (Figure 7.5).

The subsidence rate has been simply calculated from the thickness divided by the time elapsed. It is always an average and, as we will see, important periods of time can be represented by surfaces as well as in rock volumes (Chapter 1). Thus, these calculated subsidence rates are an average and as such might well be underestimates for the maximum rate at any one time. The basin had a ramp-style geometry, this means that the profile dipped gently into deeper water, with no shelf/slope break and no abyssal plain.

## 7.4　Review of the sedimentology of the Book Cliffs

The sedimentary succession of the Book Cliffs was deposited in a variety of different depositional environments including offshore shelf, wave-dominated shoreface, river-dominated delta, coastal plain with associated coal swamps, and braided alluvial systems. The key features of these sedimentary systems as they apply to the Book Cliffs are reviewed in this Section. An important consideration is that most of the changes from one depositional system to another do not happen randomly and that the different systems reflect different combinations of accommodation space and sediment supply.

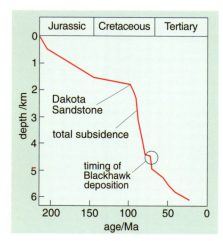

**Figure 7.5** Burial history curve for the Book Cliffs region. The plot traces the progressive burial of a horizon through time (in this case the Lower Jurassic Dakota Sandstone), calculated from the thickness of overlying strata. This can be used to calculate a subsidence rate. For the Book Cliffs strata that we are studying, a rate of 150 m/Ma has been determined. This is important for determining the mechanism controlling the geometry of the sequences (Chapter 5). We will demonstrate that the sequences in the Book Cliffs are controlled by a combination of the subsidence rate, rate of sediment supply and changes in eustatic sea-level. (Cross, 1986.)

## 7.4.1 Offshore facies association

Thick successions of fine-grained siltstone and claystone make up the Mancos Shale (Figure 7.4). These rocks are relatively soft and weather easily to give the classic 'badlands topography' of the area (Figures 7.2, 7.6). In detail, the rocks are blue-grey to grey, friable with a generally massive to very bioturbated fabric. Small-scale planar and trough cross-stratification are observed very rarely. Ammonites, belemnites and other fossils can be found and testify to an open-marine depositional setting.

**Figure 7.6** General view of the Book Cliffs near Hatch Mesa (Figure 7.1). Note that the lower parts of the cliffs are composed of blue-grey, intensely weathered shales. These are the offshore marine deposits which make up the Mancos Shale and that were deposited in quiet conditions below storm wave-base. Note also the small ridges in the foreground; these are the Prairie Canyon Member (Section 9.2.3). The cream–brown sandstones of the main cliffs comprise the shallow-marine shoreface sandstones of the Desert Member and the braided fluvial deposits of the Castlegate Sandstone. The cliff is *c.* 300 m high. *(John Howell, University of Bergen.)*

○ What does the fine-grained nature of these sediments indicate about the transport mechanism of these sediments and their position relative to storm wave-base?

● The fine grain size tells us that they were deposited from suspension in a quiet water environment below storm wave-base.

Rare occurrences of ripple cross-stratification indicate some resedimentation of material by currents on the sea-floor. The fossil content supports a marine interpretation. Extensive destruction of primary depositional features (such as planar and trough cross-stratification) by organisms bioturbating the sediment indicates that the sea-bed was a favourable place to live. Water depth was relatively shallow and the sea-floor was well-oxygenated, and so it is likely that the Mancos Shale was deposited on an open, shallow-marine shelf below storm wave-base.

## 7.4.2 Shallow-marine facies association

The coastal zone is an important site for sedimentation. Here, sediment supplied from a river to the sea is often deposited in deltas whose character is influenced by the action of waves and tides. The morphology of these deltas depends upon the interaction of the processes delivering the sediment (the river) and the processes that redistribute that sediment within the basin (waves and tides). Deltas are classified as wave-, river- or tide-dominated (Figure 7.7 overleaf), or some combination of these. The majority of shallow-marine rocks in the Book

Cliffs were deposited in wave-dominated deltas. River-dominated deltas exist within the Panther Tongue Member of the Star Point Formation and locally within the overlying Blackhawk Formation. Such systems within the Blackhawk typically pass gradationally along depositional strike into wave-dominated deltas. Depositional strike is parallel to the coastline.

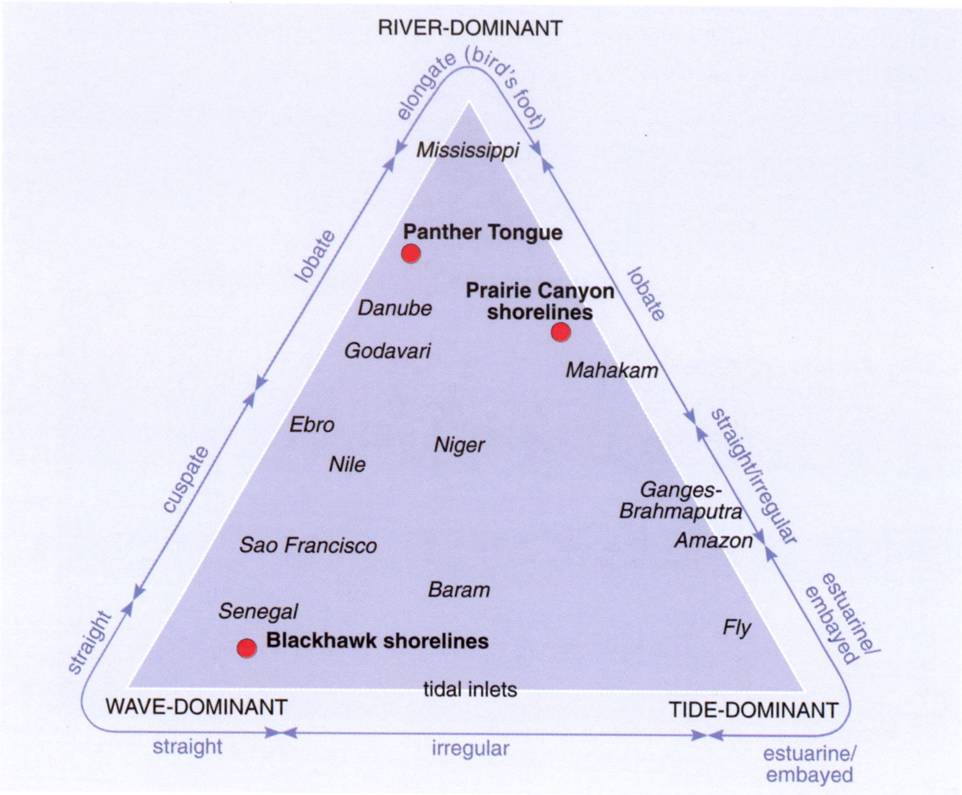

**Figure 7.7** Triangular diagram for the classification of deltas by the relative importance of the principal depositional processes (waves, rivers or tides). The deposits of the Book Cliffs are shown. Note that the majority of the Blackhawk Fm is composed of wave-dominated delta deposits. The Panther Tongue (Figure 7.4) is a river-dominated delta and the Prairie Canyon Member deposits are tidally influenced, river-dominated deltas. The Mississippi river-dominated delta is illustrated in Figure 5.8. (Galloway, 1975.)

## River-dominated delta systems

River-dominated deltas occur where fluvial systems enter a standing body of water (in this case the sea). As the flow decelerates, the river's capacity to carry sediment both as bedload and suspended load decreases, and so it deposits that sediment. The coarse-grained sediment is deposited first and progressively finer-grained sediment is laid down further into the basin. Continual addition of sediment results in the shoreline prograding and a coarsening-upward succession, called a mouthbar, being deposited (Figure 7.8). The Mississippi is a good example of a modern-day river-dominated delta (Figure 5.8 and Chapter 6).

Two additional factors are important in marine basins such as the Western Interior Basin. First, the mixing of freshwater and saltwater causes clay particles to flocculate (stick together) and they are deposited earlier than they would otherwise be. Secondly, as the seawater in the basin is denser than the incoming river water, a plume of water and suspended sediment from the river is carried far out into the basin.

In vertical successions, mouthbar deposits show a coarsening upward-profile which records the progradation of the delta. Individual beds of sandstone record a period of sediment being flushed in from the hinterland and contain either planar cross-stratification or ripples which indicate the current-induced depositional process (Figure 7.8b). Siltstone beds record quieter periods of deposition. The

(a)

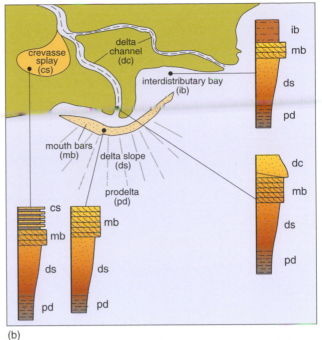

(b)

seaward

clinoform

20 m

(c)

**Figure 7.8** Features of river-dominated deltas which occur when a river system supplies sediment to the coast faster than wave and tidal processes in the basin can redistribute it. (a) The modern Godavari delta (India) shows the lobate geometry and abundant distributary channels typical of a river-dominated delta. (View is c. 160 km across.) (b) Basic facies model for a river-dominated delta. Note that the succession coarsens upward and that the principal depositional elements are the mouth bar and channels. (c) Example of a river-dominated delta succession from the Panther Tongue Member in the Book Cliffs showing a coarsening-upward profile from offshore shales to mouthbar sandstones. Note that there are surfaces within the sandstone bed that are more steeply inclined than the regional bedding; these surfaces are seaward-dipping clinoforms (Section 4.5). ((a) NASA; (b) Nichols, 1999; (c) John Howell, University of Bergen.)

mouthbar succession is commonly capped by fluvial or distributary channels that cut into the delta plain as the system progrades. Bioturbation is generally low as a result of the influx of muddy freshwater. The freshwater suppresses the full marine salinity and the resultant water chemistry is brackish, which is less supportive of bottom-dwelling fauna than normal marine water conditions.

Individual sandstone beds dip in a seaward direction (Figure 7.8c), typically at about 3°–5° more than the regional dip, and they have a lobate lateral geometry (Figure 7.8b). The delta system as a whole typically has an elongate or lobate geometry which reflects sediment input at a single point (Figure 7.8a). This is also recorded in the palaeocurrents, which show a broadly radial pattern. Individual lobes from different distributary channels and even different delta systems commonly overlap or interfinger.

## Wave-dominated shoreface delta systems

Wave-dominated shoreface delta systems are the most common delta type in the Book Cliffs stratigraphy. Sediment introduced by fluvial systems is remobilized by fairweather and storm wave processes and redistributed along the coast (Figure 7.9a). Consequently, the deposits of these systems are elongate, shore-parallel and predominately contain sedimentary structures that record wave processes, such as wave ripples and hummocky cross-stratification (HCS).

○ Where and how does HCS form?

● Above storm wave-base, due to the oscillatory motion of waves during storms. Storm wave-base is the water depth to which average storm waves can affect the sea floor, and is typically about 15–40 m. HCS is generally well preserved below fairweather wave-base which is the water depth above which the sea-bed is affected by everyday waves and is typically about 5–15 m but see discussion on amalgamated HCS below.

During storms, sand is eroded from beaches and mouthbars and redeposited as HCS sands in a sheet-like blanket across the inner shelf. During fairweather periods between storms, mud and silt are deposited from suspension between storm wave-base (SWB) and fairweather wave-base (FWWB) and cover the sheet of sand laid down during the storm. Above fairweather wave-base, the sea-bed is constantly agitated and little or no fine-grained sediment (clay and silt) is deposited. The sand-grade sediment is gradually reworked onto and along the beaches by fairweather wave processes and by longshore drift until the next storm (Figure 7.9).

A typical vertical succession through a wave-dominated shoreface in the Book Cliffs passes from offshore siltstones (deposited below storm-wave base) into interbedded siltstones and hummocky cross-stratified sandstones of the offshore transition zone (representing the area between storm and fairweather wave-bases; Figure 4.3) and upward into amalgamated HCS beds of the lower shoreface (Figure 7.9). The amalgamated beds represent deposition by storms in water depths shallower than fairweather wave-base, so any fine-grained material was not preserved between storm events. They are overlain by a zone of trough cross-stratified sandstone called the upper shoreface, representing movement of the sand on the sea-bed during fairweather conditions and by planar-laminated sandstone beds of the foreshore. The top of the foreshore deposits often contain roots from the vegetation that grew at the top of the beach in the backshore (Figures 7.9 and 7.10a).

**Figure 7.9 (opposite)** Features of wave-dominated deltas (also sometimes called shoreface systems). These deltas occur when wave processes in the basin rework the sediment introduced by rivers. Shorelines are elongate in a coast-parallel direction. Deposits coarsen-upward and contain HCS and wave ripples produced by storms. Trough cross-stratification and low-angle planar cross-bedding are the product of shoaling and breaking waves in the shallow waters of the upper shoreface. This type of shoreline system is the most common in the Book Cliffs, and some modern examples can be seen in Figure 7.3. (a) 3-D block diagram; (b) typical graphic log of a wave-dominated delta succession (typically c. 20–30 m thick.). Bio-index refers to the intensity of bioturbation: 1 = low, 6 = high. For the Book Cliffs succession, the tidal range was typically c. 2–4 m, FWWB c. 5–15 m, and SWB c. 15–40 m.

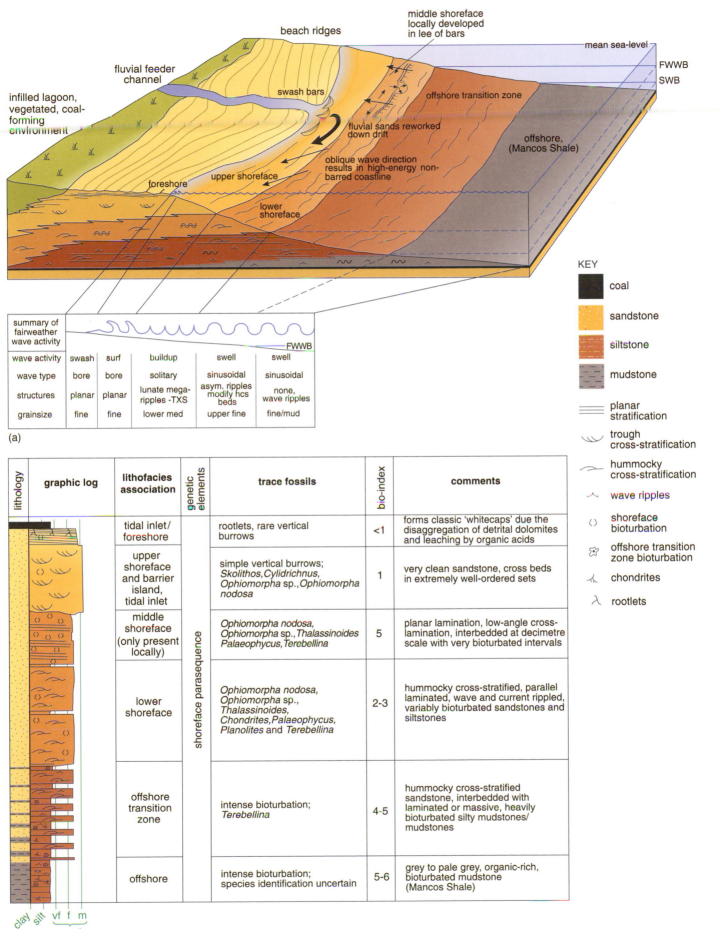

(a)

| lithology | graphic log | lithofacies association | genetic elements | trace fossils | bio-index | comments |
|---|---|---|---|---|---|---|
| | | tidal inlet/ foreshore | | rootlets, rare vertical burrows | <1 | forms classic 'whitecaps' due the disaggregation of detrital dolomites and leaching by organic acids |
| | | upper shoreface and barrier island, tidal inlet | shoreface parasequence | simple vertical burrows; *Skolithos,Cylidrichnus, Ophiomorpha* sp., *Ophiomorpha nodosa* | 1 | very clean sandstone, cross beds in extremely well-ordered sets |
| | | middle shoreface (only present locally) | | *Ophiomorpha nodosa, Ophiomorpha* sp.,*Thalassinoides Palaeophycus,Terebellina* | 5 | planar lamination, low-angle cross-lamination, interbedded at decimetre scale with very bioturbated intervals |
| | | lower shoreface | | *Ophiomorpha nodosa, Ophiomorpha* sp., *Thalassinoides, Chondrites,Palaeophycus, Planolites* and *Terebellina* | 2-3 | hummocky cross-stratified, parallel laminated, wave and current rippled, variably bioturbated sandstones and siltstones |
| | | offshore transition zone | | intense bioturbation; *Terebellina* | 4-5 | hummocky cross-stratified sandstone, interbedded with laminated or massive, heavily bioturbated silty mudstones/ mudstones |
| | | offshore | | intense bioturbation; species identification uncertain | 5-6 | grey to pale grey, organic-rich, bioturbated mudstone (Mancos Shale) |

clay  silt  vf  f  m
sand

(b)

(c)

(b)

(a)

clay
silt
sand — vf f m

lithology

(e)

(d)

As the wave-dominated delta system progrades, it will always show a coarsening-upward vertical profile. The thickness of individual facies associations (e.g. the offshore transition zone) can be used to give an estimate of the typical water depths to the various wave-bases. In the Book Cliffs succession, storm wave-base was around 15–40 m water depth and fairweather wave-base was around 5–15 m.

○ Assuming that Figure 7.9 represents a typical coastal section in the Western Interior Seaway, decide whether the system was likely to have a large (macrotidal), medium (mesotidal) or small (microtidal) tidal range.

● The thickness of the foreshore deposits, which were deposited in the intertidal zone give an indication of the tidal range which in this case is *c.* 2–3 m, i.e. microtidal to lower mesotidal. This interpretation of a relatively low tidal range is supported by the lack of tidal flat and tidal creek deposits that are typical of coastlines with higher tidal ranges.

○ Suggest two reasons why the Book Cliffs shallow-marine systems are wave-dominated and not river-dominated?

● First, the subtropical latitude favours large seasonal storms, which destroy the fluvial mouth bars and redistribute the sediment along the coast. Secondly, the newly uplifted mountain belt probably had a number of small rivers draining off it. This produced many small deltas, which were easily reworked. A more established drainage system would feed a single large river-dominated delta that could locally overprint the wave processes.

**Figure 7.10 (opposite)**   Examples of offshore transition zone, shoreface and foreshore facies from the Book Cliffs. Graphic log is *c.* 20–30 m long. (a) Upper shoreface trough cross-stratified sandstones and planar-stratified foreshore deposits from Bear Canyon (sandstone is *c.* 6 m thick). (b) Highly bioturbated middle shoreface facies from Gentile Wash. Note that this facies is not always present in the succession (pencil shows scale). (c) Offshore transition zone heterolithic deposits from Woodside Canyon (pole is 1.2 m long). (d) Low-angle dipping, planar-laminated foreshore facies from Woodside Canyon (divisions on pole at 20 cm intervals). (e) Upper shoreface to offshore transition zone facies from Woodside Canyon. Localities are shown on Figure 7.1. ((a)–(e) John Howell, University of Bergen.)

### 7.4.3   Coastal plain facies association

The coastal (or delta) plain is the low-lying area behind the shoreline, just above the high water mark, that sits on top of older shoreface and delta deposits. In the Book Cliffs the coastal plain includes two key depositional components, fluvial channels and associated overbank deposits, and coal-forming peat swamps (Figures 7.11, 7.12).

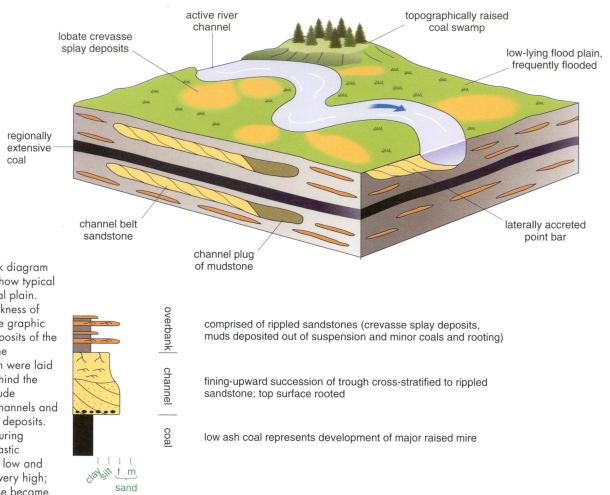

**Figure 7.11**   Block diagram and graphic log to show typical features of the coastal plain. On average, the thickness of deposits shown in the graphic log is c. 5 m. The deposits of the non-marine part of the Blackhawk Formation were laid down in the area behind the shorelines. They include meandering fluvial channels and associated overbank deposits. Peats accumulated during periods when siliciclastic sediment supply was low and the water table was very high; following burial, these became coals.

overbank — comprised of rippled sandstones (crevasse splay deposits, muds deposited out of suspension and minor coals and rooting)

channel — fining-upward succession of trough cross-stratified to rippled sandstone; top surface rooted

coal — low ash coal represents development of major raised mire

The fluvial channel deposits are typically 1–4 m thick, the grain size fines upward and the sandbodies exhibit lateral accretion surfaces (Figure 7.12b), indicating a meandering river system. Sandbodies can be up to 1 km wide, while the width of the individual channels at any point in time, calculated from the channel abandonment plugs, was in the order of 100 m.

○ Why are the channel sandbodies wider than the estimated width of the channels?

● Meandering channels deposit sand by lateral migration of point bars, where sand is deposited over a long period and produces very wide sandbodies termed channel belts. When studying the channel belt, it may be difficult to estimate the original channel width unless the remains of an oxbow lake or abandoned channel are found, formed through meander cut-off or avulsion, and later filled with mud (an abandonment plug). Abandonment plugs provide a snapshot in time of the size of the meandering river.

(a)

(b)

(c)

(d)

(e)

**Figure 7.12**   Examples of non-marine facies from the Blackhawk Formation. (a) Channel and minor overbank deposits in Woodside Canyon. Note the combination of channel-fill deposits, planar-bedded overbank sandstones, mudstones and coals. (Image *c.* 15 m high.) (b) Meandering channel deposits overlying thick coal. Note the clearly defined lateral accretion surfaces dipping towards the left. The channel belt is approximately 2 m thick; from Deadman Canyon (pole is 1.2 m long). (c) Bleached sandstone representing a mature fossil soil (palaeosol) (camera lens cap for scale). (d) Vertical root structures in overbank deposits; from Willow Creek (pencil for scale). (e) View of single-storey channel fills and overbank fines from Willow Creek (cliff is *c.* 80 m high). Localities are shown on Figure 7.1. ((a)–(e) *John Howell, University of Bergen.)*

Overbank deposits occur laterally adjacent to, and interbedded with, the channel sandbodies. The overbank deposits consist of dark- to pale-grey laminated mudstones interbedded with fine-grained, rippled, sheet sandstone beds. Evidence of rooting is common throughout (Figure 7.12d). These sediments were deposited during periodic floods when the river burst its banks. It can be difficult to determine whether the mudstones, which were deposited from suspension, represent occasional flooding of the overbank area or whether the channels, protected by levées were topographically elevated above permanently subaqueous interchannel areas. The rippled sandstone sheets represent rapid deposition of sand in crevasse splay lobes. Amalgamation of these splays formed the channel margin levées.

The coastal plain deposits of the non-marine Blackhawk Formation contain numerous coal horizons. There are at least 15 coal seams that are thicker than 1 m and some are mined. Whilst the coastal plain was heavily vegetated, it was not always a good coal-forming environment. To accumulate significant volumes of peat (to eventually form the coal), two factors are required: first, a high water table to stop oxidation of the plant matter as it dies; and secondly, a paucity of siliciclastic sediment supply. High water tables are a common feature of modern coastal plains, especially in areas of high rainfall. The prolonged maintenance of a high water table, required to produce a thick coal seam, is aided if base level is rising (e.g. during relative sea-level rise). Siliciclastic material in coal is called ash (the refractory material left in the fire grate after burning). Economic coals contain less than 5% ash, and coal cannot form at ash contents above 20%. Most Book Cliffs coals contain 4–7% ash. A mechanism is required to prevent the siliciclastic material (from either the rivers or the sea) from contaminating the peat swamp. There are two, partly linked mechanisms. First, in areas of high rainfall the peat swamp will grow above the level of the coastal plain, maintaining its own perched water table. This is called a raised mire. The modern raised mires of Borneo sit up to 10 m above the surrounding coastal plain. The second mechanism by which siliciclastic material is excluded is by storage up river during times of rapidly rising relative sea-level (discussed further in Section 8.3).

### 7.4.4  Barrier island, lagoon and estuarine facies associations

Both the river- and wave-dominated shoreline systems described from the Blackhawk Formation are progradational. This means that the amount of sediment supplied was greater than the accommodation space available to store it and the shoreline moved in a seaward direction (Section 4.2). Within the Blackhawk, there is also limited evidence for transgressive shorelines deposited when accommodation space was being created faster than the sediment could fill it. These two distinct types of shoreline occupy different sequence stratigraphical positions in time.

#### *Barrier islands and lagoons*

Barrier islands form on wave-dominated, micro- and lower mesotidal shorelines when relative sea-level is rising. Barrier islands are narrow, sand-dominated strips that separate the open marine environment from a sheltered lagoon (Figure 7.13). Depositional processes on the seaward side of modern barriers are comparable to shoreface systems except that the lagoon isolates the system from much of the fluvial sediment supply. During storms, sediment is washed from the front of the barrier, over into the sheltered lagoon and preserved as landward-dipping, current-rippled to planar-laminated 'washover fans' (Figure 7.13b).

(a)

**Figure 7.13** Barrier island processes, models and products: (a) photograph of modern barrier island/inlet system from South Carolina, USA (c. 30 km of coastline shown in image). (b) 3-D facies model for a transgressive barrier island system. (c) Schematic cross-section through a barrier island showing processes operating during landward migration of the barrier as a result of relative sea-level rise. This process is discussed further in Section 8.4. ((a) Chris Wilson, Open University; (b) and (c) Reinson, 1992.)

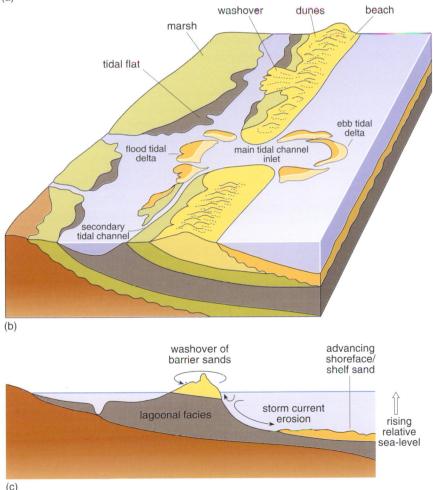

(b)

(c)

Changing salinities and anoxic conditions in the lagoon result in the preservation of mudstones with abundant organic-carbon matter and a limited diversity of trace fossils. Gradual relative sea-level rise causes the landward migration of the barrier by continued erosion on the seaward side, with washover fan deposition in the lagoon. Consequently, the barrier migrates in a landward direction and leaves very little behind (Figure 7.13c). Locally, the barrier may be cut by tidal inlets. Water flows through these channels during the daily tidal cycles and deposits small flood tidal deltas in the lagoon and ebb tidal deltas on the seaward side. The ebb tidal deltas will generally be reworked by the storms but the flood tidal ones have a good preservation potential. Lateral migration of the inlet can sometimes also be preserved within the barrier succession. Once the transgression ceases, the landward migration of the barrier will stop. The lagoon will be filled by river-derived sediment and a new progradational shoreface will be initiated, fed by the fluvial system. Because of the constant cannibalization of the barrier during transgression, preserved examples are limited and often appear linked to the initiation of the next shoreface system.

### Estuaries

We have seen that in most cases in the sediments preserved in the Book Cliffs the fluvial systems that supply sediment to the coast feed progradational shorelines (either shorefaces or deltas). There are examples, however, where river valleys become flooded during rising sea-level and form estuaries. Estuaries are complex depositional systems, largely because of the interplay between fluvial, tidal and wave processes.

○ Given we have observed that the tidal range in the Western Interior Seaway was low (Section 7.4.2), and that in open marine systems, tides are not believed to have played a significant role, why are tidal processes important in the estuaries?

● Because in estuaries, the funnel shape amplifies the tidal forces. This, combined with the fact that the system is sheltered from the waves, results in an increase in the relative importance of tidal processes (Figure 7.14).

○ List up to five criteria observable in the rock record that indicate tidal depositional processes?

● The sedimentological features indicative of tidal processes include: mud-draped ripples, rhythmically laminated heterolithic deposits, bidirectional cross-bedding, reactivation surfaces and tidal bundles.

Tidal deposits are typically heterolithic (interlaminated and interbedded, sand and mud) because the depositional energy is frequently changing during the tidal cycle (Figure 7.14). Tidal flows will deposit beds of cross-stratified sand and, while the tide turns, mud will settle from suspension and drape the sand. Estuarine depositional systems are complex. Key depositional sub-environments include tidally influenced meandering channels, bayhead deltas (where the river debouches into the estuary), tidal flats, salt marshes and high-energy tidal bars. The estuarine systems in the Book Cliffs succession contain large meandering tidal point bar deposits, tidal flats, shell beds and minor estuarine deltas (Figure 7.14). The recognition of these estuarine deposits is very important for understanding the sequence stratigraphical interpretation of the area.

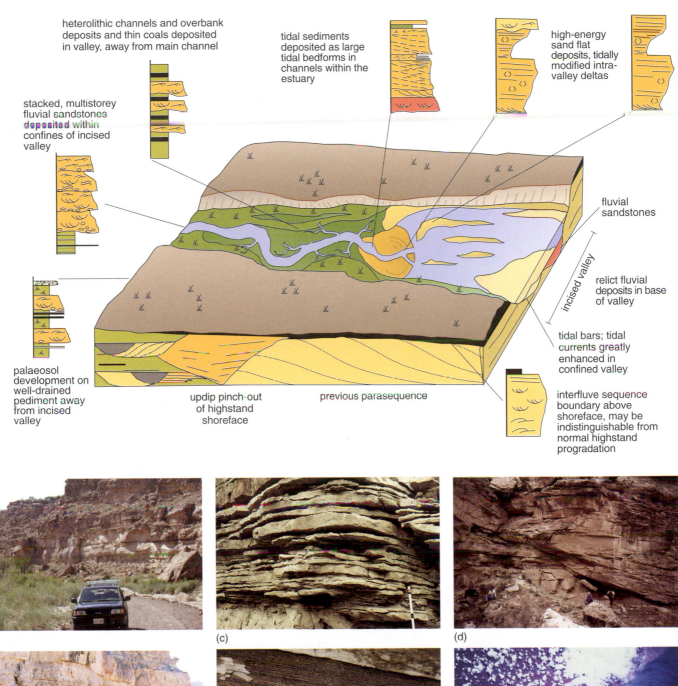

heterolithic channels and overbank deposits and thin coals deposited in valley, away from main channel

tidal sediments deposited as large tidal bedforms in channels within the estuary

high-energy sand flat deposits, tidally modified intra-valley deltas

stacked, multistorey fluvial sandstones deposited within confines of incised valley

fluvial sandstones

relict fluvial deposits in base of valley

palaeosol development on well-drained pediment away from incised valley

updip pinch-out of highstand shoreface

previous parasequence

tidal bars; tidal currents greatly enhanced in confined valley

interfluve sequence boundary above shoreface, may be indistinguishable from normal highstand progradation

incised valley

(a)

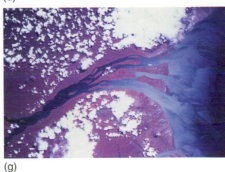

**Figure 7.14** Features of estuarine facies associations. (a) 3-D facies model for the Book Cliffs estuarine deposits and mini-graphic logs (each of *c.* 10 m) showing typical vertical successions. (b) Erosive-based tidally influenced estuarine channel system cut into older shoreface deposits, Desert Member, Tusher Canyon. (c) Bayhead delta deposits, Soldier Creek (scale bar = 30 cm). (d) Large-scale tidal bedforms (inclined heterolithic stratification), Tusher Canyon (people for scale). (e) Interbedded tidal channel, tidal flat deposits and a thin coal, Woodside Canyon. (cliff *c.* 10 m high)) (f) Detail of mud-draped sandstones typical of tidal deposits within an estuarine environment, Straight Canyon (pencil for scale). (b)–(f) Localities are shown in Figure 7.1. (g) Aerial view of small, modern estuary; the Essequibo, Guyana, South America (image *c.* 60 km from top to bottom). *((b) Stephen Flint, University of Liverpool; (c)–(f) John Howell, University of Bergen; (g) NASA.)*

### 7.4.5 Braided fluvial systems

The Castlegate Sandstone comprises a thick (up to 180 m) succession of braided fluvial deposits characterized by multiple channels, which are the product of continual switching of the position of the active channel within a broad channel belt. The key depositional element in such systems is the in-channel bar, which mainly migrates downstream, but some lateral and upstream accretionary migration occurs (Figure 7.15). There are no overbank deposits in braided fluvial systems.

Braided systems are common where the gradient of the fluvial system is high, when the discharge is highly variable and when the system is dominated by coarse-grained bedload. A key observation in the Book Cliffs is that there is an abrupt vertical change from meandering fluvial systems in the Blackhawk Formation to braided fluvial deposition in the overlying Castlegate Sandstone. This change occurs across a sharp surface which is mappable over tens of kilometres. This observation is very important in the sequence stratigraphical interpretation of the Book Cliffs, as we will see in Section 9.2.

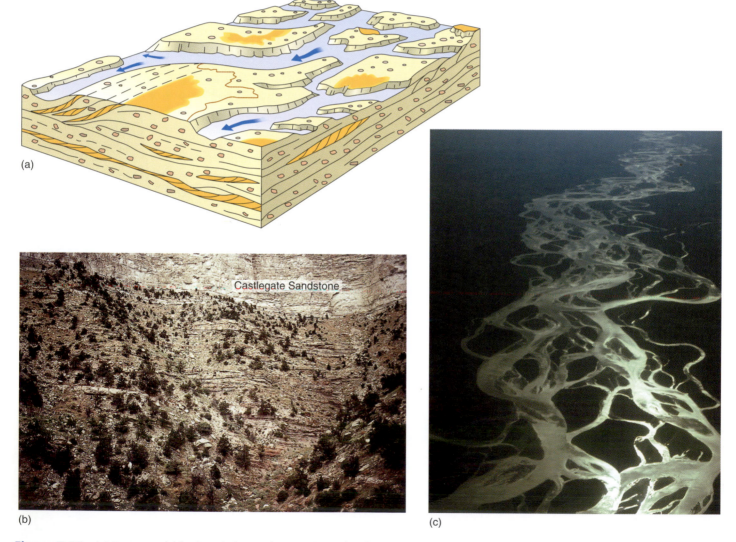

**Figure 7.15** (a) Facies model for braided river deposits. Note that the main depositional elements are in-channel bars that migrate down current. (b) Ancient sand-rich braided river deposits of the Castlegate Sandstone abruptly overlying interbedded floodplain siltstones and meander belt sandstones, Willow Creek, northern Book Cliffs (Figure 7.1). The vertical section in the photograph is c. 200 m thick. (c) Aerial view of a modern-day braided river in Siberia (image is c. 30 km long). ((a) Ramos et al., 1986; (b) John Howell, University of Bergen; (c) Bob Spicer, Open University.)

## 7.5   Summary

- The shallow-marine and alluvial sedimentary rocks of the Book Cliffs are Cretaceous in age and were deposited in a foreland basin.

- Sediment was shed from a newly uplifting mountain belt (the Sevier Orogenic Belt) the remains of which now run down the west coast of North America.

- The abundance of coals within the Book Cliffs succession indicate that the climate was warm and tropical.

- The sediments were deposited in a series of coastal plain, wave- and river-dominated deltas and offshore settings.

- Wave-dominated delta successions include offshore, transition zone, shoreface and beach deposits. The shoreface is composed of sandstones with HCS and towards the top, trough cross-stratification, indicating storm events and shoaling waves respectively.

- River-dominated deltas show a similar coarsening-upward profile to the wave-dominated systems; however, the dominant structures are current-derived and beds dip seaward (clinoforms).

- Deposits of transgressive shorelines are also locally observed, including estuarine and barrier island systems.

- The coastal plain is comprised of fluvial channels, overbank deposits and coals. Coals require specific conditions of rising water table and limited siliciclastic sedimentation to be preserved.

- The Castlegate Sandstone was deposited as a series of braided river deposits.

## 7.6   References

### Further reading

BROWN, J. E., COE, A. L., SKELTON, P. W. AND WILSON, R. C. L. (1999) *Surface Processes*, Block 4 of S260 *Geology*, The Open University, 238pp. [This provides an excellent background to the sedimentology required to make sequence stratigraphical interpretations.]

WALKER, R. G. AND PLINT, G. A. (1992) 'Wave- and storm-dominated shallow marine systems', in WALKER, R. G. AND JAMES, N. P. (eds) (1992) *Facies Models: Response to Sea Level Change*, Geological Association of Canada, 219–238. [This and other chapters are very good background to facies similar to those we encounter in the Book Cliffs.]

### Other references

KAMOLA, D. L. AND VAN WAGONER, J. C. (1995) 'Stratigraphy and facies architecture of parasequences with examples from the Spring Canyon', in VAN WAGONER, J. C. AND BERTRAM, G. T. (eds) *Sequence stratigraphy of foreland basin deposits,* American Association of Petroleum Geologists Memoir No. 64, 27–54.

PEMBERTON, S. G., VAN WAGONER, J. C. AND WACH, G. D. (1992) 'Ichnofacies of a wave-dominated shoreline', in PEMBERTON, S. G. (ed.) *Applications of ichnology to petroleum exploration*, SEPM Core Workshop No. 17, Society of Paleontologists and Mineralogists, 339–382.

VAN WAGONER, J. C. (1995) 'Sequence stratigraphy and marine to non-marine facies architecture of foreland basin strata, Book Cliffs, Utah, U.S.A.', in VAN WAGONER, J. C. AND BERTRAM, G. T. (eds) *Sequence stratigraphy of foreland basin deposits,* American Association of Petroleum Geologists Memoir No. 64, 137–224.

# 8 The parasequences of the Book Cliffs succession

## John A. Howell and Stephen S. Flint

As discussed in Chapter 4, parasequences are the basic building blocks of most sequences. In the marine succession of the Book Cliffs, the parasequences represent the shallowing-upward successions laid down by phases of delta or shoreface progradation. These are bounded by flooding surfaces expressed as hiatuses that form during periods of relative sea-level rise and thus mark the sudden upward change from shallow- to deeper-water facies.

○ Outline five reasons why the rocks in a typical Book Cliffs parasequence record the shallowing-upward progradational part of the cycle but the transgressive portion is recorded only as a surface, i.e. why is the preserved part of the depositional cycle asymmetrical?

● (i) During transgression, the shoreline becomes a barrier system (Section 7.4.4) which migrates landward and leaves nothing behind except for a transgressive erosion surface.

(ii) Combined subsidence and eustatic sea-level rise mean that the resultant rate of relative sea-level rise is typically much faster than the rate of relative sea-level fall so sedimentary systems have less chance to become established during the flooding (Section 4.2).

(iii) The coastal plain is very low lying and small relative sea-level rises lead to large landward displacements of the shoreline (Section 7.4.3).

(iv) Many of the mechanisms for generating relative sea-level rise, such as pulsed tectonic subsidence or the melting of polar ice or compaction, occur very quickly (Sections 5.1, 5.2).

(v) Finally, once the transgression is initiated, sediment is preferentially deposited in the coastal plain and the shoreline becomes sediment starved (Section 7.4.4). This sediment starvation accelerates the transgression.

Parasequences contain a variety of different facies and can show considerable variation in both depositional dip and strike directions. In Sections 8.1 and 8.2, we will examine examples of parasequences in both wave-dominated and river-dominated settings from the marine portion of the Book Cliffs, paying special attention to the vertical and lateral distribution of facies within the parasequences and how to identify and map them. In Section 8.3, we will consider the time-equivalent deposits in the non-marine environments.

## 8.1 Parasequences in wave-dominated coastal depositional systems

In wave-dominated coastal systems, a single parasequence represents the deposits of a phase of progradation from offshore transition zone–shoreface–foreshore–backshore progradation. As the coastal succession builds out, this set of linked environments will all eventually prograde past any fixed reference point. A complete parasequence (Figure 8.1) is bounded at its base by a marine flooding surface overlain by offshore transition zone shales. A coal commonly caps the parasequence.

**Figure 8.1**   Exposure photograph and graphic log of the third marine parasequence up from the base of the Sunnyside Member (denoted Sunnyside PS3 (or S3 in Figures 7.4c and 10.1)) in Woodside Canyon (Figure 7.1). Note the sheet-like geometry, shallowing-upward trend and bounding flooding surfaces. See Figure 7.9 for key. *(John Howell, University of Bergen.)*

There are at least 38 wave-dominated parasequences in the Blackhawk Formation. The following examples come from the works of Ciaran O'Byrne and Diane Kamola, who logged and mapped out examples in the Grassy and Spring Canyon members respectively.

### 8.1.1   Complete vertical succession

Some of the best examples of complete parasequences occur in the Spring Canyon Member in Gentile Wash. Many of the features of these parasequences are summarized in the graphic log and photograph (Figure 8.2a,b overleaf). This locality also has the advantage that Exxon Production Research drilled a well nearby so it is possible to compare exposure and subsurface expressions of the parasequences (Figure 8.2c).

Towards the base of the parasequence in the lower part of the succession (not shown in Figure 8.2) is an interval of blue-grey siltstone and claystone, representing offshore deposits. These pass upward into interbedded siltstones and HCS sandstones of the offshore transition zone. The HCS sandstone beds represent storm deposition and are interbedded with siltstones deposited during fairweather periods. There is an upward increase in the proportion of sandstone beds.

○   What does the upward increase in sandstone beds represent?

●   The upward increase in sandstone beds represents a gradual decrease in the water depth.

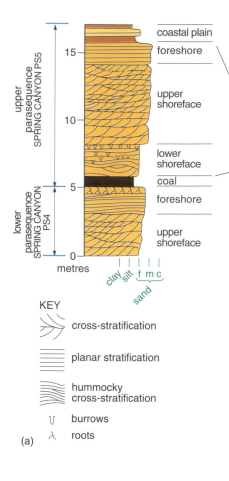

(a)

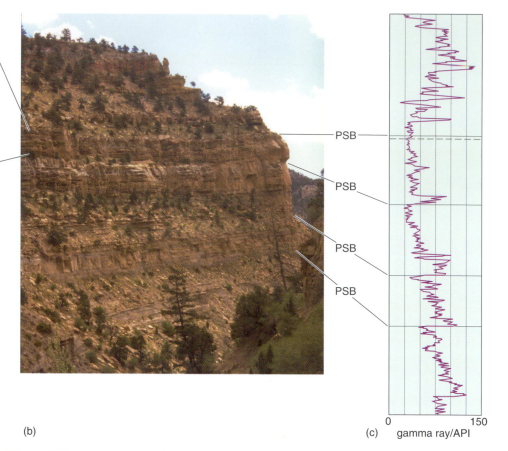

(b)

(c) gamma ray/API

**Figure 8.2** Data from Gentile Wash, northern Book Cliffs (Figure 7.1). (a) Logged section from the upper parasequence (Spring Canyon PS5). Note that the shaley lower part of the parasequence is missing here because this locality is close to the landward pinch-out of this parasequence (see text). (b) Exposure of coarsening-upward parasequences of the Spring Canyon Member. PSB = parasequence boundary (flooding surface). (c) Gamma-ray log response (see Box 3.2) from nearby borehole. ((a) and (c) Van Wagoner et al., 1990; (b) John Howell, University of Bergen.)

As more sediment is deposited, the position of the depositional surface (i.e. sea-bed) becomes shallower and progressively smaller storms are recorded in the succession. The heterolithic offshore transition zone passes into the lower shoreface which comprises amalgamated HCS beds. In this particular section, the lower shoreface is relatively thin (6 m); in most cases, it is 8–20 m thick.

The amalgamated HCS beds of the lower shoreface are locally overlain by an intensely bioturbated interval up to 3 m thick, consisting of fine-grained sandstones with extensive *Ophiomorpha*, *Skolithos* and *Terebellina* burrows, interbedded with weakly planar-bedded sandstones on a 10 cm scale. These intervals are locally present between the upper and lower shorefaces within many of the Book Cliffs parasequences and are thought to represent more sheltered conditions between bars on the shoreface. The overlying upper shoreface comprises of very well-sorted, trough cross-stratified, medium-grained sandstones.

○ What bedform and process does the trough cross-stratification of the upper shoreface represent?

● The trough cross-stratification represents the onshore migration of sandbars in the surf zone, during fairweather periods.

Sets of cross-stratification are typically 20–40 cm thick, well organized and only rarely bioturbated. It is also important to note the increase in grain size which occurs across a sharp surface that separates the lower from the upper shoreface. Palaeocurrents are generally oriented in a landward direction although a fairly high degree of variability exists.

Planar cross-stratified sandstones overlie the upper shoreface deposits. Laminations dip in a seaward direction at a very low angle (<5°). These represent the foreshore facies, deposited in the intertidal zone. The top surface is frequently rooted and coals commonly overlie the foreshore deposits. The majority of the sediments within the Book Cliffs parasequences are pale- to dark-brown in colour. However, the tops of the upper shoreface and the foreshore deposits are typically a characteristic white colour. These 'white caps' reflect the leaching of iron-rich dolomite (which breaks down to give the red-brown colour elsewhere in the section) by organic acids associated with the coals.

○  What does the succession record in terms of changing water depth?

●  The succession records a gradual decrease in water depth as sediment is added to the coastline.

A single parasequence is comprised of a series of facies associations overlying one another, each of which would have lain adjacent to one another within the depositional model. The succession within the parasequence obeys Walther's Law. From any point in the parasequence, the thickness of the overlying deposits is a proxy for the water depth (minus the effect of compaction). That is to say that from the base of the lower shoreface on the graphic log (Figure 8.2a) there is 4 m of lower shoreface, 7 m of upper shoreface and 2 m of foreshore. Consequently, the base of the lower shoreface was deposited in approximately 13 m of water.

## 8.1.2  Parasequences along a depositional dip profile

The Grassy Member (Figure 7.4) consists of four parasequences that represent the highstand systems tracts of two sequences. The lower two parasequences, Grassy parasequences 1 and 2 (GPS1 and 2), crop out between Horse Canyon and Coal Canyon in the south-east (Figure 7.1). The extensive exposure clearly illustrates the facies changes that occur along a proximal to distal depositional profile (Figure 8.3 overleaf), from the up-dip proximal pinch-out of the marine facies (where they pass into the non-marine section of the parasequence), to the distal, down-dip termination of the offshore transition zone. The down-dip changes in facies within the parasequence also affect the stratal expressions of the flooding surfaces that bound the parasequences.

The offshore transition zone deposits of the lower parasequence (GPS1) pinch-out between Coal Canyon 1 and the next canyon to the east, Stub Canyon (Figure 7.1). Basinward (east) of this point, it is very difficult to identify the parasequence boundary because there is no detectable facies change indicating changes in water depth.

The offshore transition zone deposits of GPS1 prograded as far as Coal Canyon (Figures 7.1, 8.3) and between there and Tusher Canyon the flooding surface at the base of GPS2 is marked by the superposition of offshore shale on to offshore transition zone. Further up-dip, between Tusher and Gray Canyon (Figures 7.1, 8.3) the flooding surface is represented by offshore shale on top of lower shoreface deposits. The offshore shale at the base of GPS2 thins landward and eventually pinches out just west of Gray Canyon.

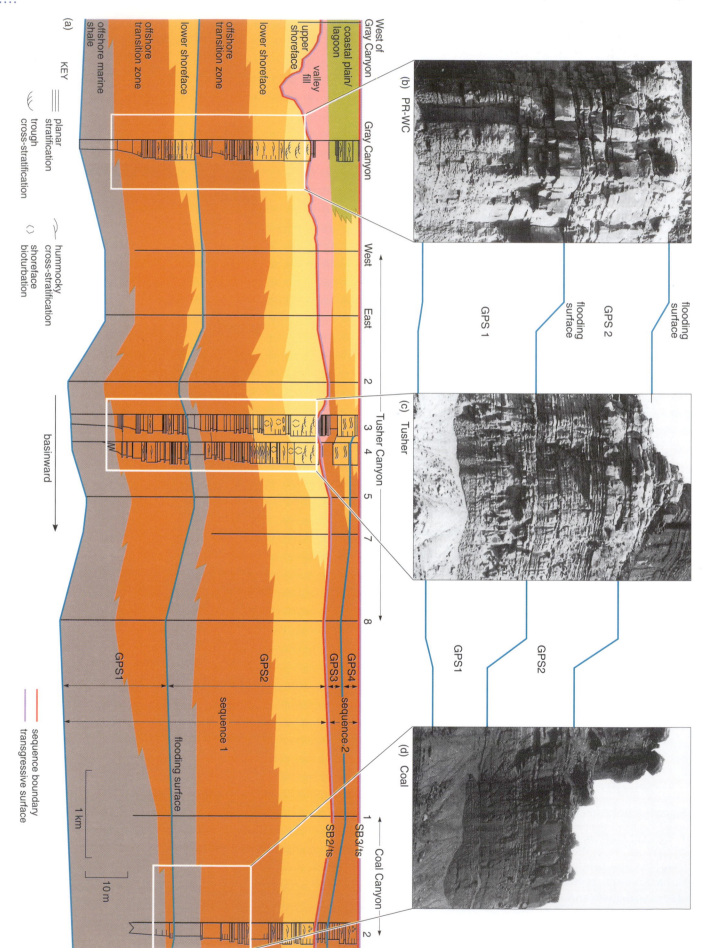

○ What can you deduce about the amount of change of water depth along the dip profile from these field relations?

● The relations indicate that, although there was an increase in relative sea-level to produce the flooding surface, the water depth was still close to storm wave-base (c. 40 m).

By the Middle Mountain sections (Figure 7.1), the water depth increase associated with the parasequence boundary formation is less than the depth to fairweather wave-base (c. 10 m) and the flooding surface is represented by lower shoreface deposits overlying upper shoreface and foreshore deposits of GPS1.

Further up-dip still, the expression of the parasequence boundary is very subtle. Upper shoreface facies of GPS2 overlie the upper shoreface and foreshore deposits of GPS1. In such cases, it can be very difficult to identify the flooding surface because the facies are all very similar. The following criteria can help:

● Juxtaposition of the two shoreface sandbodies results in anomalously thick deposits compared with other shorefaces in the same depositional system (same basin topography and same wave regime).

● In many of the examples studied from the Book Cliffs, the flooding surfaces are marked by carbonate-cemented horizons. These are believed to result from the recrystallization of lags of shelly debris deposited during the transgression (Figure 8.4).

● In many cases, the top of the parasequence may be marked by evidence of subaerial exposure, such as rooting, or even a coal. In such cases, the superposition of marine shoreface deposits is clear indication of a flooding surface.

● In many systems, pebble lags may form, concentrated by wave erosion during the transgression and formation of the flooding surface, though no pebble lags are found in the Book Cliffs succession.

GPS1 continues landward as a marine shoreface for a further 12 km to the area just south of Lila Canyon (Figure 7.1). Landward of the up-dip pinch-out of the marine facies in the upper parasequence, the flooding surface is paradoxically marked by the deposition of non-marine rocks over marine rocks. To understand this relationship, it is important to remember the incremental addition of accommodation space that generates each parasequence and to consider each parasequence as a series of linked depositional systems. We will discuss this point in more detail in Section 8.4.

**Figure 8.4**   Up-dip expression of a parasequence boundary. Unlike Figures 8.2 and 8.3, the parasequence boundary is not represented by a marine shale flooding surface here, but by a 30 cm-thick carbonate-cemented layer between two vertically amalgamated parasequences (GPS1 and GPS2). The carbonate cemented layer represents a recrystallized shell bed deposited during transgression. *(Stephen Flint, University of Liverpool.)*

**Figure 8.3 (opposite)**   (a) Cross-section constructed from graphic logs through the Grassy Member of the upper Blackhawk Formation in the Green River area (Figure 7.1). Grassy parasequences (GPS1–4) exhibit a progradational stacking pattern. Note how the offshore shale (grey) above the flooding surface at the base of GPS2 thins landward, from 8 m thick in Coal Canyon to zero in Gray Canyon. The vertical black lines indicate the positions of other graphic logs not shown here but used to construct the cross-section. Grain size varies from clay to fine-grained sand. The three photographs (b)–(d) show the landward (left) to basinward (right) change in the vertical profile of GPS1 and GPS2 as a function of position on a depositional dip profile. Note that GPS 1–4 are denoted G1–G4 in Figures 7.4c and 10.1. *(O'Byrne and Flint, 1995.)*

We have traced the boundary between two parasequences regionally and observed the way in which it changes in a depositional dip direction. The most important feature to remember is that within the marine system the parasequence boundary is always represented by a marked increase in water depth, interpreted from the facies. These changes are typically easier to see in more down-dip settings where they are marked by sudden changes in lithology, from sandstone to siltstone. In the up-dip section, the same event may be represented by cemented horizons or other subtle features.

The carbonate-cemented zones represent original colonization of the flooding surface by calcareous benthic fauna (typically bivalves) followed by subsequent dissolution of the shell material and reprecipitation as a local calcium carbonate cement during burial.

This illustrates a fundamental cornerstone of sequence stratigraphy: *that the events are correlated, not the facies*. Facies above the flooding surface change depending on where it is examined along the depositional profile, yet at all points it represents the same depositional hiatus.

One of the key features of wave-dominated shorefaces is that they are very continuous and uniform in a depositional strike direction (i.e. along the coast). There are, however, some changes which can occur along strike and these are best illustrated by re-examining the Spring Canyon Member.

Although wave processes are responsible for moving and depositing sediment in shoreface systems, the majority of the sediment is still introduced to the coast by fluvial systems (Figure 7.9). Exposures of the Spring Canyon Member behind the town of Kenilworth in the north-west (Figure 7.1) illustrate the geometries and facies of a deltaic parasequence close to the fluvial input point (ancient river mouth). Instead of the upward-coarsening succession from offshore siltstone through HCS beds into trough cross-stratified beds of the upper shoreface, the section at this locality (Figure 8.5) contains slightly different facies.

In this case, there is still an upward increase in the proportion of sandstone beds and as before, they thicken upward, typically to 40 cm thick. Instead of HCS, beds are typically planar laminated, or more rarely have current ripples and they are sharp based. Bioturbation is present but less well developed than elsewhere in the same parasequence. Individual sandstone beds dip in a basinward direction (Figure 8.5a,b); these are clinoforms (Section 4.5). Laterally (along depositional strike) the individual beds pinch out (over a distance of 2.5 km) and interfinger with flat-lying HCS beds.

The succession behind the town of Kenilworth represents a river-dominated mouth bar that sits within a shoreface succession. This mouth bar represents a point in the basin where the fluvial conditions were locally dominant although overall the system was a wave-dominated shoreface. Much of the sediment that was introduced by the river was re-mobilized by storms and redeposited along the coast. The key observation is that sediments deposited by both wave and fluvial processes can co-exist laterally within the same parasequence.

(a)

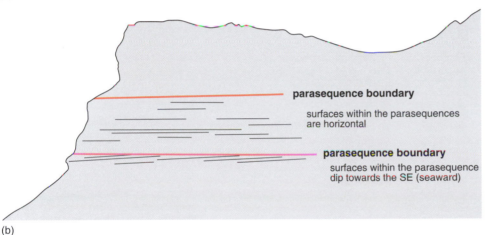

parasequence boundary

surfaces within the parasequences
are horizontal

parasequence boundary

surfaces within the parasequence
dip towards the SE (seaward)

(b)

(c)

**Figure 8.5**   Shoreface parasequence close to fluvial input point from the Spring Canyon Member behind the town of Kenilworth. (a) Photograph of the Kenilworth face showing the low-angle clinoform surfaces in the mouth bar of the Spring Canyon 4 parasequence. Cliff is c. 100 m high. (b) Line drawing of (a) to show interpretation based on the work of Diane Kamola. (c) Detailed photograph of current ripples (as opposed to HCS) in mouth bar parasequence at Kenilworth. ((a) and (c) John Howell, University of Bergen.)

## 8.2   Parasequences in a river-dominated coastal depositional system

In addition to the wave- and locally river-dominated deltas of the Blackhawk Formation, the Book Cliffs succession also includes excellent examples of river-dominated deltas in the underlying Panther Tongue Member (Figure 7.4). The lobes of this delta system are relatively small in comparison to modern-day deltas such as the Mississippi. They do, however, illustrate the key differences between river- and wave-dominated systems (Figures 7.8, 7.9).

In vertical succession, parasequences in the Panther Tongue (Figure 8.6 overleaf) appear superficially similar to the wave-dominated examples discussed previously. The lower portion is composed of marine siltstone and passes upward through an interbedded sandstone and siltstone succession into a sand-dominated upper interval. There is an upward increase in both the proportion and thickness of individual sandstone beds.

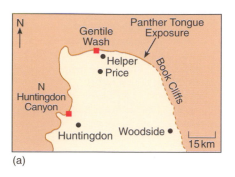

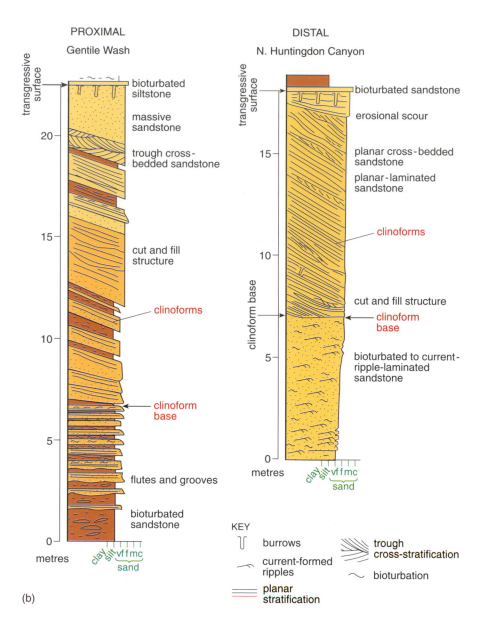

**Figure 8.6** (a) Location of graphic logs in (b). (b) Two graphic logs through the Panther Tongue river-dominated delta deposits, showing the distal thinning over 40 km from Gentile Wash (close to the position of Figure 7.8c) in the north, to the Huntingdon area further south (down depositional dip). Note the dominance of clinoforms and absence of HCS. Angle of clinoforms exaggerated. ((b) Posamentier and Morris, 2000.)

Individual sandstone beds are typically planar-laminated or more rarely current-rippled. Beds have sharp, erosive bases and, where they overlie mudstones sole marks are common. Normal grading is common within beds and the upward transition from planar-laminated to ripple-laminated records the waning of current flow. Siltstone beds record longer periods of deposition of suspended load sediment in between episodes of sandstone deposition. The sandstones represent pulses of sediment, potentially related to periods of high discharge through the fluvial feeder system. The sedimentary structures are consistent with deposition from turbidity currents and the beds are interpreted as delta-front turbidites. These result from sediment-laden, high-density river currents producing underflows that hug the sea-bed and do not mix with the overlying marine water. The beds form clinoforms that dip in a basinward direction at a slope of 3–5° (Figures 7.8c, 8.6), and record the original depositional dip on the front of the mouth bar.

The Panther Tongue parasequences are up to 25 m thick in the most proximal exposures around Gentile Wash (Figure 7.1) and comprise coarsening-upward associations of sandstones and silty claystones. The sandstones are dominated by current ripples and planar laminations with some rare HCS. Successions are more heterolithic than the Blackhawk shoreface deposits. They thin down the depositional dip profile, which in the Panther is towards the south. These river-dominated delta parasequences are thicker than the wave-dominated shoreface parasequences of the Blackhawk Formation because there was no significant reworking of sand above storm-wave base, owing to the low wave energy of the basin at this time.

Within the Panther Tongue in the Spring Canyon area (Figure 7.1), the sandstones and siltstones with clinoforms are replaced by a sandstone that is massive or contains soft sediment deformation. These massive sandstones occur in a channel-shaped body 15 m thick and 750 m wide. The margins are erosive and truncate the clinoform beds. It is interpreted as a distributary channel that was feeding sediment to the delta.

Laterally, the river-dominated deltaic parasequences are significantly different from their wave-dominated counterparts. Whereas the shoreface parasequences are represented by shore-parallel, laterally continuous sandstone bodies that extend for tens to hundreds of kilometres, the river-dominated parasequences are lobe-shaped, typically 5–10 km wide. Mapping of these lobes shows that, following river avulsion, the new delta lobe typically occupies the topographically lower part of the previous interlobe areas. These geometries indicate that significant care must be taken correlating parasequences in a river-dominated deltaic setting when using limited subsurface borehole data because there would be a tendency to correlate separate lobes as a single sandbody.

○ List up to five general differences between a river-dominated deltaic parasequence and a wave-dominated shoreface parasequence.

● River-dominated deltaic parasequences are more heterolithic, have a greater range of sedimentary structures reflecting the interaction of unidirectional currents and waves, may have thin-bedded turbidites at their bases and tend to dip more steeply in a basinward direction. On a larger scale, their geometry is more lobate than wave-dominated parasequences.

## 8.3   Parasequences in the non-marine environment

It is important to consider the parasequence as a series of linked depositional systems. Each parasequence contains a number of different facies deposited within the same time interval. We have studied the marine portion of the parasequence so now we shall correlate this up-dip (landward) and consider what the equivalent is in the marginal and non-marine environments at this time. In the marine setting, flooding surfaces that juxtapose deeper-water facies over shallow-water deposits are relatively easy to identify. As we trace the surface up-dip, it becomes more difficult to delineate. If non-marine deposits are overlain by marine deposits, the boundary clearly marks a flooding surface; however, when no marine rocks are present within a vertical section then the identification of parasequence boundaries is harder.

○ Why will parasequence boundaries be present in the non-marine system?

● The alluvial system in the coastal plain is connected to the sea. Changes in relative sea-level result in changes in water table, changes in the gradient of the fluvial systems and changes in the amount of fluvial accommodation that is available to preserve sediment (Section 4.1). All of these changes are potentially identifiable in the rock record.

In the Book Cliffs and other well-exposed successions, the easiest way to delineate parasequences in marginal and non-marine environments is physically to trace individual parasequence boundaries from marine strata into the up-dip equivalent non-marine strata. Examples from the Spring Canyon Member illustrate the main facies transitions observed between marginal marine and non-marine parasequences (Figure 8.7). An example from the Sunnyside Member is discussed at the end of this Section, to highlight one of the main departures from the general pattern. The orientation of canyons that expose the Spring Canyon parasequences makes it relatively easy to trace the parasequence boundaries from the marine facies into their non-marine equivalents. The trough cross-stratified sandstones of the upper shoreface lie adjacent to landward-dipping, rippled and planar-laminated sandstones, interbedded with unbioturbated dark-coloured mudstones (Figure 8.7) interpreted as lagoonal deposits.

The sandstone beds were deposited within the lagoon as washover fans and flood-tidal delta lobes (see Figure 7.13). Locally, rippled sandstone beds dip in a seaward direction and are interpreted as small bayhead deltas that form at the head of a bay or estuary (shown as an intravalley delta on Figure 7.14). Channel-shaped sandstones with mud-draped trough cross-stratification are rare but, where seen, are interpreted as tidal channel deposits. Rooted sandstones and siltstones, together with extensive coals, cap the lagoonal deposits. Thus each parasequence comprises the sediments that record the initiation and filling of a lagoon.

The individual lagoonal systems are typically less than 1 km wide in a seaward to landward sense. On the landward side, in the non-marine facies of the coastal plain, it is more difficult to identify individual parasequences. However, the thick coal seams that overlie the shoreface and lagoonal deposits are continuous into the non-marine realm and serve as useful markers.

Between these coal seams are 10–15 m-thick successions of fluvial channel and overbank deposits. There is typically a higher proportion of overbank deposits in the lower part of the non-marine parasequence with an increase in the proportion of fluvial channel deposits upwards. There may be poorly developed coals and carbonaceous siltstones within the non-marine parasequence; however, the laterally extensive, thick low-ash coals always form the parasequence boundaries (Figure 8.8 overleaf).

○ What is meant when we refer to ash in coal?

● Ash in coal geology refers to siliciclastic material that remains after the coal has burnt. Low ash therefore refers to coals that are almost entirely organic material with only a low percentage of siliciclastics.

Interpretation of the origin of the coals and thus their stratigraphical setting is essential to understanding and correlating the parasequences into the non-marine environment. Two models currently exist:

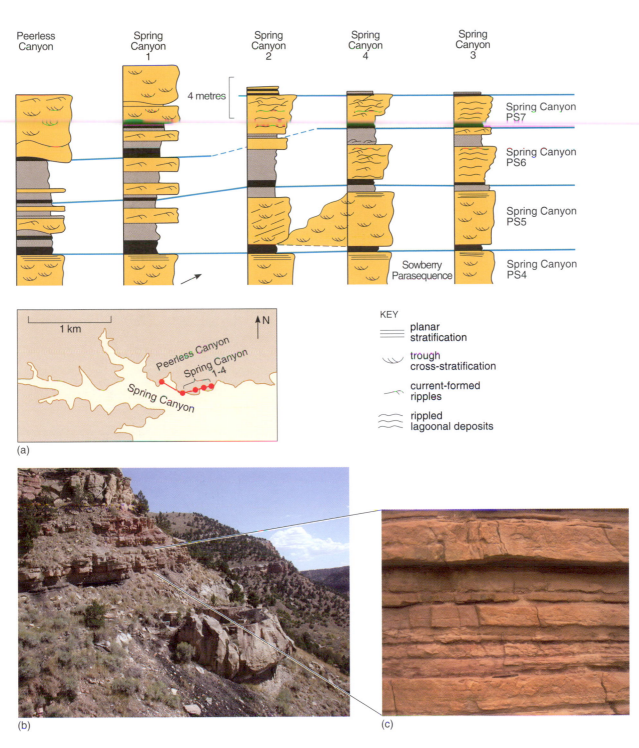

**Figure 8.7**   (a) Graphic log correlation of the Spring Canyon Member parasequences, showing the landward facies change in the Spring Canyon PS5 from barrier-island sandstone on the right-hand side to heterolithic lagoonal deposits on the left. Map shows locations used in line of section. (b) Exposure of the lagoonal deposits including channels and interbedded sands and shales (cliff c. 15 m high). (c) Detailed view of mud-draped, rippled tide-influenced lagoonal sediments (image c. 1.5 m across). ((a) Kamola and Van Wagoner, 1995; (b) and (c) John Howell, University of Bergen.)

**Figure 8.8**  Example graphic logs and photographs of non-marine parasequences in the Kenilworth Member. (a) Graphic log at Coal Creek showing Kenilworth parasequences 2, 3 and 4 in their non-marine facies. Kenilworth PS1 is still marine at this locality. The parasequence boundaries are coals, related to water table rise, landward of marine flooding. Note the increase in sandstone content upward in each parasequence.

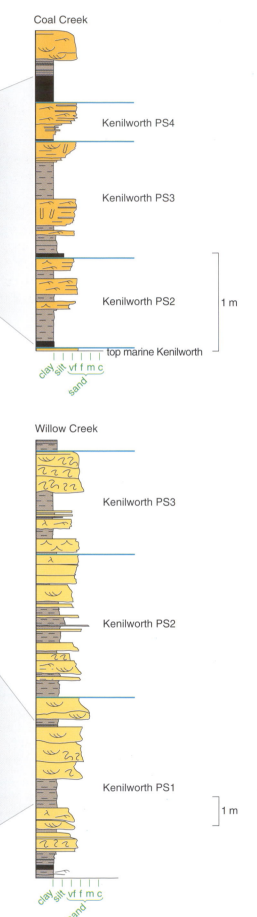

Coal Creek

Kenilworth PS4

Kenilworth PS3

Kenilworth PS2

1 m

top marine Kenilworth

clay silt vf f m c
sand

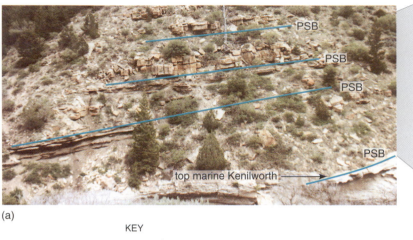

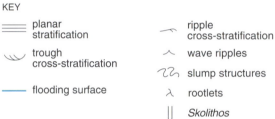

(a)

PSB
PSB
PSB
PSB
top marine Kenilworth

**KEY**

≡ planar stratification

⌣ trough cross-stratification

— flooding surface

⌒ ripple cross-stratification

∧ wave ripples

↝ slump structures

λ rootlets

∥ *Skolithos*

**Figure 8.8 (continued)**  (b) Detail of the Kenilworth non-marine parasequences further landward at Willow Creek. This section was deposited on the landward side of coastal peat mires. Here, there is an upwards increase in the proportion of channel deposits within each parasequence, and parasequence boundaries are represented by the sudden increase in proportion of fine-grained overbank deposits. The photograph shows detail of the upward increase in sandstone in a coal seam-bounded non-marine parasequence.
*((a) and (b) John Howell, University of Bergen.)*

(b)

Willow Creek

Kenilworth PS3

Kenilworth PS2

Kenilworth PS1

1 m

clay silt vf f m c
sand

## Model 1: The base of the coal represents the base of the parasequence

Although the coastal plain may be covered in luxuriant vegetation, the necessary conditions for coal formation of (a) high and rising water table and (b) low siliciclastic input are not fulfilled during the times when the parasequence is prograding, because sand is being supplied through the river to drive progradation. The water table is static and overbank flooding commonly emplaces silt and sand into the peat swamp. However, as the next parasequence starts to form, relative sea-level rises and this forces the water table to rise. Coastal mires become waterlogged and the potential to preserve the organic debris as peat and ultimately as coal is enhanced. Secondly, the raising of relative sea-level reduces the gradient of the fluvial systems, making it more difficult for rivers to continue supplying sand to the coastline and therefore into the peat swamp via overbank flooding. This material is trapped in more landward positions and the coastal region becomes more sediment-starved, thus satisfying condition (b) above. In this model, the bases of the coals are interpreted to represent the non-marine expression of the marine flooding surface.

## Model 2: The top of the coal represents the top of a parasequence

Many modern deltas in tropical climates such as the Baram delta of northern Borneo have large raised mires developed on their tops. In a raised mire, the accumulation of peat raises the floor of the bog above the regional depositional surface. A locally high water table is maintained by frequent heavy rainfall. The raised nature of the mire excludes any siliciclastic input into the system and consequently the coals developed from them have very low ash content. This model requires that the coals be initiated as low-lying coastal swamps behind the shoreline. As the parasequence progrades, the peat swamp on the coastal plain evolves into a raised mire, maintaining its own water table and excluding siliciclastic input. However, peat accumulation and preservation is terminated by a rise in sea-level which either directly floods the mire or, in more up-dip locations, the associated rise in base level allows significant amounts of siliciclastic sediment to invade the swamp. Consequently, the top of the coal marks the top of the parasequence. While the precursor peats to a coal may accumulate above base level in a raised mire during progradation, ultimate preservation is dependent upon generation of accommodation space by subsequent base level rise at the next parasequence boundary.

Intuitively, most people would think of coals as being linked to regression (static base level or base level fall) rather than the start of transgression (base level rise). The crucial argument is that of preservation, so that even raised mires have to (a) initiate as low-lying mires, when the water table (and base level) is high and (b) require generation of accommodation space to ensure their long-term preservation. Consequently, we interpret the majority of widespread, low-ash coals within the Book Cliffs as parasequence boundaries, formed in the non-marine environment during episodes of base level rise and transgression at the coastline (model 1 above). Some workers have extended these ideas to suggest that coals are usually associated with transgressive systems tracts. However, the exact base level interpretation of coals has to take into account the balance between sediment supply and accommodation space, which provides for coals accumulating in both the transgressive and highstand systems tracts. The raised mires of the modern Baram delta are an example of highstand systems tract coal-forming environments.

In the Sunnyside Member (Figure 7.4), which includes a particularly thick (up to 5 m) and very laterally extensive coal seam, the up-dip equivalent of an entire marine parasequence is represented by coal. The marine shoreface thins and pinches out into a siliciclastic coal split (Figure 8.9). A coal split is where a continuous coal seam is split laterally into two seams by a layer of non-coal material.

The Sunnyside coal split contains the marine portion of the parasequence and the coal contains the non-marine portion of the parasequences and the flooding surface. This is possible because peat compacts to as much as one-tenth of its original thickness to produce coal. Five metres of coal represent up to 35 m of non-marine sediment.

**Figure 8.9**  (a) View of Sunnyside PS3 and the adjacent parasequences. The photograph shows where Sunnyside PS3 is represented by a coal overlain by a marine shoreface (thin, pale-coloured band) that thickens seaward (to the right), forming a split in the Sunnyside coal. (Cliff is *c.* 50 m high.)
(b) Interpreted formation of the split.
(i) Relative sea-level rise results in transgression into a margin of raised mire and deposition of a shoreface sandbody. (ii) The top of the new parasequence is colonized by coal-forming plants, producing a thick peat. (iii) Early burial-related compaction is much greater for the peat than the sand, resulting in backward rotation of the shoreface sandstone body into the peat. The peat then becomes coal during later burial. ((a) *John Howell, University of Bergen.)*

(a)

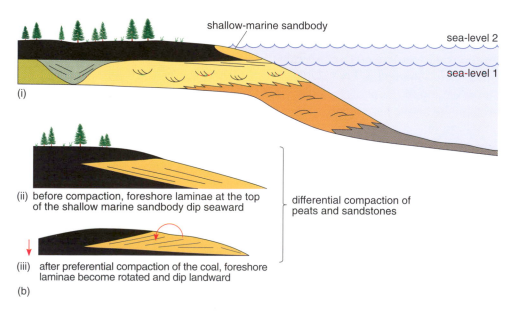

(b)

# 8.4  Model for the formation of parasequences within the Book Cliffs succession

Parasequences within the Book Cliffs succession represent the deposits of shoreface or delta progradation. This progradation is terminated by a rapid rise in relative sea-level which floods the low-lying delta top and provides the extra accommodation space into which the next parasequence progrades. In this Section, we shall consider a series of time steps that show what happens in both the marine and non-marine portion of the parasequence (Figure 8.10).

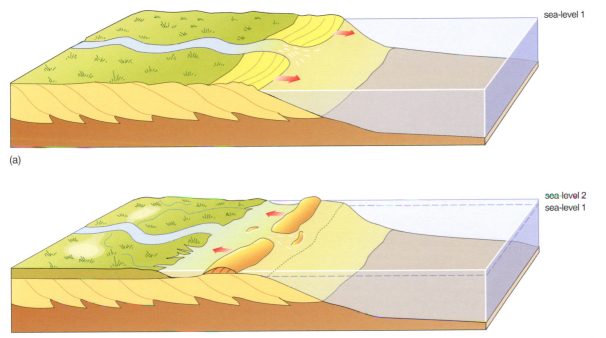

(a)

(b)

## Progradation of the first parasequence (Figure 8.10a)

Relative sea-level is stable; most sediment is passed through the fluvial system and added to the front of the shoreface. The shoreline progrades basinward. The rate of basinward movement decreases as the water depth in front of the shoreface or delta increases. Raised mires may co-exist in the coastal plain.

## Relative sea-level starts to rise (Figure 8.10b)

Relative sea-level rise causes flooding of the coastal area and the strandplain becomes a transgressive barrier island system with a lagoon on the landward side. The rise in relative sea-level reduces the fluvial gradient and sediment starts to accumulate in the coastal plain rather than at the coastline. As relative sea-level rises, the water table is raised and promotes the accumulation and preservation of coal precursors.

## Storage of sediment in the coastal plain (Figure 8.10b continued)

Deposition of sediment in the coastal plain facilitates the transgression due to decreased sediment supply and the barrier island migrates in a landward direction. Barrier migration typically leaves nothing behind except for a lag of coarse-grained material winnowed by the wave erosion (termed a ravinement lag).

**Figure 8.10**  (a) and (b) Parasequence formation takes place in several stages, controlled by the balance between the rate of generation of accommodation space (through relative sea-level change) and rate of sediment supply. Arrows mark the migration directions of facies belts. See text for explanation of each stage.

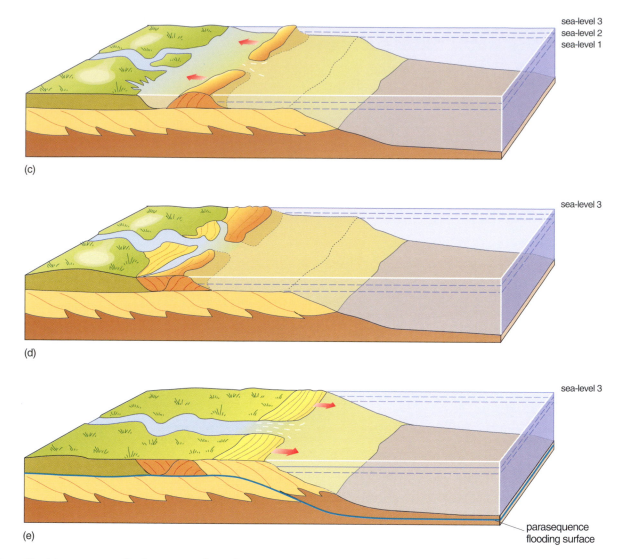

sea-level 3
sea-level 2
sea-level 1

(c)

sea-level 3

(d)

sea-level 3

parasequence
flooding surface

(e)

**Figure 8.10 (continued)** (c)–(e)
Parasequence formation takes place
in several stages, controlled by the
balance between the rate of
generation of accommodation space
(through relative sea-level change)
and rate of sediment supply. Arrows
mark the migration directions of
facies belts. See text for explanation
of each stage.

### Relative sea-level stops rising (Figure 8.10c,d)

After relative sea-level stops rising, the accommodation space generated in the
coastal plain is filled and fluvial channels start cutting into one another and
amalgamating. The lagoon system becomes filled with sediment and the barrier
island is preserved at its final, most landward point.

### Progradation of second shoreface parasequence (Figure 8.10e)

Sediment is now resupplied through the fluvial systems to the shoreface, which can
then start to prograde again. The coastal plain is now an area of sediment by-pass.

This model explains the observed stratal relationships within the parasequences in
the Blackhawk Formation. It also allows us to consider the various linked
depositional systems within the parasequence in terms of sediment balance. The
concept of sediment balance states that if sediment is being deposited within one
part of the depositional system it is not available for another and also that sediment
will travel no further than it needs to before being deposited. Consequently, during
shoreface progradation the sediment must by-pass the more up-dip fluvial system
and, during transgression and base-level rise, sediment is deposited in the fluvial
system and is not available for the marine system. This concept is also known as
*volumetric partitioning* and is summarized in Figure 8.11d (overleaf).

○ Taking into account sediment balance, if we start with a transgressive barrier island–lagoon system, and start progradation, how will the sediment balance shift?

● While the barrier is still transgressing, most of the sediment is aggrading on the coastal plain. In order to start progradation, the rivers have to carry sediment out first to fill the lagoon and then to add sediment to the shoreface. If more sediment is added to the coast than can be redistributed by longshore processes, the shoreface will prograde. At this time, there is negligible storage of sediment in the rivers and they simply act as transport conduits feeding the shoreline.

Whilst we understand what goes on during the formation of a parasequence, we have still to consider the driving mechanism that causes the initial increase in accommodation space at the parasequence boundary. Understanding the duration and frequency of the parasequences can provide an insight into their driving mechanism. The total age span of the formation, as estimated by comparing biostratigraphical information with a numeric timescale, is c. 3.5 Ma. There are at least 38 parasequences and 8 sequence boundaries in the Blackhawk Formation. As an approximate calculation, we have assumed that these 46 stratigraphical events are of equal duration. Thus, the average frequency of parasequence boundary formation is every 70–80 ka. For further discussion of the timing of the events in the Book Cliffs succession, see Chapter 10.

In the original work on parasequences, van Wagoner considered three possible causes for parasequence boundary formation (see also Section 4.2). These are discussed here in the context of the Book Cliffs:

*Pulsed, tectonically driven subsidence.* This mechanism is a possibility for the Book Cliffs succession, which was deposited in a subsiding, tectonically active basin. Periodic thrust sheet movement in the mountain belt to the west may have resulted in pulsed loading and increments of subsidence of the basin floor. However, the documented time-scales for crustal responses to loading are much greater than 70–80 ka.

*Eustatic sea-level changes.* Small-scale, high-frequency changes in global sea-level could easily have caused the flooding surface that we have observed. During some periods of Earth history, waxing and waning of continental ice sheets have caused rapid sea-level rises but slow falls because the ice sheets have decayed faster than they formed. However, the existence of glacial ice sheets during the Cretaceous 'greenhouse world' is a hotly debated topic as there are no recorded glacial sedimentary rocks of this age and other data indicate that global temperature was too warm for ice-caps to form (Figure 5.4).

*Delta lobe switching and ongoing subsidence.* When rivers avulse to a new location, they build a new delta lobe. The old lobe becomes starved of sediment and ongoing subsidence causes a relative rise in sea-level and a flooding surface at that point (Section 5.3). This mechanism is possible for the river-dominated parasequences of the Panther but less likely for the wave-dominated ones of the Blackhawk. In wave-dominated systems, where the sediment can be reworked hundreds of kilometres along the coast, the position of the fluvial channel is of much less importance. Furthermore, the relationship between the coastal plain and shoreface deposits described above indicates stepped sea-level changes rather than continued subsidence.

Whilst any of these causes may have been responsible for the Book Cliffs parasequences, a relative sea-level (tectonic or eustatic) driving mechanism is favoured for the Blackhawk flooding surfaces for the reasons outlined above.

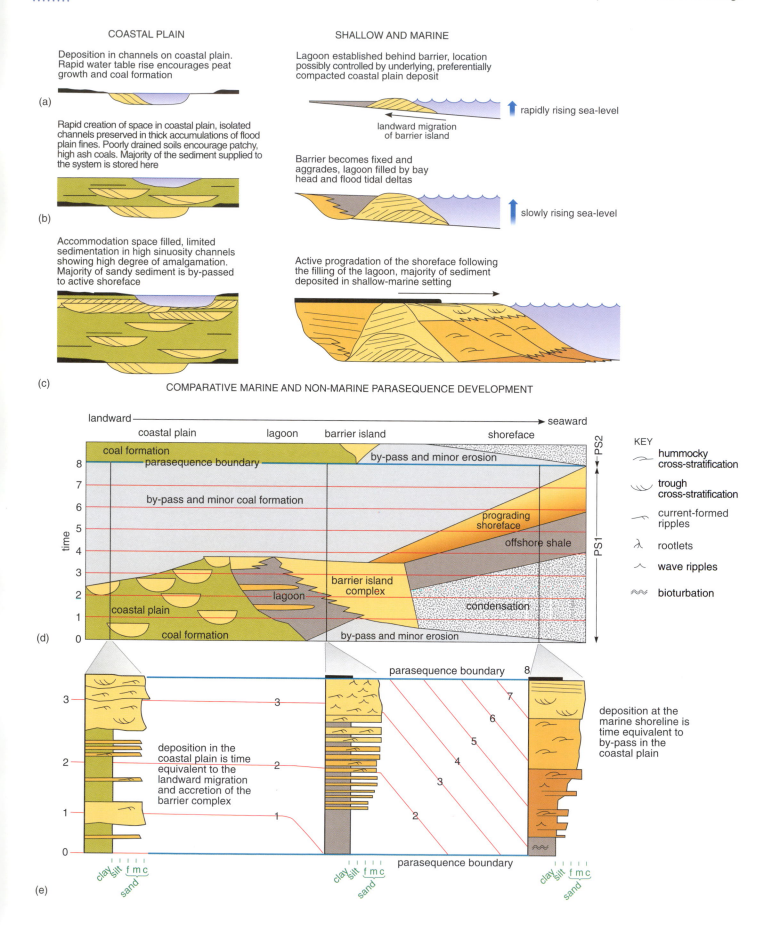

## 8.5   Parasequence stacking patterns

Parasequences commonly occur in sets that exhibit distinctive stacking patterns of progradation, retrogradation or aggradation (Section 4.2; Figure 4.7). The majority of parasequences in the Book Cliffs occur in progradational sets, although examples of retrogradational and aggradational sets are also found.

### 8.5.1   Progradational stacking pattern

The lower two parasequences of the Grassy Member that we examined previously show a clear progradational stacking pattern (Figure 8.3). This is shown by the fact that the down-dip pinch-out of the upper shoreface facies (i.e. the point it prograded to prior to flooding) of Grassy Parasequence 2 lies more basinward than that of the Grassy Parasequence 1. Consequently, through time the shoreline moved basinward. This geometry implies that the long-term rate of sediment supply was greater than the rate of creation of accommodation space (Section 4.2). The parasequence stacking patterns in the upper part of the Blackhawk are more markedly progradational than in the lower Blackhawk (Figure 7.4c; compare Aberdeen parasequences 1–3 with Desert parasequences). Following the same logic, this suggests that the balance of sediment supply versus accommodation space creation changed through time.

### 8.5.2   Retrogradational stacking pattern

A retrogradational stacking pattern occurs when the long-term rate of accommodation space creation is greater than the supply of sediment (Section 4.2). Although the individual parasequences still prograde, successive parasequences do not prograde as far into the basin as earlier parasequences. The uppermost parasequence in the Kenilworth Member (K6 on Figure 7.4c) can be clearly seen to have prograded less distance than those below it, thus signifying retrogradational parasequence stacking.

○   Why do individual parasequences still coarsen-upward (i.e. shallow-upwards) in a retrogradational parasequence set?

●   A parasequence represents a phase of delta or shoreface progradation. Individual parasequences within a parasequence set each record the basinward migration of the shoreline. In a retrogradational parasequence set, successive parasequences do not prograde as far as the underlying ones. So, when we examine a set of these units, individually they all prograde (coarsen and shallow upward) but in a retrogradational set long-term space creation is outpacing sediment supply so the succession as a whole will fine-upwards (and deepen).

**Figure 8.11 (opposite)**   Schematic representation of volumetric partitioning in space and time for the Book Cliff parasequences. It illustrates the spatial change in the volumes of sediment being deposited in the non-marine, lagoonal, shoreface and offshore facies of a single parasequence as illustrated in Figure 8.10. (a) Relative sea-level rise causes the barrier island to move landward; sediment is deposited in the lagoon created behind the barrier and on the coastal plain. (b) During landward migration of the barrier island in a period of relative sea-level rise, all the sand is stored in the coastal plain. (c) As the shoreface starts to prograde again, sediment supply fills the available accommodation space, and the coastal plain again becomes a by-pass area and the main sand volume is partitioned into the coastal zone; hence a change in volumetric partitioning or 'mass-balance'. (d) The same relationships expressed in a chronostratigraphical diagram (Section 4.3.1). Note that the time during which the majority of non-marine strata are deposited in the parasequence corresponds to the thin layer of offshore marine shale at the base of the shoreface. The majority of the shoreface sand is slightly younger than the non-marine component of the parasequence. (e) Graphic logs showing idealized parasequences from a coastal plain, shallow and marginal marine setting and their relationship to (d). For (d) and (e), the numbered red lines are the time lines.

### 8.5.3  Aggradational stacking pattern

Within the Spring Canyon Member, successive shoreface parasequences (SC4 to SC7) prograded to approximately the same position (Figure 7.4c). This defines an aggradational stacking pattern and indicates that the supply of sediment is equal to the long-term accommodation space available (Section 4.2).

Parasequence set boundaries are represented either by thick shale sections marking longer duration/higher magnitude flooding surfaces than parasequence boundaries or are coincident with erosional unconformities (sequence boundaries).

## 8.6  Summary

- In the Book Cliffs succession, the parasequences are well developed. Each parasequence represents a phase of delta or shoreface progradation bounded by flooding surfaces. For each parasequence an increment of space was created and the shoreline prograded and filled that space. Internally, parasequences show an upward shallowing of the facies that records the progradation of the shoreline. The flooding surfaces bounding the parasequences are marked by an increase in the depositional water depth of the facies above the surface from those below it. The character of the rocks recording this increase varies along a depositional profile.

- Within the Book Cliffs, the formation of a flooding surface is associated with the transition of the progradational shoreline system into a transgressive barrier island. As the barrier migrates landward, most of the siliciclastic sediment is deposited in the coastal plain. Once the shoreline stops migrating landward, the lagoon is infilled and a new shoreface prograded. Parasequence boundaries in the non-marine environment are often represented by coals.

- Individual parasequences stack into sets that record the longer-term migration of the shoreline, controlled by the larger-scale balance between the rate of sediment supply and accommodation space creation. In the Book Cliffs, most parasequence sets are progradational although some aggradational and rare retrogradational stacking is also seen. The parasequence sets in the lower part of the Blackhawk are less progradational than those in the upper part.

## 8.7  References

### *Further reading*

KAMOLA, D. L. AND VAN WAGONER, J. C. (1995) (see other references for Chapter 7). [Very descriptive work on the parasequence concept and how it applies to the stratigraphy of the Book Cliffs.]

O'BYRNE, C. J. AND FLINT, S. S. (1995) 'Sequence, parasequence and intraparasequence architecture of the Grassy Member, Blackhawk Formation, Book Cliffs, Utah, USA', in VAN WAGONER, J. C. AND BERTRAM, G. (eds) *Sequence stratigraphy of foreland basin deposits,* American Association of Petroleum Geologists Memoir No. 64, 225–257. [Good description of parasequences in the Grassy Member.]

VAN WAGONER, J. C., MITCHUM, R. M., CAMPION, K. M. AND RAHMANIAN, V. D. (1990) (see other references for Chapter 4). [Original work that defined parasequences including examples from the Book Cliffs.]

# 9   Sequences and systems tracts in the Book Cliffs

**John A. Howell and Stephen S. Flint**

In addition to parasequences, the Book Cliffs is an ideal place to study the larger-scale systems tracts and sequences that are built of the parasequences (Chapter 4). Sequence boundaries have a number of different expressions, several of which can be demonstrated with examples from the Book Cliffs.

The formation of a typical sequence in the Book Cliffs can be summarized as follows:

1   *Relative sea-level fall:* Fluvial systems that were feeding sediment to the old highstand systems tract (HST) deltas have increased energy, and incised valleys are cut typically along the courses of the old river channels. The shoreline moves rapidly seaward by the process of forced regression. During forced regression, sediment is by-passed along the valleys and deposited in more distal parts of the basin, as the falling stage systems tract (FSST). Nothing is deposited on the old coastal plain because there is no accommodation space in that area. Long-term subaerial exposure and low water table leads to the formation of mature and well-drained palaeosols on the interfluves.

2   *From its low point, relative sea-level starts to rise slowly*: FSSTs are now abandoned and transgressed. Rivers cease to incise. The shoreline starts to move up the depositional profile and the lowstand systems tract (LST) is deposited.

3   *Continued relative sea-level rise*: This leads to the generation of accommodation space in the incised valleys; these are transgressed, become estuaries and start to fill with a complex mosaic of fluvial and tidal facies that represent the lower part of the transgressive systems tract (TST). The areas between the incised valleys (interfluves) are still subaerially exposed and the site of continued soil formation, although the soils now start to become more waterlogged as the water table rises.

4   *Further relative sea-level rise fills the valley and floods the former coastal plain interfluves*: Locally, a retrogradationally stacked set of parasequences may be deposited on the newly created marine shelf area; this forms the upper TST.

5   *Point of maximum rate of relative sea-level rise*: At this point, accommodation is being created at the most rapid rate, easily outpacing the rate of sediment supply. At or after this time, condensed marine deposition occurs on the shelf, accompanied by the maximum marine incursion (maximum flooding surface). Up-dip landward manifestations include the highest water tables, formation of deposits that will form thick coals and aggradation of thick, muddy floodplain deposits.

○   How might the balance between relative sea-level rise and sediment supply affect the timing of the maximum flooding surface along the coastline?

●   The point where the maximum flooding surface ceases to form will occur later in parts of the basin that have lower rates of sediment supply. This is because with less sediment it takes longer for the space being created by the relative sea-level rise to be filled. Conversely, highstand progradation will occur earlier in areas that are supplied with more sediment.

6    *Decreasing rate of relative sea-level rise:* Eventually, the rate of sediment
     supply becomes greater than the rate at which accommodation space is
     being created and the sedimentary system starts to prograde again. These
     progradational sediments form the highstand system tract (HST).

Continued progradation of the HST is terminated by the next relative sea-level
fall, which initiates the formation of the next sequence boundary.

Sequences exist on a number of scales, and can be classified in terms of their
duration. In the Book Cliffs succession, high-order, higher-frequency sequences
are nested within a lower-order, lower-frequency sequence or group of sequence
sets (Chapter 5). Biostratigraphical data correlated with a numeric time-scale
show that the entire 500 m-thick stratigraphical interval from base Panther
Tongue to base Castlegate Sandstone represents approximately 4.5 Ma. The
sediments deposited over this interval are interpreted as a low-order sequence
and can be subdivided as follows: the Panther Tongue represents the falling
stage and lowstand sequence set; the retrogradational Storrs Member is the
transgressive sequence set and the Blackhawk Formation represents the
highstand sequence set. The base of the thick, braided fluvial Castlegate
Sandstone is interpreted as the next low-order sequence boundary (Figure 7.4c).

Within this low-order sequence, there are most of nine higher-frequency
sequences, which are especially well developed within the upper part of the
Blackhawk Formation (Figure 7.4c). Volumetrically, most of the Blackhawk
Formation is comprised of progradationally stacked parasequences, which form
HSTs. Also present are estuarine-incised valley fills and some rare TST
parasequences.

There are two reasons why the HST deposits make up most of the volume of the
sedimentary record within the Blackhawk. (i) The eight sequence boundaries
within the succession are interpreted to represent significant periods of time
when either by-pass or erosion was occurring within the Blackhawk. The
surface is therefore the only record of protracted periods when relative sea-level
was falling or low and deposition occurred in basinal areas. (ii) Within the
Blackhawk sequences, the TSTs are poorly developed because the overall rate
of sediment supply was very high.

Consequently, the actual period of time between deposition of the LST and the
point at which progradation was able to overtake accommodation space creation
was relatively short. The result in this succession is either a single TST
parasequence (e.g. Kenilworth Member) or the whole TST being represented by
only the valley fill and an overlying marine shale.

Section 9.1 is a tour through some of the best examples of sequence boundaries,
incised valleys, FSSTs, LSTs, TSTs and HSTs in the Book Cliffs case study.
Figure 7.4c provides a useful summary of the lithostratigraphical nomenclature.

# 9.1    Sequence boundaries

○ Which of the following features is *not* a result of relative sea-level fall? Explain your answer.

(i)    Subaerial exposure of previously subaqueous deposits.

(ii)   Incision of valleys by fluvial downcutting and headward erosion.

(iii)  Deposition of marine shales on top of coals.

(iv)   Change from poorly drained, boggy coastal plain conditions to well drained, mature soil-forming conditions.

● The deposition of marine shales on top of coal (iii) results from a relative rise in sea-level, not a fall. (iv) could result from either a relative sea-level fall (i.e. a forced regression) or from a normal regression due to the rate of sediment supply outpacing relative sea-level rise (Section 4.2).

The stratal expression of a sequence boundary will vary in both depositional dip and strike senses, depending upon the location within the basin. We will look at a variety of these expressions in the shallow-marine, coastal plain and offshore settings from the Book Cliffs succession (Figure 9.1).

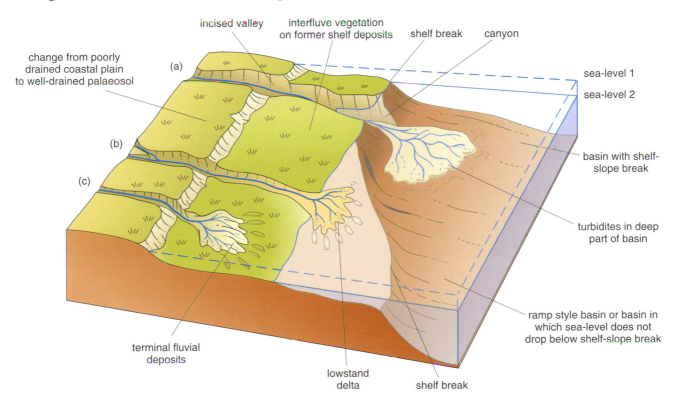

**Figure 9.1**    Schematic diagram, showing the effects of a relative sea-level fall in a basin with a continental shelf of varying width and depth. Cliff running across near the left-hand edge of the Figure shows the position of the former highstand shoreline. (a) In areas with a narrow, relatively shallow shelf, relative sea-level can fall to the shelf edge, subaerially exposing the entire shelf, rivers incise valleys out to the shelf edge and supply sediment to either shelf edge deltas or the heads of existing canyons. Gravitational processes take over and most of the sediment is remobilized within turbidity currents and deposited as basin floor fans. The sequence boundary is expressed by the erosional base of the incised valleys, the coeval interfluve surface on which a palaeosol will develop and the base of the submarine fan. (b) Effects of the same relative sea-level fall when, due to the shape of the basin, the shelf edge is not subaerially exposed. The result is a FSST and LST delta on the mid- or outer shelf (this is the case for most of the sequence boundaries in the Book Cliffs where the basin had a ramp-style topography). FSST and LST deltas may remain attached or become detached from the preceding HST, depending on the rate of sea-level fall and amount of sediment supplied. In case (c), the very low gradient of the shelf prevents the rivers carrying sediments to the new shoreline and the relative sea-level fall results in attached terminal fluvial deposits. In this case, no FSST or LST shorelines exist. This model was proposed by John Van Wagoner for the Castlegate Sandstone in the Book Cliffs.

Sequence boundaries are typically easiest to identify in the area around the shoreline of the previous HST. Within the Kenilworth Member (Figure 7.4), a progradational succession of parasequences is overlain by an erosional surface marked by an extensive drainage network of valleys separated by intervalley highs (Figure 9.2). This erosional surface is interpreted as a sequence boundary.

(a)

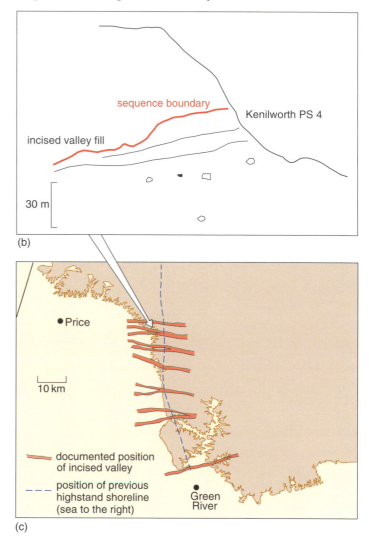

(b)

(c)

**Figure 9.2** Kenilworth Sequence Boundary. (a) Exposure photograph and (b) line drawing interpretation showing incision on the Kenilworth Sequence Boundary on the south side of Bear Canyon (Figure 7.1). The white- to brown-coloured sandstones at the top of Kenilworth PS4 are incised into by a surface cutting down to the right-hand side of the photo. Poorly exposed tidal heteroliths overlie the sequence boundary within the incision. (c) Map showing distribution of incised valleys associated with the Kenilworth Sequence Boundary. ((a) John Howell, University of Bergen; (c) Taylor and Lovell, 1995.)

The valleys are up to 16 m deep and 2 km wide (Figure 9.2c), and contain both fluvial and heterolithic tidal deposits that are interpreted as estuarine-incised valley fills. In many cases, the lower portion of the valley fill succession is fluvial and overlies lower shoreface or offshore transition zone facies of the underlying parasequence.

○ What facies are missing from the normal wave-dominated succession (Figure 7.9) at the sequence boundary? What are the implications of applying Walther's Law here?

● Middle shoreface, upper shoreface and foreshore facies are missing from the normal Book Cliffs progradational succession because the offshore transition sediments are directly overlain by fluvial and tidal deposits. Walther's Law tells us that only the facies which lie side by side in nature can exist in vertical succession unless there is a break in sedimentation. As we would never see fluvial systems next to offshore transition zone deposits in a single depositional environment, we know that there has been a break in the succession.

The surface of non-deposition and no fluvial erosion on the intervalley high areas is the interfluve sequence boundary. The interfluves between the valleys are up to 30 km wide and associated with subaerial exposure and palaeosol formation. Within areas that were subaerially exposed prior to the relative sea-level fall (i.e. on the coastal plain), the interfluves may be difficult to identify. The key observation is a change in soil type from poorly drained and immature (representing high relative sea-level) to well-drained and mature (representing subsequent falling relative sea-level). Further details of these palaeosols are beyond the scope of this book.

In areas that were submarine prior to relative sea-level fall, the interfluve surface will be marked by the juxtaposition of palaeosols on lower shoreface or offshore deposits. However, establishment of this basinward shift can be complicated by the fact that thin palaeosol horizons are easily removed by transgressive erosion during the next relative sea-level rise. Evidence indicating the existence of a soil can nonetheless be found, including plant roots, haematite and phosphate clasts, bone horizons and dolomite-cemented layers. These features are consistent with development of a subaerial, oxidized palaeosol and subsequent flooding during the later relative sea-level rise.

Up-dip, in the area dominated by alluvial and upper coastal plain deposition (i.e. landward of any marine influence in the preceding HST), the sequence boundaries are marked by a change in fluvial style. The most obvious example of this is the major sequence boundary at the base of the Castlegate Sandstone. Across this surface there is a transition from meandering fluvial systems with extensive overbank siltstones and mudstones interbedded with coals to braided fluvial deposits (Figure 7.15). The ratio of sandstone to mudstone below the surface is approximately 2 : 3 and 9 : 1 above the surface. This surface represents a major change in fluvial style resulting from an increase in the river gradient and energy of the fluvial system, driven by a relative sea-level fall.

## 9.2   Falling stage and lowstand systems tracts

### Forced regression and falling stage systems tracts

The Book Cliffs succession contains several units that are interpreted to represent forced regression (Section 4.2.2). The preservation and identification of falling stage systems tracts (FSSTs) is one of the methods of distinguishing forced regressions from normal regression (Section 4.3). However, as we saw in Section 4.3, the nature of the FSST is highly variable and in many cases it may be absent.

○   What controls the nature of the FSST?

●   The nature of the FSST is a function of the rate of relative sea-level fall and the rate of sediment supply. Factors such as the topography of the depositional slope and the magnitude of the relative sea-level fall are also significant but to a lesser degree.

We can use a wide set of criteria for recognition of forced regression. These include:

1   Evidence of long distance regression (i.e. further than other parasequences).

2   The presence of sharp-based shoreface or mouth bar deposits (indicating missing facies).

3   Presence of progressively lower-relief clinoforms in a basinward direction (indicating that the shoreline was building out into progressively shallower water, as relative sea-level fell).

4   Absence of coeval coastal plain strata.

5   Presence of a zone of separation between the underlying HST and overlying FSST and LST (i.e. that the FSST and LST are detached from the previous HST.

6   Presence of a seaward-dipping upper bounding surface (suggesting downstepping; i.e. successive shorelines lie topographically lower along the basin profile).

7   Maintenance of higher average grain size further seaward than in normal parasequences (indicating lack of accommodation space in the coastal plain for coarser grains).

8   Foreshortened stratigraphy, where the measured thickness of the parasequences is considerably less than estimates of palaeowater depth, suggesting coeval relative sea-level fall).

Several forced regressions and FSSTs are recognized within the Book Cliffs succession. It is worth remembering in the following discussion that the Cretaceous of the Western Interior Seaway was deposited on a ramp type margin similar to that shown in Figure 4.9b and part (d) of Figures 4.11 and 4.12.

### 9.2.1   Aberdeen and Kenilworth falling stage systems tracts

Within the Book Cliffs succession, FSST deposits have been interpreted in association with both the lower Aberdeen and the Kenilworth sequence boundaries (respectively shown as the LASB and KSB in Figure 7.4c). Elsewhere in the succession, FSST deposits appear to be absent, although some of the muddy heterolithic deposits of the Prairie Canyon Member of the Mancos Shale (see Section 9.3.2) may include FSSTs and LSTs.

The Aberdeen and Kenilworth examples exhibit different geometries that are interpreted to result from different patterns of relative sea-level fall. The FSST overlying the Aberdeen Sequence Boundary is a single parasequence (A2). A2 has prograded much further seaward and lies further down the depositional profile than A1 (Figure 7.4c). Furthermore, there are no coastal plain deposits associated with A2. All these aspects suggest that the parasequence was deposited whilst relative sea-level was falling.

○   Why should the lack of coastal plain deposits suggest deposition during relative sea-level fall?

●   As we saw in Section 8.4, accommodation space is generated in the coastal plain when relative sea-level rises and the fluvial profile is raised. When relative sea-level falls, the fluvial system is incising and consequently no accommodation space is available for non-marine deposition.

The FSST overlying the Kenilworth Sequence Boundary is rather different and has been the focus of a number of different interpretations. Here, we shall look at the data and then briefly discuss these different models (Figure 9.3).

*Observations*: There is a shoreface sandbody towards the top of the Kenilworth Member (K4 and K5 in Figure 7.4c; 4 in Figure 9.3a; 6–8 in Figure 9.3b; 4 and 5 in Figure 9.3c)) that differs from those above and below by extending much

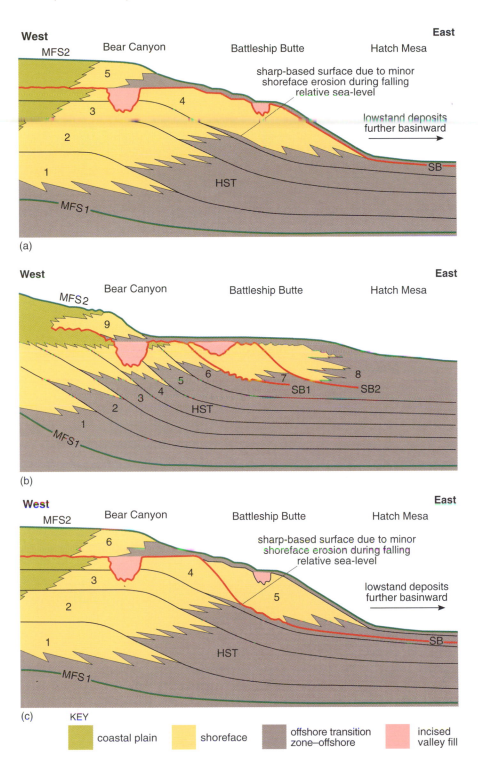

**Figure 9.3** Alternative interpretations of the stratal architecture observed in the Kenilworth Member. (a) Progradational parasequences (1–4) comprise the HST. The sequence boundary overlies parasequence 4. LSTs exist further basinward. (b) An alternative interpretation of the same data by Simon Pattison, who proposed a greater number of HST parasequences (six in total) and regarded parasequence 4 of (a) as a composite 'attached LST'. This model requires two sequence boundaries, both lying within parasequence 4 of model (a), and in this interpretation there are no LST deposits further basinward. (c) In this model, the sandstone body parasequence 4 of (a) is divided into two: a lower unit (4) which is part of the HST of the underlying sequence, and a FSST (5) which represents deposition soon after the onset of relative sea-level fall(s). *((b) Pattison, 1995.)*

further into the basin than either the underlying Kenilworth or overlying Sunnyside parasequences. A thin package of coastal plain sediments lie landward of the shallow-marine deposits.

*Interpretation 1*: In the original work on this unit, Taylor and Lovell interpreted the entire sandbody (4 in Figure 9.3a) as a single shoreface parasequence deposited during the upper part of the HST. They placed the Kenilworth Sequence Boundary on top of the unit. This interpretation does not account for why the unit extends so far into the basin, has limited coastal plain deposits and is sharp-based over such a significant distance.

*Interpretation 2*: Subsequent work by Pattison separated the sandstone body into three genetic units (6, 7 and 8; Figure 9.3b). The first of these units he termed a HST parasequence, the second two were interpreted as an 'attached LST' (equivalent to an attached FSST). He concluded that two sequence boundaries (SB1 and SB2) ran though the sandstone body and that there were no further LST deposits further basinward. This interpretation is somewhat over-complicated and does not explain the occurrence of an incised valley close to the pinch-out of the sandbody (7). Furthermore, the depths of the two incised valleys elsewhere at this level imply a greater magnitude of relative sea-level fall than the position of the proposed 'attached lowstands' would here.

*Interpretation 3*: The preferred interpretation here, as shown in Figure 9.3c, is that the lower part of the sandstone body represents the last HST parasequence of the underlying unit. The upper part of it represents the FSST which is sharp-based and prograded into the basin under conditions of forced regression during falling relative sea-level. Further relative sea-level fall detached the shorelines and pushed them further out into the basin. Later FSST and LST shorelines are interpreted to exist further into the basin.

### 9.2.2 Sand-rich forced regression shoreline deposits (the Panther Tongue)

The major sequence boundary at the base of our study interval is overlain by sandy river-dominated delta parasequences of the Panther Tongue (Figures 7.4, 7.8, 8.6), which have been interpreted as a forced regressive deposit. Specific lines of evidence include a downstepping stacking pattern and the lack of coeval coastal plain deposits. Biostratigraphical estimations of water depth for the basal muddy deposits are 75–100 m, yet the measured thickness of the Panther is 25 m maximum. This is consistent with the 'foreshortened stratigraphy' (criterion 8 in the numbered list near the start of Section 9.2) for a forced regression, assuming the depositional water depth estimation is correct. The Panther clinoforms in the northern, proximal area are 15 m high and thin basinward to 10 m, satisfying criterion 3 above (Figure 8.6). The exact position of a major sequence boundary at the base is unclear because the Panther Tongue does not exhibit a markedly sharp base (therefore it does not fit criterion 2). The master sequence boundary may run as a correlative conformity in the Mancos Shale just below the sandstone. In this interpretation, the Panther would become sharp-based up-dip to the north, but the Panther is not exposed at this level in this area.

### 9.2.3 Muddy, heterolithic shoreline deposits (Prairie Canyon Member of the Mancos Shale)

The offshore marine strata of the Mancos Shale include a number of heterolithic deposits called the Prairie Canyon Member. In the past, these have been identified as 'offshore bars' and 'tidal sand ridges'. The deposits, which typically weather proud of the soft Mancos Shale, are 5–25 m thick, more sand-rich than the surrounding siltstones and grey to brown in colour (Figures 9.4, 7.2a and 7.6).

At least five units have been mapped from both the outcrops and wells drilled behind the cliffs (Figure 7.4). Correlations demonstrate that these units are most likely FSST and LST deposits that were laid down while the sequence boundaries in the Blackhawk Formation were being formed and sand was by-passed through incised valleys to a point further into the basin where accommodation space existed (Figure 7.4c). The deposits lie at least 40 km basinward of the preceding highstand shorelines.

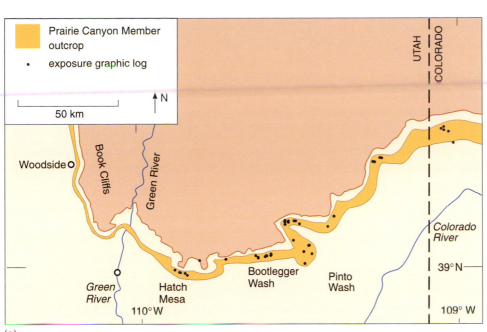

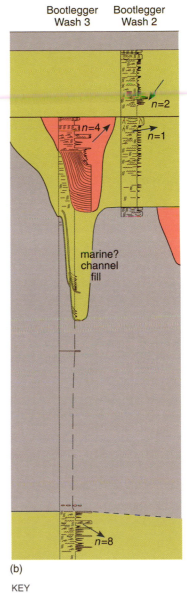

**Figure 9.4** Characteristics of the Prairie Canyon FSST and LST deposits. (a) Outcrop distribution of Prairie Canyon deposits in the eastern Book Cliffs. Their stratigraphical position is shown in Figure 7.4. (b) Graphic log showing multiple horizons of tidal channels and muddy coarsening-upward parasequences within the Mancos Shale. *(Hampson et al., 1999.)*

Within the five individual units, facies associations include:

1. Small-scale coarsening-upward units, up to 5 m thick. These parasequences contain thinly bedded, current-rippled and planar-laminated sandstone beds that become thicker and more abundant upward. Individual sandstone beds are commonly wave-rippled toward their tops. These units are interpreted as small, river-dominated delta mouth bars. The coarsening-upward profile indicates progradation and the presence of wave ripples confirms a water depth above storm wave-base. Mapping out of these units demonstrates a lobate geometry. Between these lobes, comparably thick successions of siltstones, similar to the Mancos Shale but with a shallow-water trace fossil assemblage, are interpreted as low-energy, muddy shoreline deposits.

2. Small channel sandstones, typically less than 5 m thick, which contain fine- to medium-grained sandstone with abundant coaly debris. These channels may be either massive or contain inclined heterolithic stratification (tidal cross-stratification). They trend east–west and feed the mouth bars. Channels are laterally separated by the interfluve soil horizons described above.

The shallow-water sedimentary structures and trace fossil assemblages are good evidence for these not being offshore deposits. Furthermore, the presence of tidal cross-stratification, fluvial channels and rooted interfluves supports the FSST and LST interpretation (Figure 9.4b). However, it is interesting to consider why these deposits are sedimentologically different to those in the underlying HST. The highstand parasequences are sandy, wave-dominated systems while the FSST and LST parasequences are muddy, river-dominated systems with very low wave energy. A number of explanations are possible. First, during falling and low relative sea-level, the width and therefore the wave fetch of this narrow, enclosed basin would be reduced. Secondly, the wave energy would be greatly dissipated on a very low angle ramp; and finally it is possible that at least some of the sandy material deposited was later removed back into the estuaries during subsequent relative sea-level rise. This latter process is common in Holocene estuaries.

### 9.2.4   Braided fluvial lowstand deposits (the Castlegate Sandstone)

The braided fluvial deposits of the Castlegate Sandstone overlie a major sequence boundary at the top of the Blackhawk Formation (Figure 7.15b). The Castlegate thins from over 160 m in the north-west, where it overlies coastal plain deposits, to 20 m in the south-east where it overlies marine shales (Figure 7.4). The basinward equivalent is a series of fresh water oolitic ironstones. Study of this unit by Van Wagoner led him to propose that the fluvial deposits themselves represent the LST: this is somewhat different to the other units we have just considered. The scenario he envisaged was that following the major relative sea-level fall, the high-gradient fluvial systems deposited large amounts of sand as alluvial fan to braided river systems. Because of the basin geometry, the low angle of the ramp resulted in these fluvial systems becoming terminal and not reaching the new shoreline (Figure 9.1). Freshwater lakes developed between the fan edge and the shoreline, which was now far to the east and totally sediment-starved, hence the formation of iron-rich ooids (Figure 9.5). This situation is unique to the Castlegate Sandstone and is not found in the Blackhawk Formation.

**Figure 9.5** (a) Map of the incised valley and interfluve drainage network developed on the Castlegate Sequence Boundary in the eastern Book Cliffs. (b) Schematic interpretation of the Castlegate braided terminal fluvial deposits; unlike many systems, there appears to be no lowstand delta or shoreline basinward of the incised valleys in the study area. *(Van Wagoner, 1995.)*

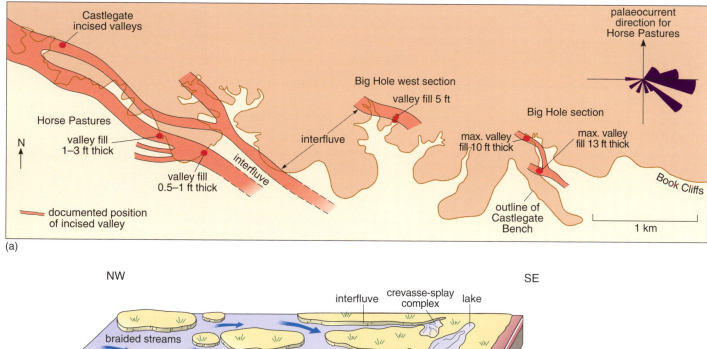

## 9.3  Transgressive systems tracts

The transgressive systems tracts are considered here to have two parts, the filling of the incised valleys (lower TST) and the re-establishment of shallow-marine shoreface systems over the filled valleys and flooded interfluves (upper TST). The parasequences in the upper TST, if developed, will exhibit a retrogradational stacking pattern.

The base of the incised valleys represents the sequence boundary and transgressive surface superimposed because, at the palaeogeographical position of the incised valleys, there are no true LST or FSST deposits. The fills of the incised valleys are, therefore, assigned to the TSTs. The sequence boundary surface at the incised valley floor occupies the equivalent amount of time as the FSST and LST deposits further basinward. Deposition and preservation of deposits in the valley can only start once accommodation space is regenerated at this position in the basin, which happens during the subsequent early rise in relative sea-level. The fill of the valley is therefore transgressive and belongs in the TST.

### 9.3.1  Lower transgressive systems tracts; incised valley fills

As we have seen above, when relative sea-level falls, valleys are cut into the old HST by fluvial systems. While the FSSTs and LSTs are being deposited, these valleys are areas through which most of the sediment is by-passed. Small pockets of fluvial sands may be deposited on the base and at the margins of the valleys during falling and low relative sea-level. As base level is falling, major deposition and preservation of valley fill is precluded, but, at this time, FSSTs and LSTs are being deposited in the more distal sections where there is available accommodation space. Thus, volumetric partitioning takes place (Section 8.4).

Subsequent relative sea-level rise floods these valleys and the relatively shallow lower portions of the valley system (basinward) are rapidly filled with sediment. The middle portion, where the deepest part of the valley was cut into the old HST shoreline and coastal plain, becomes an estuary, sheltered from the wave processes in the basin and with a topography that amplifies the small tidal range experienced along the shorelines facing the basin. The incised valleys fill with a complex array of tidal and fluvial facies. We shall look at examples from above the Sunnyside Sequence Boundary at Woodside Canyon and the Desert Sequence Boundary at Tusher Canyon (Figures 7.1, 7.4c). Further up-dip, beyond the tidal limit, the incised valleys are filled with stacked fluvial channel deposits.

### *Sunnyside incised-valley fill — Woodside Canyon*

The modern Woodside Canyon (Figure 9.6) cuts through the Book Cliffs and exposes one of the best examples of an estuarine-incised valley fill.

The underlying progradational parasequence is up to 28 m thick and marks the top of the HST of the underlying sequence (Figure 9.6 overleaf, 7.4c). The valley system is up to 20 m deep, with clear margins (Figures 9.6c, 9.7 on p.192). Mapping of the system around the modern canyons reveals three separate branches that fork upstream (Figures 9.6b, 9.7). At its maximum, the valley system is 3.5 km wide.

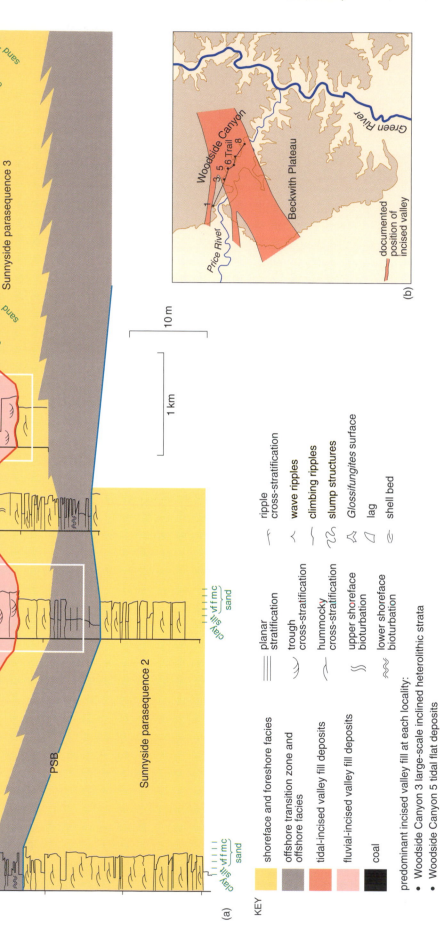

**Figure 9.6** The Woodside Canyon incised-valley complex. (a) Correlation of Sunnyside Member measured sections shows clear truncation of Sunnyside shoreface parasequence 3 by the sequence boundary and the variety in the estuarine fill of the incised valley. Note the incision on the sequence boundary; the fluvial deposits at the base of the valley fill; the predominance of tidal deposits within the valley and planar nature of the overlying flooding surface (PSB). (b) Map showing location of the correlation line in (a) and position of the incised valley with respect to the present-day Woodside Canyon. The white and pale-brown coloured shoreface deposits of Sunnyside PS3 on the right-hand side have been removed by the sequence boundary and replaced by later, poorly exposed, heterolithic tidal deposits (cliff is c. (c) Sequence boundary and margin of incised valley in Woodside Canyon. The white and pale-brown coloured shoreface deposits of Sunnyside PS3 on the right-hand side have been removed by the sequence boundary and replaced by later, poorly exposed, heterolithic tidal deposits (cliff is c.

**Figure 9.7** 3-D visualization of the drainage network produced by erosion on the Sunnyside Sequence Boundary. The main valley axis is at the base of the map and two tributary valleys from the west and north-west drain into it. The line of cross-section A–A' refers to the cross-section in Figure 9.6.

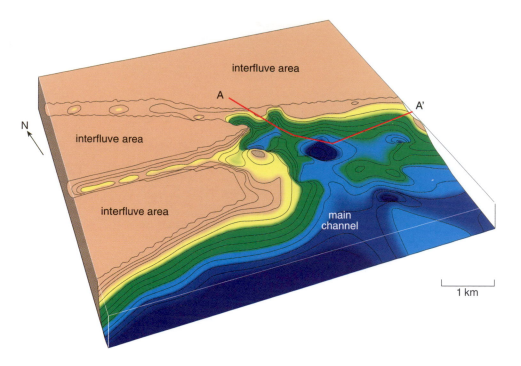

The valley system is filled by a complex array of facies. Coarse-grained, trough cross-stratified sandstones up to 4 m thick occur locally at the base (Figure 9.6a,f). These contain no marine body or trace fossils and are interpreted as fluvial sandstones, deposited locally in barforms during relative sea-level fall and low sea-level. They are overlain by up to 16 m of tidal deposits laid down during the early relative sea-level rise and form the lower TST. Tidal facies include sandy subtidal bars, large-scale inclined heterolithic stratification (which is evidence for lateral accretion in meandering tidal channels (Figure 9.6d), tidal flat deposits (Figure 9.6e), thin coals (Figure 7.14e), shell beds and intra-estuary deltas (Figure 9.6g). The distribution of these facies illustrates several discrete base levels within the estuary, which indicate that it was filled in a stepped fashion.

It is important to differentiate incised valleys, cut during falling relative sea-level and filled during subsequent transgression, from simple tidal and fluvial channels, which were active on the coastline during normal progradation. There are a number of features within the valley system discussed above which aid such a distinction:

- An upward facies transition from fluvial to tidal deposits indicates that conditions became more marine upward (transgressive). A simple progradational system would become less marine upward as the coastline moved gradually basinward.

- The existence of a number of discrete base levels within the fill. A simple system such as a fluvial or tidal channel that was prograding will only contain evidence for a single base level. Evidence for multiple base levels within the valley fill succession, such as coals, soils and emergent tidal flat overlain by subtidal bars, indicates that high-frequency parasequence-scale relative sea-level changes took place during the valley filling.

- The scale of the incised valley fill features are significantly larger both in width and depth than normal distributary channels.

- The significant amplification of tidal currents (4 m-high bedforms) would not occur on a wave-dominated, microtidal coast. Relative sea-level fall, cutting of river valleys and their subsequent transformation into funnel-shaped estuaries during relative sea-level rise is required for this degree of tidal amplification.

## Incised valleys within the coastal plain

Incised valleys may occur either as estuarine systems encased within non-marine deposits or, further up-dip, as stacked fluvial channel fills. The stacked fluvial channels were discussed when we looked at sequence boundaries, but it is worth considering one of the estuarine systems in the Sunnyside Member in more detail.

Within the Sunnyside Member, a large, tidally influenced sandbody encased within non-marine deposits has been interpreted as an incised valley (Figures 9.8, 9.9 overleaf). The unit is up to 20 m thick and comprises two coarsening-upward packages. Each is typically sand-dominated and contains both planar and trough cross-stratification.

Locally, especially towards the base, large-scale mud-draped trough cross-stratification indicates a subtidal origin (Figure 9.9c). Associated mudstones show a restricted trace fossil assemblage of *Chondrites*; these observations indicate poorly oxygenated conditions (Figure 9.9d). Fluvial channels with abundant soft sediment deformation locally cut the tops of the two coarsening-upward packages. The whole unit, including the fluvial channels, is capped by a 3 m-thick, very extensive coal horizon. The sandbody is exposed along the cliff line between Alrad and Pace Canyons (Figures 7.4, 9.8a) and can be mapped in the subsurface as a blocky sandstone on borehole logs. Fossil wood with *Teredolites* borings was noted in this unit in a core from an oil well 30 km up-dip from the underlying highstand shoreline deposits.

○  What is the significance of the *Teredolites* borings?

● The boring bivalve *Teredolites* can only survive in waters of marine salinity, and bored wood is commonly washed into modern estuaries by the incoming tide, but is very rare in fluvial channels. The implication of the *Teredolites* borings is that the wood was in contact with marine waters.

The evidence leads us to the interpretation of these sandbodies as a large estuarine complex. The significant factor is that the valley network is cut into coastal plain deposits and overlain by a coal. This indicates that the valley system was cut and, during the subsequent transgression, tidal-estuarine conditions were forced up the valley network beyond the position of the former HST shoreline (Figure 9.8a,b). This is common in modern estuarine systems. The presence of fluvial deposits cutting into the top of the succession indicates that the valley filled to capacity before the top of the lower TST had been deposited.

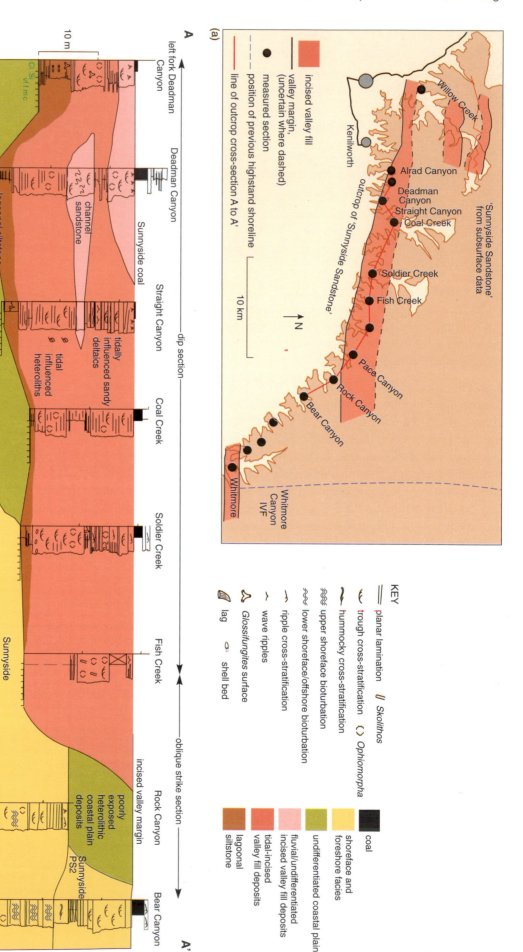

**Figure 9.8** Detail of the incised valley complex overlying the Lower Sunnyside Sequence Boundary: an incised valley complex in a non-marine setting. (a) Map showing the distribution of incised valleys above the Lower Sunnyside Sequence Boundary in the northern part of Book Cliffs. Position of previous highstand shoreline shown in blue. (b) Cross-section along the axis of one of the incised valleys, showing the predominance of sandy tidal deposits.

(a)                                            (b)

(c)                                            (d)

**Figure 9.9** Photographs of the incised valley complex overlying the Lower Sunnyside Sequence Boundary in a non-marine setting. (a) The Sunnyside Sandstone in Straight Canyon; note the two distinct sandbodies and the recessively weathered, heterolithic unit in the centre (cliff *c.* 20 m high). (b) Coarsening-upward succession of tidally influenced deltaic deposits at the top of the valley fill succession in Soldier Creek (Figure 7.1) (person shows scale). (c) Large-scale cross-bedding with organic carbon-rich drapes. Cross-bedding of this magnitude provides evidence for meso- to lower macrotidal conditions and represents amplification of the tidal wave within an estuary (cliff *c.* 6 m high). (d) Smaller-scale, heterolithic tidal cross-stratification (image is *c.* 1 m top to bottom). *((a)–(d) John Howell, University of Bergen.)*

### 9.3.2   Upper transgressive systems tracts

As the rate of relative sea-level rise continues to increase, the incised valley systems are completely filled and the interfluves (former shelf and lower part of the old highstand coastal plain) are flooded. This flooding surface develops diachronously up the depositional profile. Marine conditions are returned to the more distal/seaward part of the fluvial profile while in the up-dip section the valleys are still active sites of fluvial deposition. Deposition of the upper TST continues up to the maximum flooding surface. The Blackhawk Formation differs from the classic model in having poorly developed upper TSTs, represented by single parasequences or flooding surfaces alone.

The newly uplifted Sevier Mountain belt and the tropical climate resulted in very high overall sediment supply to the Blackhawk shoreline. Consequently, the upper TST is typically not well developed because there was insufficient time for a TST shoreline to form. Marine shale tongues commonly represent a combination of the flooding surface above the interfluves/incised valley fills, and the upper TST. However, there are two examples: the Kenilworth and Desert Members, which show the backstepping parasequence stacking pattern that the model predicts for the upper TST. The cross-section and maps from the Kenilworth Member (Figure 7.4c) clearly show the progradational parasequence set of the lower HST (parasequences K1 to K4; Figure 7.4c) and the overlying sequence boundary and associated incised valleys. A single shoreface parasequence (K6; Figure 7.4c) overlies the valley-fills. It has a very limited progradational extent (7 km) compared to the other parasequences we have seen (mean 20 km) and is also notably thicker (up to 30 m). This parasequence is overlain by the maximum flooding surface, above which is the progradational HST parasequence set of the Sunnyside Member (S1–S3; Figure 7.4c). K6 is therefore interpreted as the upper part of a TST.

Landward of the shoreface deposits in this Kenilworth TST, the time equivalent coastal plain succession contains abundant coals and a low proportion of channel sandstones to overbank deposits. This architecture is typical of coastal plain deposits laid down under conditions of rapidly rising base level. Rapid sediment accumulation results in overbank deposits not being reworked by channels, which in turn are small and of low energy. Coals are developed and preserved because of the very high water table and excess accommodation space (Section 8.3).

## 9.4   Maximum flooding surfaces

The Blackhawk Formation maximum flooding surfaces lie within the thick tongues of Mancos Shale (Figures 7.2, 7.4c). In the offshore, the deepest-water deposits that represent the maximum flooding surface are not always easy to identify where exposed. As it is not always possible to identify an exact surface, so a maximum flooding 'zone' is more usually identified, based on evidence of slight organic-carbon enrichment within the mudstones. This is often marked by an increased concentration of uranium, which is preferentially concentrated in organic matter, and which in turn is reflected in gamma-ray logs (Box 3.2). In the shallow-marine system, the maximum flooding surface is defined by the deepest-water deposits. These are typically offshore shales, overlying shoreface facies.

○   In the shallow-marine realm, how do we differentiate the maximum flooding surface from other flooding surfaces?

●   The maximum flooding surface is differentiated from other flooding surfaces because it has the most landward extent. It will also typically be overlain by the thickest shale interval and mark a change in the parasequence stacking pattern from retrogradational to progradational.

In the coastal plain system, the maximum flooding surface is very difficult to distinguish. The interval around the surface is typically represented by thick coals and mud-rich fluvial systems with a high proportion of preserved overbank deposits.

## 9.5   Highstand systems tracts

In general, the HST is characterized by aggradational to increasingly progradational stacking of parasequences, which can clearly be seen in most members of the Blackhawk (Figure 7.4c). Good examples include the lower part of the Kenilworth Member, the Sunnyside Member and the Grassy Member. The upper HST reflects a slowdown in the rate of relative sea-level rise and thus a decrease in the rate of creation of accommodation space, which leads to an increased distance of parasequence progradation (Section 4.3.1). Although the majority of the sedimentary rocks within the Blackhawk Formation are contained within progradationally stacked HST parasequence sets, much of the time represented by the Blackhawk corresponds to surfaces such as the sequence boundaries, when sediment was being deposited in FSSTs and LSTs further into the basin.

## 9.6   Summary

•   Sequences contain up to four systems tracts, although in the Book Cliffs transgressive systems tracts are often poorly developed or absent.

•   Falling stage systems tract deposits are present in the Kenilworth and Aberdeen sequences. In other sequences, the rates of relative sea-level change are interpreted to have been too rapid for the preservation of this systems tract.

- The Book Cliffs succession includes examples of sand-rich falling stage systems tract and lowstand systems tract (the Panther Tongue); muddy, heterolithic falling stage and lowstand systems tracts (the Prairie Canyon Member) and a terminal fluvial lowstand systems tract (the Castlegate Sandstone).

- The lower part of the transgressive systems tract is represented by the filling of valleys incised during falling sea-level. The deposits within these valleys are tidal–estuarine in origin. Excellent examples are seen in the Book Cliffs.

- The upper part of the transgressive systems tract is ideally marked by a retrogradationally stacked parasequence set. Examples of this can be seen in the Kenilworth and Desert sequences. In the other sequences in the Book Cliffs, they are absent: this is interpreted to be because of the high sediment supply.

- Most of the rocks in the Book Cliffs are progradationally stacked highstand parasequences. However, significant amounts of time may be represented by surfaces when no sediments were preserved.

## 9.7   References

HAMPSON, G. J., HOWELL, J. A. AND FLINT, S. S. (1999) A sedimentological and sequence stratigraphic re-interpretation of the Upper Cretaceous Prairie Canyon Member (Mancos B) and associated strata, Book Cliffs area, Utah, USA', *Journal of Sedimentary Research*, **69**, 414–433.

O'BYRNE, C. J. AND FLINT, S. S. (1996) 'Interfluve sequence boundaries in the Grassy Member, Book Cliffs, Utah: Criteria for recognition and implications for subsurface correlation', in HOWELL, J. A. AND AITKEN, J. F. (eds) *High resolution sequence stratigraphy: innovations and applications*, Special Publication of the Geological Society No.104, 207–220.

PATTISON, S. A. J. (1995) 'Sequence stratigraphic significance of sharp-based lowstand shoreface deposits, Kenilworth Member, Book Cliffs, Utah', *Bulletin of the American Association of Petroleum Geologists*, **79**, 444–462.

PLINT, A. G. AND NUMMEDAL, D. (2000) (see other references for Chapter 4).

POSAMENTIER, H. W. AND NUMMEDAL, D. (2000) (see other references for Chapter 4).

TAYLOR, D. R. AND LOVELL, W. W. (1995) 'Recognition of high frequency sequences in the Kenilworth Member of the Blackhawk Formation, Book Cliffs, Utah', in VAN WAGONER, J. C. AND BERTRAM, G. (eds) *Sequence stratigraphy of foreland basin deposits*, American Association of Petroleum Geologists Memoir No. 64, 257–277.

VAN WAGONER, J. C. (1995) (see other references for Chapter 7).

# 10 Sequence stratigraphical evolution of the Book Cliffs succession

### John A. Howell and Stephen S. Flint

In this final Chapter, we consider the large-scale architecture of the Book Cliffs succession and, based on the available data, we can speculate on the origin of the interpreted relative sea-level changes with reference to the processes introduced in Chapter 5. Figure 10.1 summarizes the sequence stratigraphical interpretation for the Blackhawk Formation.

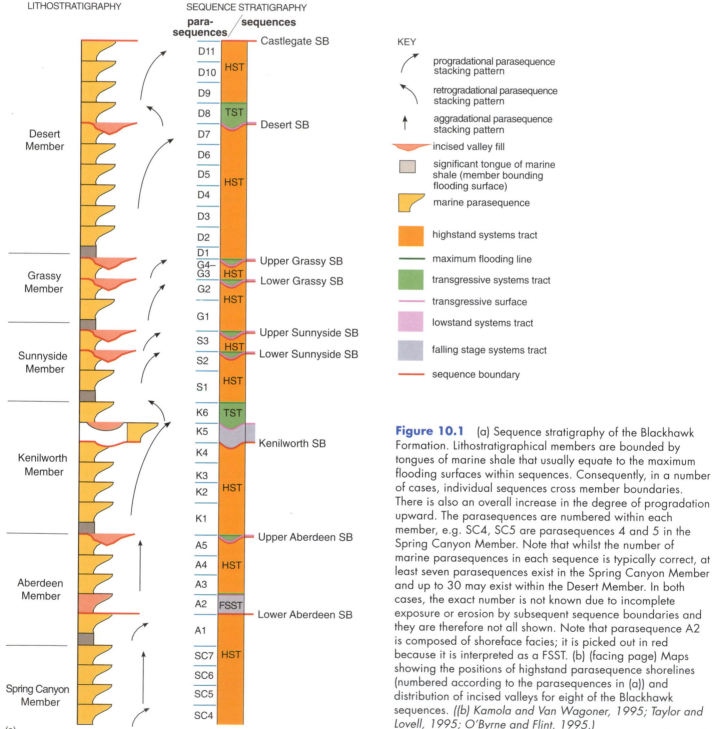

**Figure 10.1** (a) Sequence stratigraphy of the Blackhawk Formation. Lithostratigraphical members are bounded by tongues of marine shale that usually equate to the maximum flooding surfaces within sequences. Consequently, in a number of cases, individual sequences cross member boundaries. There is also an overall increase in the degree of progradation upward. The parasequences are numbered within each member, e.g. SC4, SC5 are parasequences 4 and 5 in the Spring Canyon Member. Note that whilst the number of marine parasequences in each sequence is typically correct, at least seven parasequences exist in the Spring Canyon Member and up to 30 may exist within the Desert Member. In both cases, the exact number is not known due to incomplete exposure or erosion by subsequent sequence boundaries and they are therefore not all shown. Note that parasequence A2 is composed of shoreface facies; it is picked out in red because it is interpreted as a FSST. (b) (facing page) Maps showing the positions of highstand parasequence shorelines (numbered according to the parasequences in (a)) and distribution of incised valleys for eight of the Blackhawk sequences. ((b) Kamola and Van Wagoner, 1995; Taylor and Lovell, 1995; O'Byrne and Flint, 1995.)

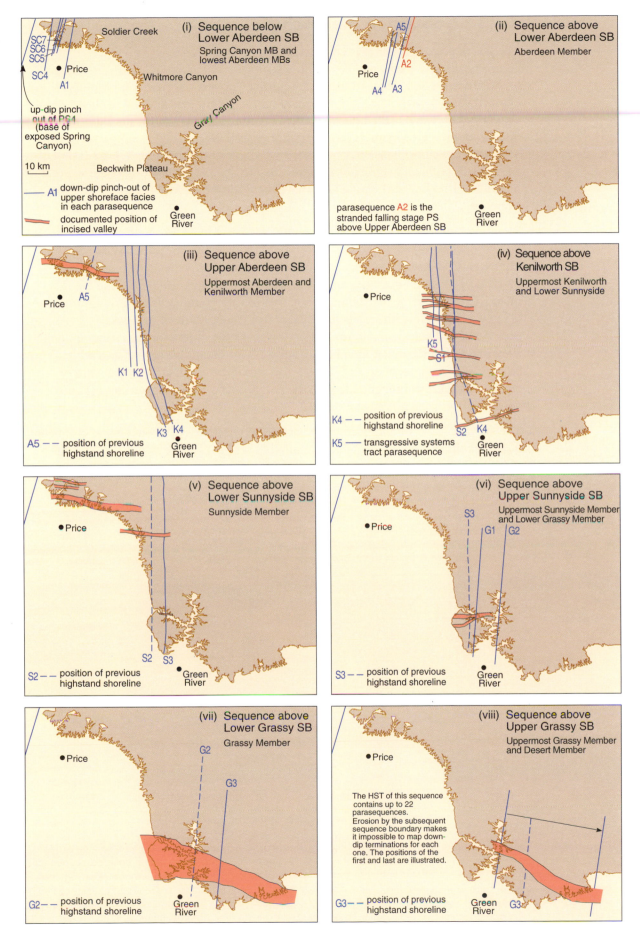

(b)

**Figure 10.1 (continued)**

## 10.1  Sequence hierarchy

The Blackhawk Formation is composed of nine high-frequency sequences that comprise the low-frequency Panther to Castlegate sequence. The early Blackhawk high-frequency sequences (those in the Spring Canyon and Aberdeen Members) are characterized by aggradational highstand parasequence sets and few well-developed sequence boundaries (Figures 7.4c, 10.1a). Younger sequences of the Sunnyside, Grassy and Desert Members have strongly progradational highstand parasequence sets, well-developed sequence boundaries and associated incised valley fill deposits. Lithostratigraphical member boundaries mapped by previous workers correspond to maximum flooding surfaces in all cases except for the top of the Spring Canyon Member.

The nine high-frequency sequences within the Blackhawk Formation were defined from the work of Van Wagoner, Taylor and Lovell, Pattison, Kamola and van Wagoner, O'Byrne, and Flint and Howell. Within these sequences, there is a systematic change in parasequence stacking patterns over time. This can be quantified by plotting the shoreline position of each parasequence through time (Figure 10.1b). This illustrates the change from aggradational to weakly progradational stacking patterns within lower members to strongly progradational stacking within the upper members (Figure 10.2). The flooding surfaces which bound individual members (which are also maximum flooding surfaces) are of significantly greater magnitude than the parasequence flooding surfaces (Figure 7.4c).

The superposition of a high-frequency relative sea-level curve on a lower-frequency curve results in a complex composite relative sea-level curve which controls the stacking patterns and characteristics of the high-frequency sequences within the low-frequency sequence (Figures 5.10, 5.13 and 10.3).

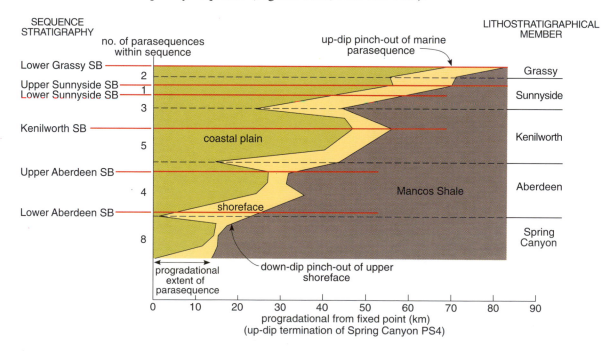

**Figure 10.2**  Summary of progradation of the Blackhawk Formation siliciclastic wedge through time. The graph shows the position of the up-dip and down-dip terminations of the upper shoreface facies within each parasequence. The lithostratigraphy and sequence stratigraphy are superimposed. Note the overall change through time from the broadly aggradational parasequences of the Spring Canyon and Aberdeen sequences to the strongly progradational stacking patterns in the Grassy. This graph is a simplification of the same pattern which can be seen in the cross-section (Figure 7.4c) and which shows the even more pronounced progradational parasequences in the Desert Member.

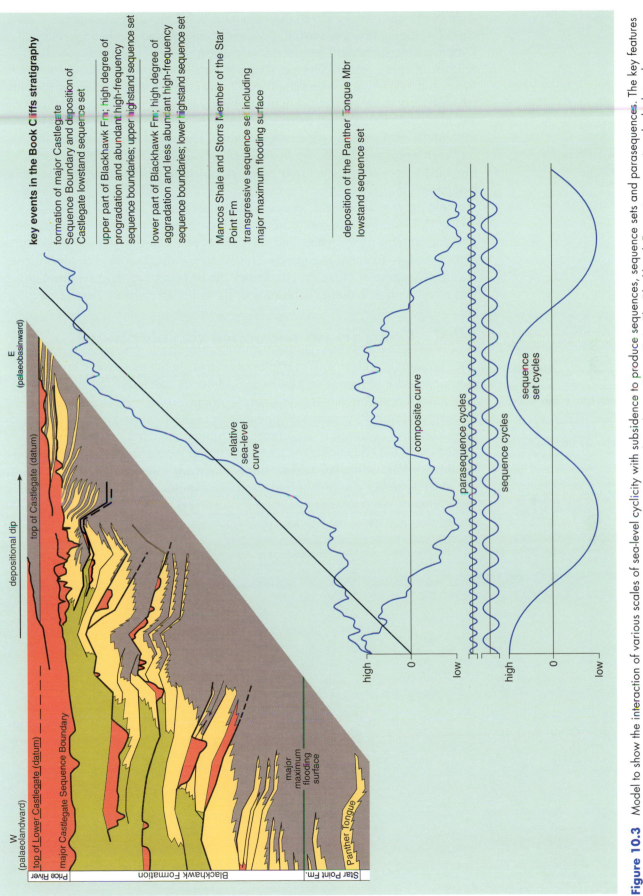

**Figure 10.3** Model to show the interaction of various scales of sea-level cyclicity with subsidence to produce sequences, sequence sets and parasequences. The key features of the geometries of the Book Cliffs succession can be seen in the summary cross-section, based upon Figure 7.4c. The Blackhawk Formation comprises the highstand sequence set of this lower-order composite sequence. (Top left: Hampson et al., 2001; bottom right: Van Wagoner et al., 1990.)

key events in the Book Cliffs stratigraphy

formation of major Castlegate Sequence Boundary and deposition of Castlegate lowstand sequence set

upper part of Blackhawk Fm; high degree of progradation and abundant high-frequency sequence boundaries; upper highstand sequence set

lower part of Blackhawk Fm; high degree of aggradation and less abundant high-frequency sequence boundaries; lower highstand sequence set

Mancos Shale and Storrs Member of the Star Point Fm transgressive sequence set including major maximum flooding surface

deposition of the Panther Tongue Mbr lowstand sequence set

It should be stressed, however, that care must be taken when making large-scale chronostratigraphical correlation of sequences using stacking patterns alone. Differential basin margin subsidence or sediment supply variations can and do produce coeval aggradational and progradational stacking patterns along-strike. In Section 5.5, it was shown that the FSST and LST of individual sequences stack progressively in an aggradational to progradational pattern to form a falling stage and lowstand sequence set; transgressive sequence sets are characterized by a retrogradational stacking pattern of individual sequences, and highstand sequence sets have a progradational pattern although less so than the falling stage and lowstand sequence set. The lower part of the Blackhawk low-frequency sequence comprises poorly developed sequence boundaries together with aggradational stacking patterns at both the sequence and parasequence scale (Figures 7.4c, 10.3). Well-developed unconformities and more pronounced progradational stacking patterns of sequences and parasequences are developed within the upper part of the low-frequency sequence. This change is interpreted to occur because during deposition of the lower Blackhawk the low-order relative sea-level change was more rapid than during deposition of the upper Blackhawk Formation.

## 10.2   Controls on sequence development

The literature is replete with examples of both eustatic and tectonic driving mechanisms for sequence development, each being tailored to the authors' preferences; however, direct unequivocal evidence for the exact driving mechanism (or more usually the combination of mechanisms) is often lacking. A more objective approach is to compare the timing and duration of stratigraphical features with the known rates of potential driving mechanisms. In the case of the Blackhawk Formation, the existing time framework is limited. Figure 10.4 shows the general rates of common processes that cause changes in accommodation space. Below, we discuss in turn the relevant processes with respect to the chronostratigraphy of the Blackhawk Formation.

**Figure 10.4**   Rates of processes that control stratal architecture in sedimentary basins. The estimated rate for the formation of a Blackhawk Sequence Boundary was calculated from the total number of stratigraphical events (HST, TST parasequences, valley fill successions and lowstand shorelines)/duration of the Blackhawk to give the approximate duration to each event. A rate was then calculated from the magnitude of relative sea-level fall (valley depth)/duration of the event. All other rates from a variety of published sources. *(Based on a diagram by Tom McKie (unpublished).)*

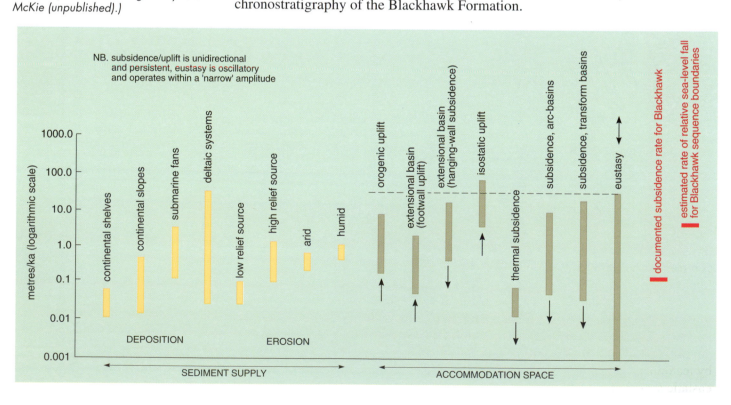

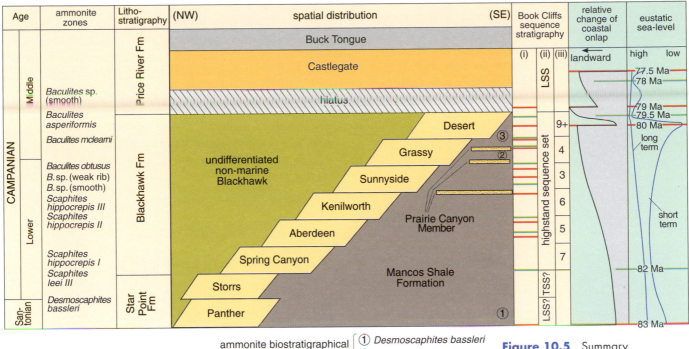

The ammonite biostratigraphy provides three correlation points to the globally recognized biozones. It should be noted that whilst the ages of the base and top of the Blackhawk are well constrained biostratigraphically, the exact ages of member boundaries are not known and are displayed schematically (Figure 10.5). It is interesting to consider the relation between the 'Mesozoic – Cenozoic Cycle Chart' (see Section 4.6 and Figure 10.5) and the Blackhawk stratigraphy. The sequence boundary predicted at '80' Ma from the Mesozoic – Cenozoic Cycle Chart has been correlated to the Desert Sequence Boundary (ammonite zone *Baculites mclearni*) by John van Wagoner and the relative sea-level fall at '79' Ma has been correlated by van-Wagoner with the Castlegate Sequence Boundary. The intervening 79.5 Ma flooding surface from the cycle chart would, in this interpretation, correspond to the maximum flooding surface in the Desert Member. Comparison of the biostratigraphical data with the geological time-scale indicates an age of approximately 83 Ma for the Panther Tongue, which would tie in with the pronounced 83 Ma sequence boundary on the Mesozoic – Cenozoic Cycle Chart and would, in addition, tie the Storrs Member maximum flooding surface to the 82 Ma maximum flooding surface.

Positive correlation of observations from the Book Cliffs with the Mesozoic – Cenozoic Cycle Chart suggests that eustatic sea-level was a major control on certain parts of the Book Cliffs stratigraphy, but the proposed correlation is only as good as the absolute age dates. There are, however, within the limits of age control, some clear mismatches with the Mesozoic Cenozoic Cycle Chart. Interestingly, the Mesozoic – Cenozoic Cycle Chart predicts that the eustatic sea-level change associated with the Desert Member Sequence Boundary should be of greater magnitude than the Castlegate Sequence Boundary, but field evidence proves that the basinward shift of facies at base Castlegate is much greater. One possible interpretation is that the Castlegate Sequence Boundary was enhanced by tectonic processes in the Book Cliffs area and is thus a product of both eustatic sea-level fall and local tectonics.

ammonite biostratigraphical markers found in the Mancos Shale Formation
① *Desmoscaphites bassleri*
② *Baculites mclearni*
③ *Baculites asperiformis*

**Figure 10.5**   Summary chronostratigraphy for the Book Cliffs succession, including biostratigraphical zones; ammonites found within the Mancos Shale Formation; summary of the sequence stratigraphical interpretation showing (i) sequence boundaries (orange), maximum flooding surfaces (green), (ii) sequence sets and (iii) number of parasequences in each member; and in the green columns relative sea-level and onlap curves from the Mesozoic – Cenozoic Cycle Chart (Section 4.6) for Campanian and Upper Santonian. LSS = falling stage and lowstand sequence set; TSS = transgressive sequence set. *(Left: Kamola and Van Wagoner, 1995; middle: Young, 1955; right: Haq et al.,1988.)*

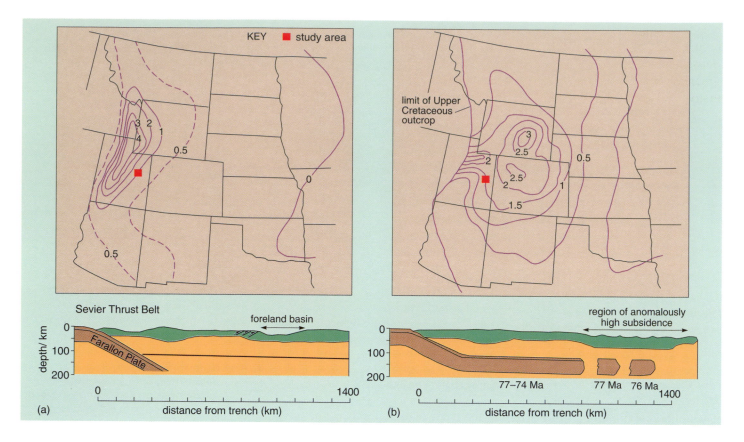

**Figure 10.6** Plate tectonic evolution of western North America during the Cretaceous and early Cainozoic. (a) Sevier structural style; narrow, asymmetric foreland basin caused by loading by thin-skinned thrust sheets. (b) Laramide structural style, initiated in the latest Cretaceous, in which shallower subduction resulted in broader subsidence, followed by a series of intrabasinal uplifts that formed the modern Rocky Mountains. These uplifts segmented the marine Western Interior Basin into a series of smaller, non-marine basins during the early Tertiary. The contours are in km and indicate the depth of the sedimentary basin. The Blackhawk Formation was deposited during the transition from structural style (a) to (b). *(Cross, 1986.)*

The tectonic subsidence that contributed to creating accommodation space for the Blackhawk Formation is related to two distinct phases of deformation in the thrust belt to the west. The first phase of flexural loading (Sevier orogeny) began with the deposition of the Dakota Sandstone (*c.* 97 Ma) and was characterized by thin-skinned tectonics and rapid subsidence in a narrow N–S trending foreland basin, immediately in front of the uplifting mountain belt (Figure 10.6). Thin-skinned tectonics means that the crust was being shortened and uplifted via a series of low-angle thrust faults that did not affect the lower Palaeozoic and Precambrian metamorphic basement, but were restricted to the late Palaeozoic and Mesozoic cover. At *c.* 80 Ma, there was a transition to slower subsidence that covered a much broader area, followed by uplift (with comparatively little deformation) over a huge area (Figure 10.6). The change in structural style occurred during the deposition of the Blackhawk Formation.

Using the available biostratigraphical dates, the average subsidence rate for the Blackhawk Formation has been calculated (Figure 7.5). A subsidence rate of 0.014 cm/yr (140 m/Ma) has been calculated. This rate, when plotted onto Figure 10.4, is average for the active subsidence mechanisms shown. It is a much lower rate of accommodation space generation than the maximum possible through thermal and glacio-eustatic sea-level change. Uncertainties on the match with the Mesozoic – Cenozoic Cycle Chart, and the absence of continental ice-caps in the 'greenhouse world' of the Cretaceous, have lead several authors to interpret the relative sea-level fall that resulted in the deposition of the Castlegate Sandstone as tectonic in origin. The evidence is strong for a tectonic component to the long-term accommodation space history (at sequence set scale), but tectonic mechanisms for generating the higher-frequency cycles are more difficult to envisage. Changes in thrust propagation rates and development of local imbricate structures within thrust belts bounding foreland basins have been suggested as mechanisms for controlling stratigraphical cyclicity in the basin-fill. Early papers related movement of basin-bounding thrusts directly to progradation of siliciclastic tongues into the basin. The link between these processes is now appreciated to be more complex, with a 'lag-time' required for erosion and sediment release between thrust movement and sedimentary high-frequency response.

The initial basin response to thrust sheet advance is one of deepening and production of a flooding surface, before the 'catch-up' of sediment supply. It has also been appreciated that thrust fault dynamics are complex and there is a hierarchy of imbricate structures which move at different frequencies and different rates. Many authors have interpreted a causal link between the shortening history of the thrust belt and cyclicity in the basin fill. The main problem at this time is the paucity of data to link rates of fault movement quantitatively with the estimated duration of high-frequency cycles (parasequences, parasequence sets and high-frequency sequences).

There has been much discussion on the influence of asymmetric foreland basin subsidence patterns on sequence architecture. Thrust loading (the load on the crust produced by uplift of a fold-thrust belt) produces higher rates of foreland basin subsidence adjacent to the thrust belt (Figure 10.7 overleaf). This large-scale geometry was used by Posamentier and Allen to propose that sequence boundaries characterized by incised valleys will not be developed close to the thrust belt, owing to the rate of subsidence being equal or greater than the rate of sea-level fall. The Blackhawk Formation dataset indicates that this hypothesis is oversimplified. In the Book Cliffs, which can be considered as a single datapoint within a basin of over 1000 km width, the Blackhawk Formation shows a temporal evolution from high to low accommodation space through the highstand sequence set. This is expressed as greater parasequence progradation distances and more closely spaced, more deeply incised valleys up section. This may be related to a gradual decrease in subsidence rate with time, towards the Castlegate Sequence Boundary.

A recent discussion on the sedimentology and sequence stratigraphy of the Desert Member and Castlegate Formation clearly highlights how different driving mechanisms can be interpreted from the same data. Yoshida and others prefer a tectonic driving mechanism for high-frequency sequence development. This is based on evidence including changes in provenance within the Castlegate, changes in palaeocurrents to run parallel to basement structural trends and certain angular relations across sequence boundaries. These are consistent with similar relationships documented from the Pyrenean foreland basin and directly related to deformational processes.

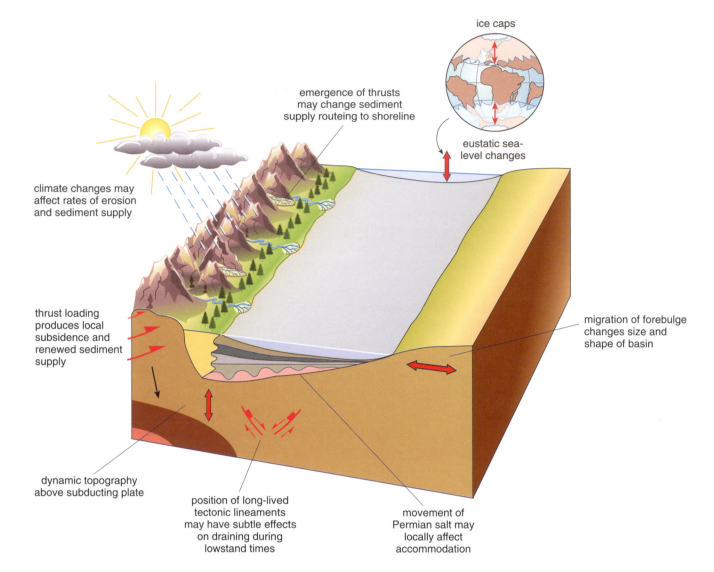

ice caps

emergence of thrusts
may change sediment
supply routeing to shoreline

eustatic sea-
level changes

climate changes may
affect rates of erosion
and sediment supply

thrust loading
produces local
subsidence and
renewed sediment
supply

migration of forebulge
changes size and
shape of basin

dynamic topography
above subducting plate

position of long-lived
tectonic lineaments
may have subtle effects
on draining during
lowstand times

movement of
Permian salt may
locally affect
accommodation

**Figure 10.7** Schematic summary diagram to indicate the controls on sequence development in the Book Cliffs. Note that there are complex feedback relationships between several of the variables, thus making it difficult to isolate the effect of a single variable. However, the rate of change of accommodation space is the fundamental control on both the volume and type of sediment preserved in the rock record and on the partitioning of time between rocks and surfaces.

## 10.3 Milankovich processes and palaeoclimate

Changes in solar insolation due to highly predictable oscillations in the Earth's orbit (Milankovich cycles, Box 2.2) and resultant high-frequency cycles of climate change have been proven to have controlled late Cainozoic glacial cycles (Section 5.1). These episodes of glacial advance and retreat produced eustatic sea-level changes on the tens of thousands to hundreds of thousands of years frequencies.

The oxygen-isotope record of seawater from marine rocks has allowed a precise measurement of palaeoclimatic fluctuations in the Milankovich frequency band back through the middle–late Tertiary, allowing definitive tying of well-dated stratigraphical cycles to glacially driven eustatic sea-level changes (Section 3.2).

Milankovich processes are the most widely accepted driving mechanism for high-frequency sea-level changes and are therefore an obvious mechanism to propose for generation of high-frequency flooding surfaces and sequence boundaries in the Blackhawk Formation. Research workers have identified *c*. 100 ka duration Milankovich cycles over a distance of 1500 km across the Western Interior Basin, from siliciclastic cycles in the west to carbonate cycles in the east. This correlation effectively precludes thrust belt tectonics as the origin of the cycles because the eastern section of the basin was not directly affected by the local subsidence variations caused by thrust movements (Figure 10.7).

The main objection to glacially driven eustatic sea-level changes is the paucity of evidence for development of continental ice-caps during the Cretaceous 'greenhouse' period. The Earth's orbital dynamics are understood to have been similar during the Cretaceous and the Tertiary so it is reasonable to expect climatic cyclicity in the polar regions at this time. There is an increasing body of evidence that non-glacial orbital forcing can occur in greenhouse climatic periods, which could produce subtle sedimentological cyclicity based on changing rates of erosion, sediment release and eustatic sea-level changes (see Section 5.1.2).

It is unlikely that there was one single control on development of Blackhawk Formation stratigraphy (Figures 10.5 and 10.7). While the larger, member-bounding flooding surfaces may represent discrete episodes of increased subsidence associated with movement on individual thrust fault segments, there is no quantitative evidence for a tectonic mechanism which acts on a cyclic frequency high enough to produce parasequence flooding events (tens of thousands of years frequency). Even more difficult is the identification of a tectonic mechanism in a foreland basin setting, which can produce regional relative falls in sea-level (i.e. widespread vertical uplift of the sea-floor) on a *c*. 500 ka cyclicity, inboard of any forebulge. Unloading of thrust sheets (erosional removal) takes place relatively slowly to produce low-frequency isostatic rebound. We have to conclude that high-frequency orbitally forced sea-level fluctuations, superimposed on thrust belt-controlled lower-frequency cycles of subsidence, were probably the most important control on Blackhawk Formation stratigraphical development (Figure 10.7). Regarding tectonic drivers, it is important to distinguish between controls on the focusing of incision from controls on the changes in accommodation space that lead to incision. Inherited intrabasinal tectonic features controlled the position of incised valleys but did not drive the cyclic changes in accommodation space.

## 10.4   Summary

- The late Cretaceous succession exposed in the Book Cliffs provides a large-scale field laboratory for the practical demonstration of sequence stratigraphy methodology and concepts.

- A low-frequency unconformity-bounded depositional sequence comprises a falling stage and lowstand sequence set delta system (Panther Tongue Member), a backstepping transgressive sequence set (Storrs Member) and a 300 m-thick aggradational to increasingly progradational highstand sequence set (Spring Canyon to Desert Members of the Blackhawk Formation). This low-frequency sequence is truncated by the Castlegate Sequence Boundary and is overlain by braided river sandstones. The basinward shift in facies at this sequence boundary is approximately 100 km.

- The Blackhawk Formation highstand sequence set contains most of nine nested high-frequency unconformity-bounded sequences that stack upwards in an increasingly progradational style. These high-frequency sequences include falling stage and lowstand forced regressive shorelines (only exposed/preserved for some cases), incised valley fills, poorly developed transgressive systems tracts and well-developed highstand systems tracts.

- Sequence boundaries are regionally mappable and include both incised valley and interfluve palaeosols. Flooding surfaces cross facies boundaries landward, from marine shale, through carbonate-cemented intra-shoreface zones to coals.

- Sequence development in the Book Cliffs is interpreted to have been controlled by high-frequency glacio-eustatic sea-level cycles superimposed on lower-frequency cycles of thrust sheet emplacement and erosion. The complex interaction of these variables produced the accommodation space and sediment supply combinations interpreted from the stratigraphy. During periods of increased tectonic subsidence, parasequence stacking patterns were more aggradational, with less incision on sequence boundaries (lower Blackhawk Formation). During periods of lower subsidence, parasequence stacking patterns became strongly progradational with more erosion on sequence boundaries (upper Blackhawk Formation).

## 10.5 References

### Further reading

KAMOLA, D. L. AND HUNTOON, J. E. (1995) 'Repetitive stratal patterns in a foreland basin sandstone and their possible tectonic significance', *Geology*, **23**, 177–180. [Model for tectonic driving mechanisms for parasequence boundaries.]

### Other references

VAN WAGONER, J. C. (1998) 'Sequence stratigraphy and marine to non-marine facies architecture of foreland basin strata, Book Cliffs Utah, USA', Reply to Yoshida *et al.*, *American Association of Petroleum Geologists Bulletin*, **52**, No. 8, 1606–1618. [Discussion of possible driving mechanisms for sequences in the Book Cliffs. Published reply to the discussion paper of Yoshida *et al.*, 1998.]

YOSHIDA, S., MIALL, A. D. AND WILLIS, A. (1998) 'Sequence stratigraphy and marine to non-marine facies architecture of foreland basin strata, Book Cliffs Utah, USA: Discussion', *American Association of Petroleum Geologists Bulletin*, **52**, 1596–1606. [Discussion of possible driving mechanisms for sequences in the Book Cliffs.]

# PART 4 CARBONATES

# 11 Carbonate depositional systems

## Dan W. J. Bosence and R. Chris L. Wilson

As mentioned in Chapter 4, it is not possible to transfer an understanding of siliciclastic sequence stratigraphy directly to the interpretation of carbonate successions. There are two key reasons for this. First, carbonate sediments are produced at, or close to, where they are deposited, either as organic skeletons of calcium carbonate or as sedimentary grains directly precipitated from seawater. Secondly, carbonate sediments are susceptible to dissolution and/or cementation if they are alternately bathed in marine and fresh pore waters as sea-level rises and falls. This Chapter explores the significance of these statements, reviews the key controls on carbonate deposition, and examines some settings where carbonate sediments accumulate today.

## 11.1 Carbonate sediments are different

Siliciclastic sediments are derived from processes of weathering, erosion, transport and deposition that deliver rock debris to sedimentary basins. Rivers supply siliciclastic sediment into basins at a series of point sources along basin margins; this debris is then dispersed by bedload transport, gravity-driven flows, and in suspension. Carbonate sediments are different: they are produced in specific areas called carbonate factories, either as the skeletons of marine organisms or by direct precipitation from seawater. Most carbonates, therefore, are made close to where they are deposited rather than being delivered into the depositional basin from the surrounding land area.

From studies of present-day carbonate-producing environments and ancient limestone successions, we know that there are three main types of carbonate factories (Figure 11.1, overleaf):

1  *Warm-water carbonate factory.* Shallow-marine, tropical waters supporting rapidly calcifying communities of organisms that rely on photosynthesis for their energy (e.g. corals in association with their symbiotic algae, calcified green and red algae). Such communities today build shallow-water coral reefs. These waters are often supersaturated with calcium carbonate that can be precipitated at the sea-floor to form carbonate grains such as ooid sands or lime mud.

2  *Cool-water carbonate factory.* Shallow- to moderate-depth marine shelf environments in temperate, arctic and some tropical areas that support calcifying communities of marine benthic invertebrates such as molluscs, bryozoa, foraminifers and barnacles in association with calcified red algae. These communities (with the exception of algae) are not confined to the photic zone and may have lower rates of carbonate production than those from shallow-marine tropical waters described above.

3  *Pelagic carbonate factory.* Oceanic areas where conditions are right for planktonic organisms such as coccoliths and foraminifers to thrive. Calcified plankton grow in the shallow photic zone of the oceans and after death their skeletal debris sinks to accumulate in deep-water areas. These are called pelagic or deep-water carbonates. However, if the waters are very deep and cold, they are likely to be undersaturated with respect to calcium carbonate, the pelagic carbonate debris will dissolve and so may never reach the sea-floor.

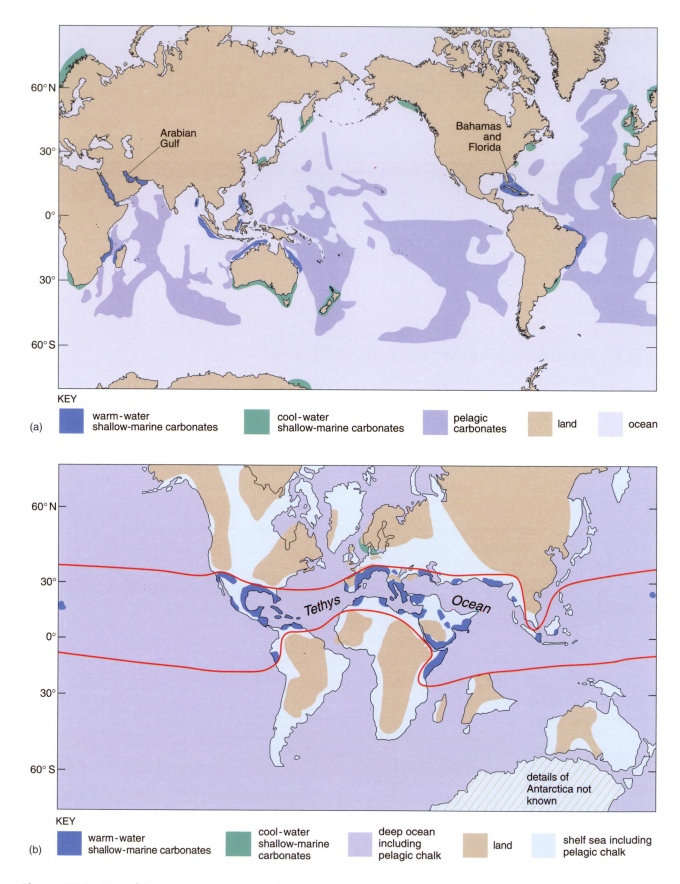

KEY

(a) ■ warm-water shallow-marine carbonates   ■ cool-water shallow-marine carbonates   ■ pelagic carbonates   ■ land   ■ ocean

KEY

(b) ■ warm-water shallow-marine carbonates   ■ cool-water shallow-marine carbonates   ■ deep ocean including pelagic chalk   ■ land   ■ shelf sea including pelagic chalk

**Figure 11.1**  Map of the world showing (a) distribution of three main carbonate factories where sediments are produced and accumulating at the present time (named localities are referred to in the text); (b) in approximately late Cretaceous times. Note that the areas of pelagic deposition are not shown separately but include much of the deep ocean and shelf sea. Not all the deposits shown are entirely contemporaneous. Red lines across (b) indicate the subtropical zone of major carbonate platform growth. ((a) Jenkyns, 1986; (b) Sohl, 1987.)

○ Figure 11.1a shows that cool-water carbonates and warm-water carbonates occur at similar latitudes on either side of the North Atlantic. What is the reason for this?

● This is because the ocean currents in the western North Atlantic bring warm, equatorial waters northwards along the American margin whilst on the African margin cool northern waters are brought southwards. Thus, cool- and warm-water carbonates can be formed at the same latitude on opposite sides of the same ocean.

○ Pelagic carbonates occur at intermediate water depths, such as the flanks of ocean ridges, but not in very deep waters. What is the reason for this?

● The reason is because at great depths the supply rate of pelagic carbonate grains sinking from the shallow photic zone is balanced by the rate of carbonate dissolution. The depth at which this occurs is known as the carbonate compensation depth, which is usually at about 3000–4000 m although in colder waters it is considerably shallower.

Figure 11.1a is not typical of the global distribution of carbonate sediments in the past, for two main reasons. The first is because during periods of high global sea-level such as the late Cretaceous (Figure 11.1b), vast areas of present-day continents were flooded, so that shallow-water carbonate factories were very extensive. The second reason is that prior to the Cretaceous coccolithophores were not common so it was not possible for deep-sea pelagic oozes to accumulate.

The rate at which the three types of factories generate calcium carbonate sediment has been obtained by measuring the thickness and age of Holocene carbonate sediment accumulating in each factory (Figure 11.2). These data demonstrate that carbonate deposition, particularly in shallow warm-water

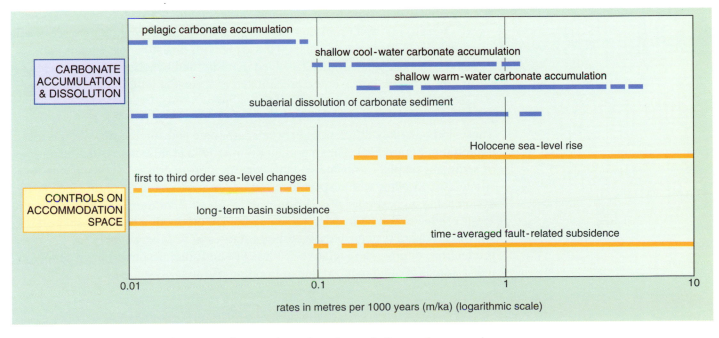

**Figure 11.2** A comparison of the rates of accumulation for pelagic, shallow cool-water and shallow warm-water carbonates, and the rate at which accommodation space may be created by glacio-eustasy, basin subsidence, and fault-related subsidence.

factories (e.g. coral reef growth), is able to outpace long-term basin subsidence and first to third order rates of sea-level rise. Only the most rapid sea-level rise at the end of glacial periods over the past million years and fault-related subsidence can outpace shallow warm-water carbonate accumulation rates. This has important implications for sequence stratigraphy.

○ Recall what happens during a relative sea-level *rise* on a continental shelf on which siliciclastic deposition occurs.

● During the resultant transgression, shorelines are pushed landwards causing sediment starvation on the more distal parts of the shelf.

○ Bearing in mind what you have just learnt about rates of carbonate deposition, what would happen as sea-level rises over a carbonate shelf?

● Because the sediments are manufactured on the shelf (and not, as is the case with siliciclastic sediments, delivered from the land), the carbonate factory usually can produce sediment fast enough to outpace or keep up with the rise in sea-level. Thick carbonate successions, therefore, develop over the shelf in contrast to the sediment starvation that occurs over siliciclastic shelves during transgressive episodes.

○ Recall what happens during a relative sea-level *fall* on a continental shelf dominated by siliciclastic deposition.

● During relative sea-level fall, there is incision of fluvial systems resulting in increased sediment supply from continental areas which is then transported into the basin.

○ Bearing in mind the composition of carbonate sediments, what additional processes different from those in siliciclastic systems might occur during relative sea-level fall in a carbonate system?

● Previously deposited marine calcium carbonate may dissolve away if acidic ($CO_2$-rich) rainwater replaces the marine pore fluids during relative sea-level fall. This will generate a karstic landscape with limestone pavements and cave systems. None of this dissolved material will be redeposited within the basin. It might, however, be precipitated as a cement within underlying sediment to turn it into a lithified limestone.

These and other key differences between carbonate and siliciclastic systems are explored in Figure 11.3. The important point to note at this stage is that in terms of sediment supply, carbonate systems can often operate in the opposite way to siliciclastic systems with respect to variations in relative sea-level. In siliciclastic systems, the highest rates of sediment supply often occur during falling relative sea-level due to fluvial incision (when carbonate production becomes shut off), whilst carbonate systems have their highest rates of sediment supply (or production) during rising relative sea-level (when siliciclastic sediment supply to marine areas is often reduced due to flooding of fluvial systems).

**Figure 11.3 (opposite)**  A comparison of the sequence stratigraphy of siliciclastic and carbonate systems. The siliciclastic system shown is a margin with a shelf break where the relative sea-level fall causes sea-level to fall below the shelf break. Note that these sketches are very generalized and this topic is treated in more detail in Chapter 12. *(Lower three panels: James and Kendall, 1992.)*

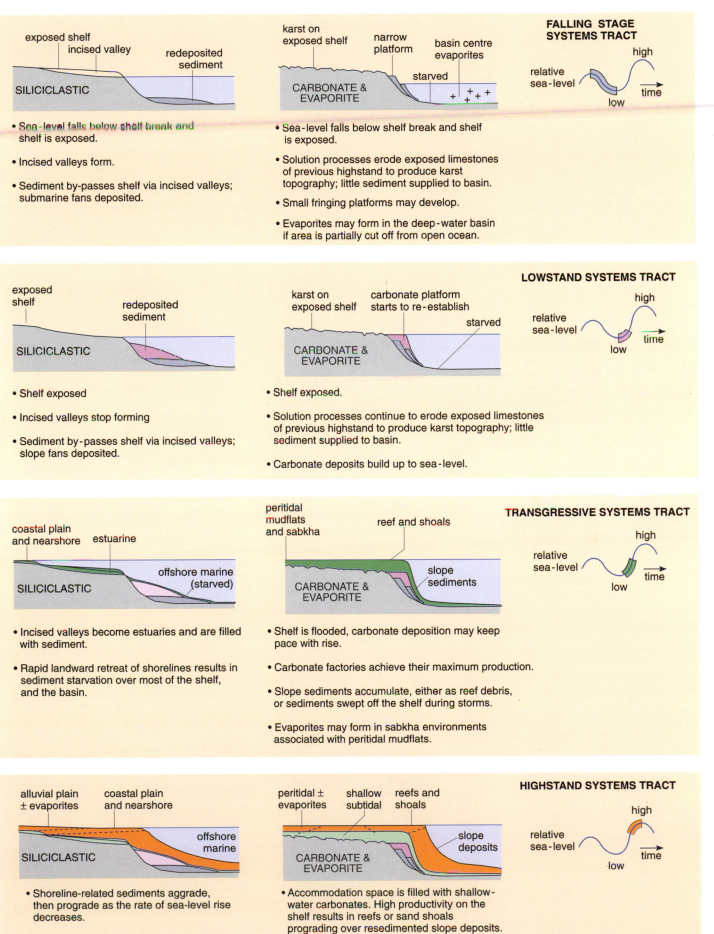

## 11.2 Carbonate platforms

The ability of carbonate depositional systems to keep pace with the continued creation of new accommodation space means that very thick carbonate successions may accumulate (Figures 11.4, 11.5). For example, the carbonate environments we see today in South Florida and the Bahama Banks south-east of Florida probably began to form during the Jurassic as thinned continental crust bordering the newly opened Atlantic Ocean began to subside. Since then, about 5 km of carbonate sediment has accumulated on the Bahama Banks (Figure 11.4) to build a carbonate platform that is totally isolated from any influx of siliciclastic sediment (apart from tiny amounts blown in by the wind). During this same period, the carbonates of the present-day Apennines (Figure 11.5) were forming on the other margin of this ocean.

**Figure 11.4** Seismic section and interpretative section across the western side of the Great Bahama Bank, showing a long history of bank aggradation and progradation. The drifts shown in dark blue are slope carbonates redistributed by contour currents. (For location of cross-section, see Figure 11.8.) *(Anselmetti et al., 2000.)*

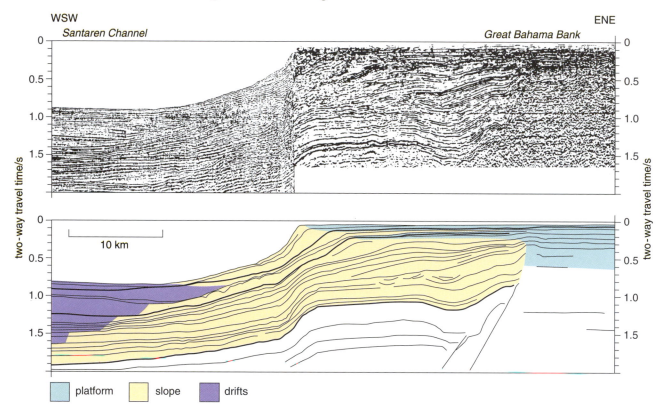

**Figure 11.5** Large-scale exposure of a Lower Jurassic carbonate platform from the Apennines, Italy, showing a thick succession of shallow-water carbonate sediment. This demonstrates that carbonate sediment production kept pace, or outpaced, subsidence in this region in the Early Jurassic. *(Dan Bosence, Royal Holloway University of London.)*

The examples in Figures 11.4 and 11.5 are both carbonate platforms that resulted from carbonate production in warm-water carbonate factories. The warm-water carbonates are by far the most common type of carbonate accumulation in the geological record.

○ What might be the reason for warm-water limestones dominating the geological record of carbonate rocks?

● Warm-water carbonate factories can accumulate carbonate sediment at faster rates than cool-water and pelagic carbonate factories (Figure 11.2).

The term *carbonate platform* is used to describe recent and ancient thick deposits of shallow-water carbonates. Such platforms exhibit a range of morphologies (Figures 11.6, 11.7 overleaf):

- a *rimmed carbonate platform* has a shelf margin rim or barrier such as a reef * or sand shoal, that partially isolates an inner platform or lagoon (Figures 11.6, 11.7a).

- a *ramp* is a platform that is gently inclined (<1°) towards an open sea or ocean (Figures 11.6, 11.7b).

- an *epeiric platform* develops when an extensive cratonic area is covered by a shallow sea (Figure 11.6).

- an *isolated platform* (sometimes referred to as a carbonate bank) is isolated from continental landmasses (Figures 11.6, 11.7a) and an atoll is a small isolated platform usually formed over a subsiding volcano.

- a *drowned platform* develops when carbonate deposition is unable to keep up with the rate of subsidence. This may occur when carbonate production rates are suppressed by inhospitable temperatures, nutrients or siliciclastic sediments and/or, when subsidence rates are increased (Figure 11.6).

CARBONATE PLATFORMS

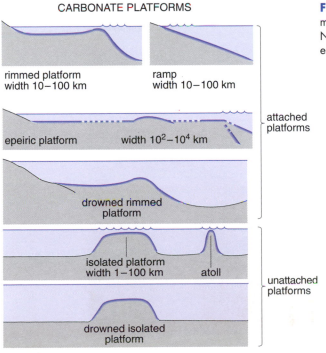

rimmed platform
width 10–100 km

ramp
width 10–100 km

epeiric platform                 width $10^2$–$10^4$ km

attached platforms

drowned rimmed platform

isolated platform
width 1–100 km            atoll

unattached platforms

drowned isolated platform

**Figure 11.6** Sketch cross-sections showing the morphologies exhibited by carbonate platforms. Note that the gradients shown are vertically exaggerated. *(Tucker and Wright, 1990.)*

\* In this book, reef describes any body of carbonate rock that was formed dominantly by growth of organisms with calcareous skeletons and possessed some topographic relief at the time of its formation.

(a)

**Figure 11.7** Landsat images of two modern carbonate platforms. These Thematic Mapper images collect light reflected from the land and sea-floor down to depths of about 40 m. (a) and the water-covered areas of (b) are true colour images whereas the land areas of (b) are false colour. (a) Caicos Bank: this is an isolated rimmed carbonate platform in the southern Caribbean (paler blue area). It is partially rimmed along its northern edge by fringing reefs and a string of islands with a Pleistocene limestone basement appearing green from the vegetation cover in this humid tropical setting. The eastern part of the platform is partially covered by ooid sand shoals. The southern and western margin of the platform is unrimmed and passes gradually into deeper waters (dark blue). (b) Southern margin of Arabian Gulf illustrating the inshore and coastal areas of a carbonate ramp in an arid climate setting. Deep (black) waters to the north gradually shallow towards the south to a series of ooid sand barrier islands that isolate a lagoonal area cut by tidal channels. The low relief sabkha (salty mudflats) area appears as a dark band and passes inland to a coastal plain with isolated aeolian sand dunes (cf. Figure 11.16). *(Harris and Kowalik, 1994.)*

(b)

Rimmed carbonate platforms, ramps and epeiric platforms are all types of attached carbonate platform that are adjacent to a nearby continental landmass, which may mean that at times land-derived siliciclastic sediments are mixed with the carbonates. In contrast, isolated platforms are unattached. Drowned platforms may be either attached or unattached.

○ Figures 11.4 and 11.6 show that carbonate platforms have flat tops. Why should this be so?

● Because carbonate deposition stops when it reaches sea-level, although it may continue to prograde at the platform margins as shown in the highstand systems tract on Figure 11.3.

Rimmed platforms are fringed by reefs or carbonate sand shoals. These absorb wave energy, and restrict the circulation of water by wave- or tide-induced currents to and from the platform interior or lagoon. Platform rims develop best along the windward margins that are most affected by the intense wave energy from storms and ocean swell, whereas unrimmed margins are characteristic of leeward margins that are less affected by storms and ocean swell.

Ramps do not develop rims because there is no major break of slope into shallow water that can be colonized by reef-building organisms. In addition, organic productivity is not as high because the wave energy is not as intense as it is along a shelf margin where the oceanic swell and storm waves are enhanced by the steep slope and shallow area of the shelf. However, the gradual shallowing of the ramp does result in strong wave action in the shoreface and intertidal zones resulting in the formation of shoreline carbonate sand bodies and beaches. They therefore develop similar morphological features to siliciclastic shorelines (see Figure 7.9), including barrier islands as shown in Figure 11.7b.

Rimmed carbonate platform slopes can be very steep because they can be built sub-vertically by reefs or originate from carbonate sediment that has become cemented by calcium carbonate on the sea-floor. Such slopes and sub-vertical submarine cliffs are therefore much steeper than uncemented siliciclastic slopes.

○ What effect might these steep slopes have on depositional sequences formed when relative sea-level falls below the platform margin?

● Because of the steep slope, the area available for carbonate production will become greatly reduced; perhaps only a fringing reef will develop on the slope. This means that not much sediment will be supplied during deposition of the falling stage and lowstand systems tracts.

Thus, the sedimentary architecture may be quite different in rimmed carbonate platforms compared to siliciclastic systems (Figure 11.3). However carbonate ramps, with their lower angles of slope, develop similar features to those in siliciclastic systems.

Studies of present-day and Phanerozoic carbonate platforms show that there is a range of factors controlling their development. These are summarized in Table 11.1 (overleaf). Many of these controls influence the sequence stratigraphical development of carbonate platforms.

**Table 11.1** Controls and subcontrols on nature and occurrence of carbonate platforms.

| Controls | Influences and effects |
|---|---|
| **Biogenic:** | |
| Evolutionary changes | Different organisms have evolved and dominated carbonate factories through geological time, changing the nature and texture of sediments produced. |
| Warm-water, shallow-marine carbonate factory | Common on today's shelves and in the past, and representing the highest-producing carbonate communities (including coral reefs). Cainozoic biotas dominated by corals and calcified green algae; Mesozoic biotas by corals, sponges and rudist molluscs; and Palaeozoic biotas by tabulate and rugose corals, stromatoporoids and bryozoa. |
| Cool-water, shallow-marine carbonate factory | Common on today's shelves and increasingly being recognized in ancient successions. Cainozoic biotas dominated by foraminifers, molluscs, calcareous red algae, bryozoa and barnacles. |
| Pelagic carbonate factory | This has evolved considerably through time with evolution of new calcifying planktonic groups such as planktonic foraminifers and coccoliths. |
| **Oceanographic:** | |
| Global greenhouse and icehouse conditions and carbonate mineralogy | Greenhouse periods: carbonate grains (particularly ooids) and cements precipitated in seawater are formed of the stable form of $CaCO_3$ that contains very low Mg levels, known as low magnesium calcite. |
| | Icehouse periods: grains and cements precipitated as unstable $CaCO_3$ (aragonite and high magnesium calcite). These are more readily dissolved than low magnesium carbonate to form secondary porosity during sea-level falls. |
| Temperature and salinity controls | Precipitated carbonate grains such as ooids and lime mud form in shallow-marine environments where temperatures and salinities are slightly elevated above normal marine values. Many invertebrates cannot survive in high salinity waters and biogenic carbonate production is reduced. Lowered salinities and colder waters lead to undersaturated waters that dissolve calcium carbonate. |
| Nutrient controls | Marine waters with high nutrients from coastal runoff or upwelling favour benthic and planktonic algae whilst open ocean waters with low nutrient levels favour reef-building corals and large benthic foraminifers. This is because corals and large benthic foraminifers are adapted to low nutrient conditions as they tightly recycle the nutrients with the symbiotic algae in their tissues. |
| Light penetration | Light penetrates approximately the upper 50–100 m of water but is greatest in the upper 10–20 m. The amount of light affects productivity of photosynthetic organisms such as planktonic and benthic algae. Also it affects invertebrates with algal symbionts that assist with calcification of corals, large benthic foraminifers, and some large bivalves. |
| Water circulation and oxygenation | Good water circulation is required for some organisms, particularly for reef builders to thrive, and for ooids to be formed. Wave and tidal currents influence the sediment texture. Well-oxygenated waters are essential for growth of all calcifying invertebrates. |

**Tectonics and climate:**

| | |
|---|---|
| Siliciclastic sediment supply | High siliciclastic sediment supply from the land inhibits carbonate production because benthic organisms become swamped by traction deposits and light cannot penetrate waters with a high suspension load. |
| Basin margin subsidence | This creates accommodation space required for thick carbonate platforms to accumulate, e.g. subsidence on passive ocean margins for attached platforms to accumulate. Rapid subsidence may drown shallow-water carbonate platforms. |
| Basin margin uplift | Uplift of hinterland area may increase siliciclastic sediment supply. Uplift of a deep shelf into shallow waters may initiate faster carbonate production. Uplift and subaerial exposure will kill off a carbonate platform. |
| Fault-related subsidence | Rapid fault-related subsidence can drown warm-water, shallow-water carbonate platforms. Submarine fault scarps may form steep-sloping platform margins. |

## 11.3   Modern carbonate depositional environments

Before exploring the sequence stratigraphical development of carbonate deposits in more detail, a basic knowledge of the different types of carbonate facies that can develop on carbonate platforms is required. This Section therefore provides a summary of several modern carbonate depositional environments representative of warm-water carbonate platforms: South Florida, the Bahama Banks, and southern margin of the Arabian Gulf (Figure 11.1a).

### 11.3.1   Climatic and oceanographic setting of Florida and the Bahamas

South Florida and the Bahamas (Figure 11.8, overleaf) experience a semi-humid, subtropical climate with a wet season from July to December (100–150 cm annual precipitation). South Florida (Figures 11.8, 11.9) is an attached carbonate platform with a large, low-lying hinterland (including the Everglades Swamps) and there is considerable freshwater runoff into the shallow-marine coastal environments that can reduce salinities to 6‰* in Florida Bay in the wet season. In contrast, warmer spring and summer temperatures and aridity may increase salinities to 60 in Florida Bay. The reef areas experience salinities of 35–38 and water temperatures varying from 18–30 °C. The adjacent Bahamas Banks are isolated platforms surrounded by deep, open-marine waters (salinity 36 and 22–31 °C). The large shallow bank top environments are seasonally arid and salinities up to 46 are recorded. Both areas are microtidal, with a range of about 0.8 m at the ocean margin, but diminishing into the platform interiors.

South Florida and the Bahamas are within the north-easterly tradewind belt and so the winds blow predominantly from the north-east in winter and from the south-east in summer. Therefore, reef-fringed margins form on the eastward-facing, windward platform margins (Figure 11.8). The more gently sloping, westward-facing leeward margin of South Florida is a carbonate ramp that grades westwards into the Gulf of Mexico.

---

* Salinity is a measure of total amount of dissolved ions in seawater expressed in parts per thousand (‰) by weight; normal seawater has a salinity of 35‰. Note that salinity is now often expressed as a number without appending the ‰ unit.

**Figure 11.8** Map of South Florida and the Bahamas showing regional distribution of major carbonate facies. In particular note occurrence of reefs on most windward margins and ooid shoals on the leeward margin of Great Bahama Bank. Red line shows position of seismic section shown in Figure 11.4. (Jones and Desrochers, 1992 based on Wanless and Dravis, 1989.)

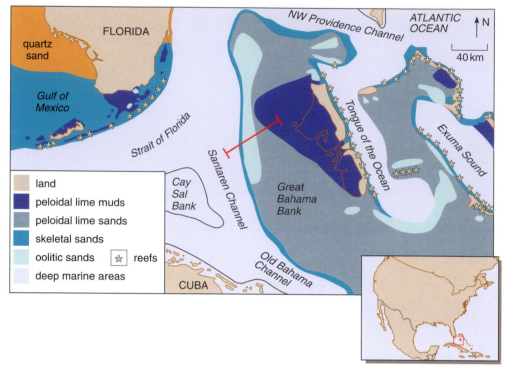

## 11.3.2 South Florida

In shallow-water settings, Holocene carbonate sediments, which are locally lithified, rest on the eroded and cemented Pleistocene Key Largo Limestone (Figure 11.9d,e). Many of the islands of the Florida Keys are lithified reefs or carbonate sand shoals that accumulated during the last Pleistocene sea-level high (c. 120 ka). The Pleistocene–Holocene boundary in the South Florida area is therefore a sequence boundary.

During the Holocene post-glacial transgression, marine waters flooded this margin, resulting in biogenic carbonate deposition on the sequence boundary at the Pleistocene–Holocene boundary. The transgressing sea gradually flooded the platform-top to reach its present-day position. Previous reef margins were flooded and new reefs colonized old reef-rimmed margins to re-establish a partially reef-rimmed margin that we see today in South Florida. The Florida Keys form a more continuous barrier and separate interior platform (or lagoonal) environments of Florida Bay from the back reef and reefal environments of the Reef Tract (Figure 11.9). The shallow coastal waters of Florida Bay have variable salinities, dissolved oxygen levels and suspended carbonate sediment loads which restrict the occurrence of fully marine organisms. Therefore, seagrass beds (e.g. *Thalassia*, Figure 11.10a overleaf) and calcareous organisms that live on them (their epibiotas including foraminifers, serpulids and bryozoa) are not so luxuriant as in the back reef area, and other benthic carbonate producers are restricted to molluscs, foraminifers and calcareous green algae (*Penicillus* and *Udotea*, Figure 11.10b). When dead, these organisms decay and the remaining shelly debris accumulates as a bioclastic mudstone- and wackestone-textured sediment (see Box 11.1 for explanation of textural limestone classification). Shells are concentrated on beaches and storm ridges as gravel-sized molluscan grainstones and packstones (Figure 11.10c). The origin of the carbonate mud is complex because lime mud may also be produced by bioerosion (organisms grazing or boring into skeletal carbonate) or by chemical precipitation. Carbonate mud production has been extensively studied in South Florida. Transects from the

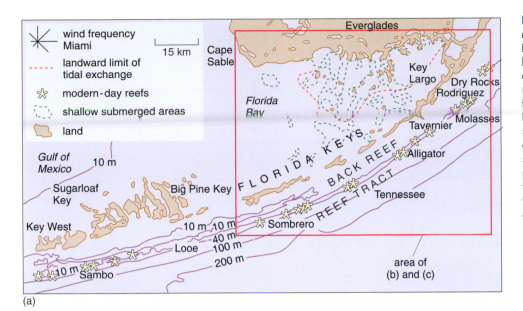

**Figure 11.9** Features of South Florida rimmed carbonate platform. (a) Bathymetry and location of the Florida Keys, a line of Pleistocene reefs; modern-day reefs grow at the shelf edge. (b) and (c) Distribution and details of sediments for part of the Keys (for *Halimeda*, see Figure 11.10i). (d) Idealized cross-section showing distribution of sediments across the shelf. (e) Generalized cross-section through a rimmed carbonate platform in the Keys showing idealized facies belts. Turquoise area shows thickness and details of Holocene carbonates. *(Wright and Burchette, 1996 based on Ginsburg, 1956 (b), (d), (e), Sellwood, 1986 (b),(c),(f), Ginsburg and James, 1974 (c), and Enos, 1977 (f).)*

open water areas to restricted inner bay environments indicate that there are changes in the calcite : aragonite ratio and geochemistry (e.g. strontium concentration) of the muds. The same changes are found in the skeletons of the benthic marine communities in these environments. This suggests that mud is produced from the breakdown of skeletons of algae and benthic invertebrates, as well as some shoreline erosion which also generates mud-sized carbonate.

**Figure 11.10** Surface and underwater photographs of carbonate producers and sediments from Florida Bay and Reef Tract.
(a) Excavated bed of seagrass (*Thalassia*) illustrating dense growth of shoots (1 cm diameter), roots and leaves that trap and bind carbonate mud, much of which is produced from organisms living on the grass blades. (b) Muddy bottom of Florida Bay with population of calcified green algae (10–15 cm high). *Udotea* is a fan-shaped alga (centre foreground) and *Penicillus* is a bush-shaped alga (centre back); other algae are non-calcified forms. (c) Florida Bay shoreline showing accumulation of molluscan shell gravel and king crab (crab *c.* 15 cm across). (d) Low altitude aerial view of Florida Bay illustrating shallow-water mud mounds that preferentially accumulate carbonate sediments through the trapping of sediment in seagrass beds. (e) Box core of intertidal microbial mat with horizontal microbial/sediment laminations and vertical desiccation cracks (scale bar = 10 cm). (f) Mud pebble conglomerate formed from erosion and redeposition of intertidal microbial facies (scale bar = 10 cm). (g) Current-washed skeletal sand in back-reef area with colony of coral *Acropora* (*c.* 30 cm across). (h) Coral (*Diploria c.* 1 m across) overturned by a recent storm to reveal bioeroded (by echinoids and parrot fish) undersurface to colony. (i) *Halimeda*, a calcareous green alga (*c.* 10 cm high). ((a)–(h) *Dan Bosence, Royal Holloway University of London; (i) Chris Wilson, Open University.*)

Within Florida Bay, muddy biogenic carbonate sediments are swept by storm waves into shallow-water mud mounds (Figure 11.10d) which are stabilized by seagrass (*Thalassia*). These mounds have distinct windward erosional margins on their north-east sides and accretionary seagrass stabilized leeward (south-westerly) margins. Mounds therefore migrate in an offshore, down ramp, direction on this leeward margin. Intermound areas are characterized by a thin cover of bioclastic (molluscan, algal, foraminiferal) packstones and grainstones or may be bare of sediment and expose the underlying Pleistocene Key Largo Limestone.

Where mounds build up to sea-level, they are colonized by mangroves, and islands form with windward beaches and storm ridges. Ponded areas within the islands have variable salinities and are colonized by microbial mats that consist of layers of blue-green algae and bacteria. These form laminated micrites (Figure 11.10e) with desiccation structures and irregular-shaped holes up to a few millimetres across known as fenestrae. These may be reworked by storms to generate mud–pebble conglomerates (Figure 11.10f). These facies are typical of intertidal to supratidal carbonate sediments throughout the geological record.

The Reef Tract forms the rimmed margin to this platform and, because this is the northernmost extent of the Caribbean reef zone, it comprises a discontinuous series of reefs. Back reef areas have skeletal grainstone sands and gravels with scattered patch reefs constructed by the framebuilding corals *Montastrea*, *Diploria*, *Acropora* and *Porites* (Figures 11.10g–h, 11.11c–d) with encrusting coralline algae and attached *Halimeda* (Figure 11.10i), alcyonarians and gorgonian soft corals. The patch reefs are bioeroded by sponges, worms, parrot fish, echinoids and molluscs (Figure 11.10h) which generate much of the skeletal sand grains around the patch reefs.

(a)

(b)

(c)

(d)

**Figure 11.11**  Surface and underwater photographs to illustrate Florida Reef tract corals and reef zones. (a) Surface view at low tide of reef flat where coral growth has stopped due to intertidal desiccation and wave erosion (foreground c. 10 m across). (b) Reef crest zone of robust branches (15–20 cm across) of *Acropora* growing towards direction of wave approach. (c) Reef framework construction by massive form of *Montastrea* coral on the reef front (fish c. 20 cm long). (d) Reef talus facies in fore reef zone comprising *Diploria* (c. 40 cm across) and broken coral branches in a skeletal gravel. ((a)–(d) Dan Bosence, Royal Holloway University of London.)

The main barrier reefs have a reef flat of *Acropora palmata* (Figure 11.11a) with a seaward spur and groove zone (Figure 11.12). Spurs are built by seaward-projected branches of *A. palmata*, and by massive *Montastrea* and *Diploria* coral heads (Figure 11.11b,c) and encrusting coralline algae. Down the windward margin of the reef, a *Millepora* terrace gives way to a deeper-water *Montastrea* and *Diploria* zone (Figure 11.11b). Lower areas of the slope are characterized by reef talus deposits (Figure 11.11d) which pass offshore to a terrace of rhodoliths (coralline algal nodules) at 30–50 m water depth.

We shall see in Section 13.1 that similar reef morphologies and coral zones are preserved in ancient reefs.

A descriptive classification of reef frameworks is shown in Box 11.1. Figure 11.12 is an idealized block diagram of a modern coral reef. Corals mainly form framestones and bafflestones with encrusting coralline algae producing secondary frameworks of bindstones. Grooves and channels through the reefs and reef talus accumulate rudstones, and floatstones characterize sheltered deeper-water fore reef and back reef areas.

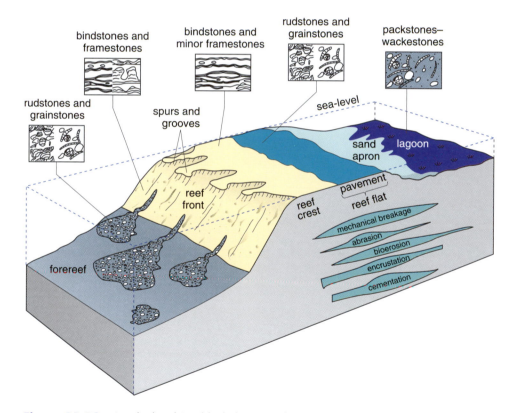

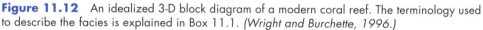

**Figure 11.12**   An idealized 3-D block diagram of a modern coral reef. The terminology used to describe the facies is explained in Box 11.1. *(Wright and Burchette, 1996.)*

## Box 11.1 Descriptive carbonate rock classification based on texture

The classification shown in Figure 11.13 is based on the texture of the carbonate rock, that is, the relationship between grains and the mud matrix (often called micrite — an abbreviation of microcrystalline (<4 μm) calcite) or the absence of matrix between grains. In modern sediments the mud is composed of aragonite or calcite (but through diagenetic processes the aragonite will change to calcite in ancient limestones). Thus, a *mudstone* contains very few coarse grains (<10%), and a *wackestone* has more than 10% grains (the texture is similar to that exhibited by siliciclastic greywackes — hence the prefix 'wacke' *). Where the sediment is grain-supported, a *packstone* denotes that the space between the grains is filled with carbonate mud, but a *grainstone* either has voids between the grains or this space has been filled by cement introduced after deposition. The terms mudstone–grainstone were proposed by Dunham and the whole classification scheme used in Figure 11.13 is usually referred to as the Dunham classification. Additional terms on the right of the diagram were added by the researchers Embry and Klovan to cover the coarse grain sizes (floatstone and rudstone) and the types of deposits associated with reefs. Although these terms were established for lithified rocks, they can be usefully used for the unlithified sediments as described in this Chapter.

○ If the recent carbonate sediments referred to in Figures 11.8–11.10 (some of which are illustrated in Figure 11.15) were to become limestones, then which terms of the Dunham classification would be used for each of the following. (*Note*: one or two terms may apply because you do not always have enough information for a precise classification.)

1   Oolitic sands (no mud)

2   Skeletal sands (no mud)

3   Peloidal lime muds

4   *Halimeda* and mollusc sand (grain-supported)

5   Mollusc and foraminiferal mud (mud-supported, >10% grains)

6   Pellet mud

7   Muddy bioclast sand (grain-supported)

● 1 Oolitic (or ooid) grainstone; 2 skeletal (or bioclastic) grainstone; 3 peloidal wackestone, packstone or mudstone; 4 *Halimeda* and mollusc grainstone or packstone (depending on the percentage mud); 5 mollusc and foraminiferal wackestone; 6 pellet (or peloidal) packstone, wackestone or mudstone; 7 bioclastic packstone.

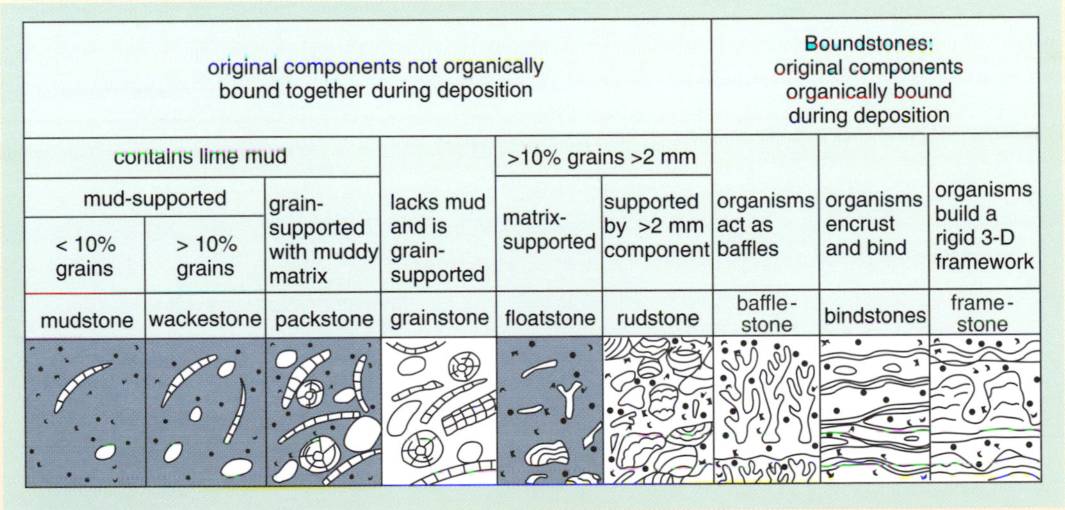

| original components not organically bound together during deposition | | | | | | Boundstones: original components organically bound during deposition | | |
|---|---|---|---|---|---|---|---|---|
| contains lime mud | | | | >10% grains >2 mm | | | | |
| mud-supported | | grain-supported with muddy matrix | lacks mud and is grain-supported | matrix-supported | supported by >2 mm component | organisms act as baffles | organisms encrust and bind | organisms build a rigid 3-D framework |
| < 10% grains | > 10% grains | | | | | | | |
| mudstone | wackestone | packstone | grainstone | floatstone | rudstone | baffle-stone | bindstones | frame-stone |

**Figure 11.13**   The Dunham classification of limestones (mudstone–grainstone) based on the textural relationship between grains, mud matrix and intergranular pore space (which may be filled by post-depositional cements) and the Embry and Klovan classification for coarse grain sizes (floatstone and rudstone) and reefal or organically bound carbonate rocks (bafflestone to framestone). *(Tucker, 1991 based on Dunham, 1962, and Embry and Klovan, 1971.)*

* Wacke was originally a miner's term for a large stone, but later adopted to mean a sandstone-like rock resulting from decomposition of basaltic rocks. It is now used for siliciclastic rocks with grains (e.g. sand or pebbles) enclosed in more than 15% mud-sized matrix or carbonate rocks with >10% grains in a mud-sized matrix.

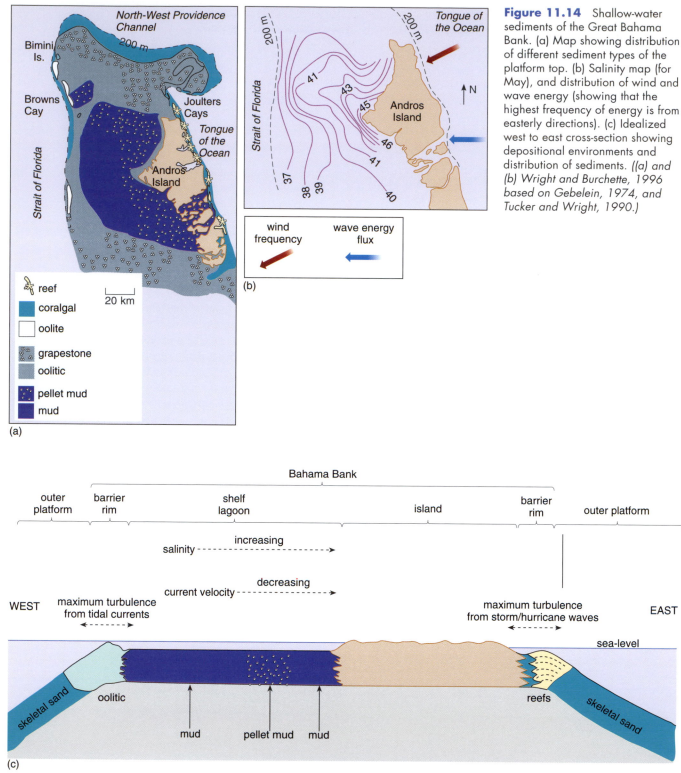

**Figure 11.14** Shallow-water sediments of the Great Bahama Bank. (a) Map showing distribution of different sediment types of the platform top. (b) Salinity map (for May), and distribution of wind and wave energy (showing that the highest frequency of energy is from easterly directions). (c) Idealized west to east cross-section showing depositional environments and distribution of sediments. ((a) and (b) Wright and Burchette, 1996 based on Gebelein, 1974, and Tucker and Wright, 1990.)

## 11.3.3 The Great Bahama Bank

The Great Bahama Bank is the best present-day example of a large, isolated unattached rimmed platform. As in South Florida, the distribution of sediments across the platform is influenced by water depth, wave and tidal energy, and variations in salinity. With the aid of what you have already learnt from the

Florida carbonate platform, explore some of these influences by answering the following questions with the aid of diagrams in Figure 11.14 and the photographs of some of the shallow-water grains in Figure 11.15.

○ What is the relationship between the distribution of coral reefs, and the prevailing wind direction?

● Coral reefs develop on the windward margins of the Bank, in other words mainly on north-east- and east-facing margins.

○ What is the distribution of the oolitic sediments, and what factors appear to control their distribution?

● Oolitic sediments occur around the leeward margin of the Bank, where there are no coral reefs, in shallow turbulent waters of slightly increased salinity (normal marine salinity is 35).

○ What is the distribution of the carbonate muds, and what factors appear to control their distribution?

● The carbonate muds occur in the shallow shelf lagoon, under conditions of calm water and relatively high salinity (and temperature); both of which contribute to precipitation of calcium carbonate from seawater.

The broad pattern of sediment distribution shown on Figure 11.14a has probably existed since Neogene times. However, the leeward margin of the Bank has prograded westwards into the Florida Strait and Santaren Channel (Figure 11.4), but the windward (eastern) side has remained as a steep reef-rimmed escarpment for much of this time.

(a)

(b)

(c)

**Figure 11.15** Photographs of shallow-water carbonate sediments from the Bahamas. (a) Ooid sand: spherical grains probably possess peloids as nuclei, and the more irregular grains have shell fragments as nuclei. (b) Grapestone grains: aggregates of shell fragments and peloids cemented by very fine-grained aragonite. (c) Peloidal and skeletal sand grains (i.e. shell fragments converted to very fine-grained aragonite by boring algae) that after collection were washed out of a mud. (Photographs are all c. 2 cm across.) *(Roger Till.)*

## 11.3.4  The southern Arabian Gulf

The Arabian Gulf is a marine foreland basin (Box 3.1) created by loading during deformation of the Zagros Mountains of south-east Iran (Figures 11.1a, 11.7 and 11.16). The steep northern margin of the Gulf is dominated by siliciclastic sediments derived from erosion of the Zagros Mountains, but the lower relief southern or foreland margin of the Gulf Coast States (e.g. Abu Dhabi and Dubai) is an excellent example of a present-day carbonate ramp. The prevailing winds ('Shamal') are onshore from the north-west which make this a storm-dominated coastline, and storm wave-base and fairweather wave-base have an important control on sediment texture (see below). Due to the restricted opening to the Indian Ocean through the Strait of Hormuz, and the arid climate setting, salinities are elevated in the Gulf to 40–45.

The Gulf is not very deep (<100 m) and marls with a pelagic carbonate component represent the basinal facies of the outer ramp environment (i.e. part of the ramp slope below storm wave-base (SWB) at about 50–60 m depth). These marls grade up the gentle ramp slope into muddy and then clean molluscan and oolitic sands at fairweather wave-base (FWWB at around 10–20 m). This zone between SWB and FWWB is known as the mid ramp, and the inner ramp is the area above FWWB together with coastal regions (Figure 11.16b). Areas of clean bioclastic sand are found in inner ramp environments and these are commonly cemented on the sea-floor to produce extensive hardgrounds. Small patch reefs (Figure 11.17a, overleaf) occur with a low diversity coral fauna because of the elevated salinities and sometimes low winter temperatures.

The inner ramp is dominated by skeletal sand beaches or ooid barriers in the Abu Dhabi area (Figures 11.7b, 11.16c, 11.17b). Aragonitic ooids are precipitated today in the relatively high salinity marine waters with a moderate tidal and wave energy regime. Ooid barrier islands are situated up to 20 km away from the shoreline sabkhas* and broad, shallow lagoons have developed behind the barriers. The ooid shoals are cut by tidal channels which lead out into ebb and flood, ooid sand deltas.

The lagoons have elevated salinities of 40–50 and so have reduced marine faunas comprising gastropods and ostracods that occur in peloidal wackestones. In the lagoons, tidal channels erode down to the lithified Pleistocene limestone and have a basal lag of lithoclasts, shelly material and peloids. The lagoonal muds are aragonitic and both the absence of calcified green algae (like those present in South Florida and the Bahamas) and compelling geochemical evidence (beyond the scope of this book) indicate that these muds were precipitated chemically from the lagoonal waters rather than having an organic origin as is the case in Florida.

The sabkhas are broad coastal plains (Figures 11.7b, 11.16a, 11.17c) that may be flooded by lagoon waters when storm winds blow landward. Intertidal areas have extensive populations of filamentous and sphere-shaped microbes on the tidal flats which, together with sediment deposited on the sabkhas when they are flooded by storms, give rise to microbial laminites or mats (Figures 11.16b, 11.17d) and small dome-shaped stromatolites (Figure 11.17e). High aridity in the area leads to a net evaporation of floodwaters and the drawing up of saline groundwaters into the sabkha. This leads to precipitation of evaporite minerals

---

* 'Sabkha' is the Arabic word for an area of low-lying salty ground.

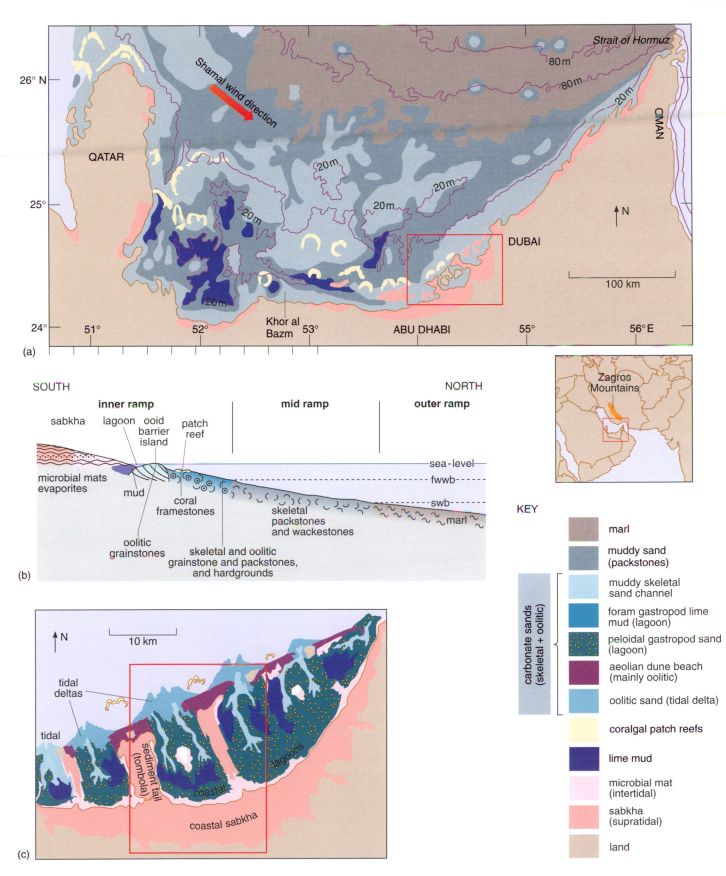

**Figure 11.16** Carbonate sediment distribution in the southern Arabian Gulf: (a) Generalized facies zones (red box indicates area of (c): red box on small map of the Middle East shows area of (a)); (b) vertically exaggerated sketch cross-section (south to north) showing depth profile and facies on ramp; (c) sediment distribution associated with barrier islands and lagoons in the Abu Dhabi area (red box shows area of Figure 11.7b). ((a) and (c) Wright and Burchette, 1996 based on Purser, 1973.)

**Figure 11.17** Sedimentary facies of the inner ramp of the southern Arabian Gulf. (a) Coral patch reef with skeletal gravels and sands. (b) Ooid sand beach on barrier island. (c) View across sabkha that has been recently flooded. (d) Microbial mats showing laminae and desiccation cracks. (e) Domed stromatolites built up from alternating microbial and sediment-rich layers. (f) Pit dug into supratidal sabkha revealing fine-grained anhydrite (with the texture of cottage cheese) precipitated within pre-existing carbonate sediments. ((a)–(f) *Dan Bosence, Royal Holloway University of London.*)

such as dolomite, gypsum and anhydrite. Fine-grained anhydrite grows within the lagoonal sediments. It forms folded layers and nodules due to expansion caused as the new mineral grows within pre-existing lagoonal sediments (Figure 11.17f).

Because of the short period of time since marine flooding of this margin, there has been little development of a stratigraphical section. However, progradation of the shoreline would be expected to produce a shallowing-upward succession from outer to mid to inner ramp facies capped by subaerial exposure of the sabkha surface.

(a) View *c.* 2 m across

(b) View *c.* 30 m across

(c) Scale bar = 1 m

(d) View *c.* 30 cm across

(e) View *c.* 70 cm across

(f) Scale bar = 15 cm

## 11.4 Summary

- Unlike siliciclastic sediments, which are transported into marine basins along their margins, carbonates are manufactured near where they are deposited in carbonate factories. There are three types of marine carbonate factories: warm-water, cool-water and pelagic. Each one is productive in a defined range of environmental conditions.

- Warm-water carbonate factories are able to keep up with all but the fastest rates of increase in accommodation space. This has very significant implications for their development during changes in accommodation space. Unlike siliciclastic systems, in which the largest rate of sediment supply to the continental shelf is often during falling relative sea-level and the smallest is during rising relative sea-level, rates of carbonate deposition are frequently lowest during falling relative sea-level and often keep up with, or outpace, rates of relative sea-level rise.

- On the basis of their morphology, carbonate platforms are classified as rimmed carbonate platforms, ramps, epeiric platforms, isolated platforms (including atolls) and drowned platforms (Figure 11.6). Platforms may be attached to shoreline or be unattached and surrounded by deeper waters.

- Carbonate platforms are most commonly formed by warm-water carbonate factories filling accommodation space to sea-level. Kilometre-thick successions occur in regions of long-term subsidence unaffected by siliciclastic supply. Long-term eustatic sea-level rise may also lead to moderately thick successions. Such platforms may drown when rates of carbonate production are reduced by decreased or increased temperatures or salinities, increased nutrients or siliciclastic supply and/or increased subsidence rates.

- In modern warm-water carbonate settings, the distribution of different sediment types is influenced by water depth, energy of environment, salinity and climate and tectonics. High-energy sediments are deposited on or near platform margins, with reefs predominating on windward sides, and sands (ooidal and skeletal) on leeward sides. Towards platform interiors, less well-sorted skeletal or peloidal sands occur, with carbonate muds being deposited or precipitated in the most sheltered areas. In the latter case, tidal flats may occur, which, in more arid settings contain gypsum and anhydrite that form just beneath the surface as pore waters are evaporated. Studies of modern and ancient carbonate successions have lead to the formulation of generalized facies models for rimmed shelves and ramps (Figure 11.18, overleaf).

- In rimmed shelves (Figure 11.18a,b), increased wave energy and limited availability of nutrients results in the formation of coral reefs and carbonate sands near the steep shelf edge. These absorb wave and tidal energy and so restrict circulation into the shelf interior, where muddy sediments characteristic of lower energy levels occur. Material shed from the rimmed shelf into deeper water is resedimented on, or at the base of, the platform slope.

- Ramps (Figure 11.18c,d) are characterized by gentle (<1°) seaward-inclined slopes over which the development of carbonate sand bodies is controlled by the position where storm wave-base and fairweather wave-base intersect the sea-floor. Barrier reefs do not develop on ramps, but patch reefs may develop in places (e.g. the Arabian Gulf). In arid climates (e.g. the Arabian Gulf), salinities are high enough for evaporites to form, particularly in the intertidal and supratidal (sabkha) zones. These do not occur in more humid climate settings (e.g. South Florida).

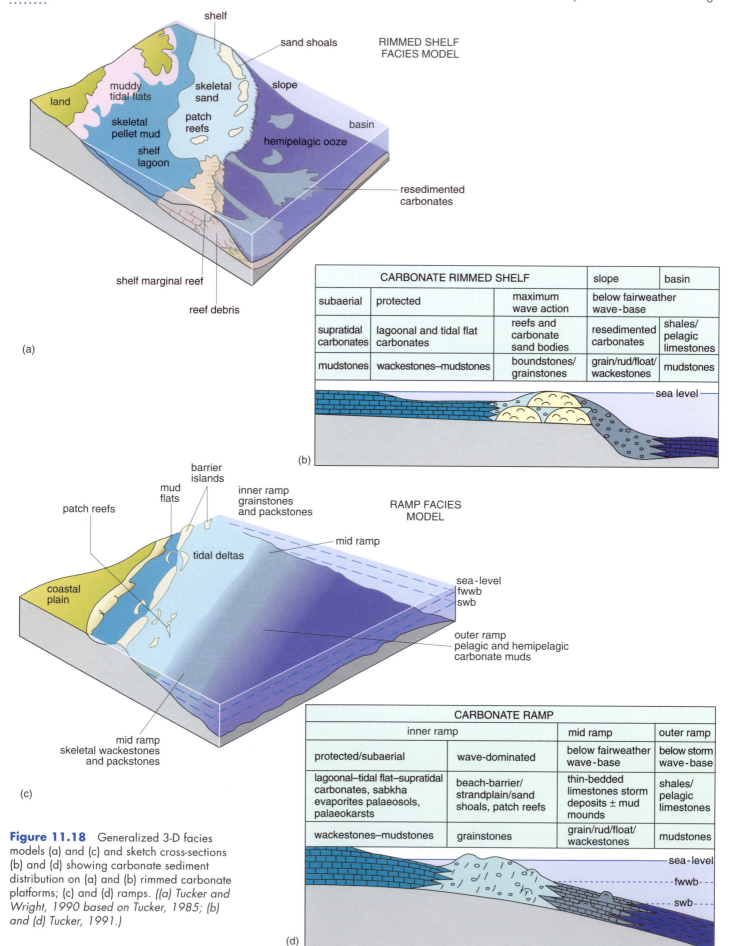

**Figure 11.18** Generalized 3-D facies models (a) and (c) and sketch cross-sections (b) and (d) showing carbonate sediment distribution on (a) and (b) rimmed carbonate platforms; (c) and (d) ramps. ((a) Tucker and Wright, 1990 based on Tucker, 1985; (b) and (d) Tucker, 1991.)

RIMMED SHELF FACIES MODEL

shelf
sand shoals
slope
muddy tidal flats
skeletal sand
land
skeletal pellet mud
patch reefs
basin
shelf lagoon
hemipelagic ooze
resedimented carbonates
shelf marginal reef
reef debris

| CARBONATE RIMMED SHELF | | | slope | basin |
|---|---|---|---|---|
| subaerial | protected | maximum wave action | below fairweather wave-base | |
| supratidal carbonates | lagoonal and tidal flat carbonates | reefs and carbonate sand bodies | resedimented carbonates | shales/ pelagic limestones |
| mudstones | wackestones–mudstones | boundstones/ grainstones | grain/rud/float/ wackestones | mudstones |

sea level

(b)

RAMP FACIES MODEL

barrier islands
mud flats
inner ramp grainstones and packstones
patch reefs
mid ramp
tidal deltas
coastal plain
sea-level
fwwb
swb
outer ramp
pelagic and hemipelagic carbonate muds
mid ramp skeletal wackestones and packstones

(c)

| CARBONATE RAMP | | | | |
|---|---|---|---|---|
| inner ramp | | | mid ramp | outer ramp |
| protected/subaerial | | wave-dominated | below fairweather wave-base | below storm wave-base |
| lagoonal–tidal flat–supratidal carbonates, sabkha evaporites palaeosols, palaeokarsts | | beach-barrier/ strandplain/sand shoals, patch reefs | thin-bedded limestones storm deposits ± mud mounds | shales/ pelagic limestones |
| wackestones–mudstones | | grainstones | grain/rud/float/ wackestones | mudstones |

sea-level
fwwb
swb

(d)

(a)

## 11.5 Further reading

BROWN, J. E., COE, A. L., SKELTON, P. W. AND WILSON, R. C. L. (1999) (see further reading list for Chapter 7).

WRIGHT, V. P. AND BURCHETTE, T. P. (1996) 'Shallow water carbonate environments', in READING, H. G. (ed.) *Sedimentary Environments: Processes, Facies and Stratigraphy* (3rd edn), Blackwell Science, 325–391. [This provides an excellent overview, including a treatment of carbonate sequence stratigraphy similar in some respects to that presented in this book.]

SCHOLLE, P. A., BEBOUT, D. G. AND MOORE, C. H. (1983) *Carbonate depositional environments*, American Association of Petroleum Geologists Memoir No. 33, 708pp. [Still the best descriptions and colour illustrations of the complete range of carbonate depositional environments and facies.]

TUCKER, M. E. (2001) (see further reading list for Chapter 1).

TUCKER, M. E. AND WRIGHT, V. P. (1990) *Carbonate Sedimentology*, Blackwell Scientific Publications, Oxford, 482pp. [An excellent advanced text for those who wish to follow up the introduction provided in this book. Its treatment of carbonate sequence stratigraphy is, however, brief, because this area of research was in its infancy at the time of publication.]

WALKER, R. G. AND JAMES, N. P. (1992) *Facies Models: Response to Sea Level Change*, Geological Association of Canada, 454pp. [Part III on 'Carbonate and evaporite facies models' contains six chapters that provide more detailed and well illustrated follow-up reading. The fact that it contains more treatment of sequence stratigraphy than Tucker and Wright (1990), yet was published only about two years later, is testament to the rapid advances in understanding that had been made over a short period of time.]

# 12 Sequence stratigraphy of carbonate depositional systems

*Dan W. J. Bosence and R. Chris L. Wilson*

Section 11.1 discussed how sediment supply in siliciclastic and carbonate depositional systems is fundamentally different so that carbonate systems respond in a different way to siliciclastic systems when subjected to relative sea-level changes. This Chapter explores some of the key controls on sequence development in carbonates, and the resultant stratal architecture. The sequence stratigraphy of rimmed carbonate platforms and ramps is then examined. The Chapter concludes by showing how numerical forward modelling can be used to explore the complex relationships between sediment supply, accommodation space and relative sea-level change. Also we show how numerical modelling can be used to bracket some of the rates of processes that control the formation of carbonate platforms.

## 12.1 Major controls on the sequence stratigraphy of carbonate platforms

### 12.1.1 Sediment supply and carbonate platform flooding

Carbonate sediment supply, or more specifically carbonate production, is proportional to the area of flooded platform top (Figure 12.1). It is logical, therefore, to expect the greatest supply of sediment in carbonate platform systems to occur during periods of elevated relative sea-level because shallow-water carbonate-secreting communities (the carbonate factories) will be able to extend over the entire upper surface of carbonate platforms. This results in the shallow-water sediments filling or even overfilling the newly created accommodation space. Shallow-marine and peritidal carbonate sediments will accumulate up to, or just above, sea-level after which the carbonate factory will shut down.

Alternatively, if surface currents are capable of transporting sediment from the platform interior to the margin then carbonate production will continue as the carbonate sediment is transported onto the sloping platform margins and therefore progradation is enhanced. This will enable the carbonate factory to keep producing shallow-marine carbonate grains to feed progradation of the platform margin. This is the case in the carbonate platforms in Florida and the Bahamas today, and the process has been aptly labelled 'highstand shedding'. In Florida, sediment is shed westwards out of Florida Bay and down the ramp into the Gulf of Mexico (Figure 11.9). In the Bahamas, the platform top has a long history of shedding sediment off the leeward margin to the west. This is dramatically shown by the westerly prograding clinoforms in Figure 11.4. This process is in direct contrast to siliciclastic systems that often have their highest rates of sediment supply during falling and low relative sea-level.

Shallow-marine warm-water carbonate factories produce sediment at greater rates (1–10 m per 1000 years) than rates of increase of accommodation space due to long-term basin subsidence and low-order sea-level changes (0.01–0.1 m per 1000 years; Figure 11.2). This means that on many platforms shallow-water carbonate sediment may infill or overfill accommodation space even when relative sea-level is rising. Such systems will, therefore, show aggradation or

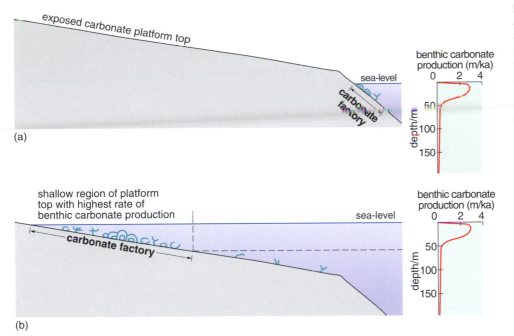

**Figure 12.1** Relationships between carbonate platform morphology, sea-level and carbonate sediment supply. The gradients shown on the cross-sections are vertically exaggerated. The graphs on the right show vertical changes in carbonate production (in metres of vertical thickness per 1000 years) with depth (in metres below sea-level). Production reaches a maximum between 10 and 30 m water depth and then decreases with increasing water depth. (a) During periods of relatively low sea-level, the limited horizontal extent of the carbonate platform results in small total amounts of carbonate being generated because of the small area of high productivity. (b) Flooding over the platform top significantly increases the horizontal extent of the productive area, and sediment production will rapidly fill all the newly created accommodation space.

progradation during both rising relative sea-level and high sea-level. Both will form regressive or shallowing-upward profiles even though relative sea-level is rising. This can also occur in some siliciclastic systems with continuous high rates of sediment supply such as river-dominated deltas. On the other hand, carbonate ramps do not have such highly productive, shallow-water habitats and appear to behave more like siliciclastic continental shelves.

## 12.1.2   Relative sea-level change and sequence development

The fact that by far the greatest amount of carbonate sediment is manufactured in water depths less than 50 m means that changes in relative sea-level have an enormous influence on sediment production (Figure 12.1 and Section 4.1). This in turn influences the geometry of strata deposited to form lowstand (LST), transgressive (TST), highstand (HST) and falling stage (FSST) systems tracts as shown in Figure 12.2. In Figure 12.2a,d (overleaf), progradation due to highstand shedding is dominant during deposition of the HST. During forced regression, a small wedge of sediment (Figure 12.2b(i)) may form due to low carbonate productivity on the slope formed by the top of the previous HST (Figure 12.2a). If production is higher, or a significant amount of physical erosion of the carbonate slope takes place, or the slope is of a lower angle, or the rate of relative sea-level change is slow, progradation with downstepping or downlapping will occur (Figure 12.2b(ii)). In all cases, the earlier sediments will become subaerially exposed and an unconformity representing the sequence boundary will start to form. A similar geometry will continue in the ensuing lowstand of relative sea-level (Figure 12.2b(i)) but the amount of sediment deposited may be small or absent in carbonate systems, especially where there is a steep margin to the platform. At this stage, the platform may be chemically eroded (i.e. dissolved) if in a humid climate and this material is removed from the succession, leaving very little trace in the stratigraphical record.

○   Bearing in mind the typical relative rates of carbonate production and rates of increase in accommodation space (Figure 11.2), what four possible geometries might be expected during a period of relative sea-level rise?

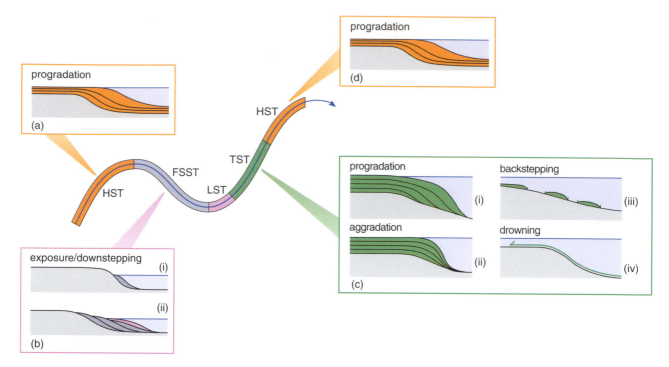

**Figure 12.2** Cartoons showing the development of carbonate depositional systems tracts through an idealized cycle of relative sea-level change. Note: (b)(ii) shows the FSST in grey and the LST in purple. See text for further explanation. *(Wright and Burchette, 1996.)*

● (1) A progradational TST will form if carbonate production rates are in excess of the rate of increase in accommodation space (Figure 12.2c(i)).

(2) An aggradational TST will form if carbonate production keeps pace with the increase in accommodation space (Figure 12.2c(ii)).

(3) If carbonate production cannot keep pace with the increase in accommodation space, a series of small carbonate accumulations will form in succession and show a retrogradational pattern of small drowned platforms along the slope of the previous highstand (Figure 12.2c(iii)).

(4) If the rate of increase in accommodation space during relative sea-level rise is so high that carbonate factories cannot become re-established, then platform drowning occurs (Figure 12.2c(iv)). The platform may then be draped by a thin layer of pelagic sediments.

The variability in sequence architecture in TSTs and HSTs is explored further in Figure 12.3. This illustrates the stratigraphical geometries that result from variable rates of increase in accommodation space in relation to carbonate production along a basin margin with an attached carbonate platform. Note that the FSST and LST (shown in purple on Figure 12.3) in each case are composed of siliciclastic rocks that were transported across the carbonate platform via incised valleys.

○ What might be controlling the variation in accommodation space along the basin margin in Figure 12.3?

● There are two components to the relative sea-level curve: eustasy and subsidence. To explain the geometries in Figure 12.3, either subsidence or eustatic sea-level would have to be decreasing from (a) to (b) to (c).

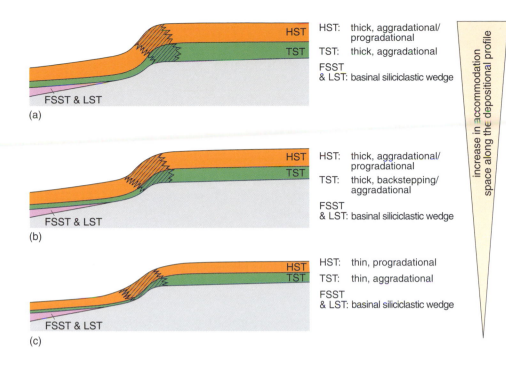

**Figure 12.3** The effects of changing the rate of creation of accommodation space in an attached carbonate platform along a basin margin. Note the FSST and LST is developed in siliciclastic facies and that carbonate production keeps pace with increasing accommodation space. Areas with oblique lines represent shelf edge facies. *(Wright and Burchette, 1996 based on Garcia-Mondéjar and Fernández-Mendiola, 1993.)*

○ How could you determine whether local subsidence or eustatic sea-level was controlling the geometry?

● By examining other basin margins around the world to see if other sections of equivalent age were the same and therefore if eustasy was the dominant component to the sea-level curve.

The sequence stratigraphical development of different types of carbonate platform is explored in detail in Sections 12.2 and 12.3.

### 12.1.3 Small-scale cycles or parasequences

As with siliciclastic sequences, shallowing-upward parasequences can form the building blocks of depositional sequences developed on shallow-water areas of carbonate platforms (Chapter 4). In many ancient carbonate platform top successions, several hundred parasequences may occur (e.g. Figure 11.5).

○ Casting your mind back to the modern environments of carbonate deposition introduced in Section 11.3, what type of facies will constitute the small-scale shallowing-upward parasequences that might be forming in the carbonate platforms of (a) South Florida and (b) southern Arabian Gulf?

● (a) In South Florida lagoons (Figure 11.9), muddy carbonates might accumulate over the sequence boundary eroded into the Key Largo Limestone. The lagoonal muds may then shallow upward to grass-bed wackestone and shoreline beaches or mangrove peats (Figure 11.10). On the reef margin (Figure 11.9), new reefs will re-establish on old reefs of the underlying Pleistocene Key Largo Limestone with an intervening sequence boundary.

(b) In the southern Arabian Gulf (Figures 11.16, 11.17), a small-scale cycle is being deposited as the sabkha shoreline progrades out to produce a shallowing-upward succession of lagoonal muds overlain by intertidal and supratidal microbial laminites and then aeolian sands.

These small, metre-scale cycles (parasequences) consist of a flooding surface overlain by shallow subtidal facies overlain in turn by shallow-marine or tidal flat facies (Figure 12.4a and Section 4.2). Each parasequence records the creation and subsequent infilling of accommodation space.

○ How is it possible to accumulate successions containing tens or even hundreds of such cycles?

● There must be a long-term continuous increase in accommodation space that is greater than the sediment supply. This is caused by subsidence or eustatic sea-level rise, or a combination of both. Each cycle, therefore, is the result of a high-frequency change in accommodation space superimposed on a low-frequency one (Chapters 4 and 5).

Systematic changes in the thickness of successive carbonate cycles are often observed: they may thicken upwards or thin upwards.

○ Because carbonate production rates are usually greater than changes in accommodation space, resulting in the accommodation space always being filled, what can systematic changes in the thickness of successive carbonate cycles tell us about changes in the longer-term changes in accommodation space?

● Thinning-up indicates a decrease in the rate of long-term addition of accommodation space, and thickening-up indicates an increase in this rate.

This interpretation is explored in Figure 12.4b, showing how cycle thickness trends characterize different systems tracts. In addition, the proportion of different facies types within the cycles will vary according to the rate of the longer-term increase in accommodation space. When more accommodation space is created during rapid relative sea-level rise, greater amounts of marine subtidal sediment would be expected than when accommodation space is limited during slow relative sea-level rise or relative sea-level fall. Similar to siliciclastic sequences, carbonate depositional sequences may be stacked into sequence sets, reflecting their position in a lower order lower frequency cycle of relative sea-level change (Chapter 5).

### 12.1.4 Dissolution and cementation

During relative sea-level fall and low relative sea-level, carbonate platform top sediments are drained of the seawater contained within pore spaces and this is replaced by meteoric water, derived from rainfall, which is $CO_2$-rich and therefore acidic. Such fluids will dissolve unstable aragonitic grains (such as ooids, molluscs, corals) and may reprecipitate this dissolved carbonate as a calcite cement. This leads to lithification of the original mixed mineralogy carbonate sediment. This early lithification means that there is not much physical erosion of sediments deposited previously, and so very little carbonate is physically eroded and transported into the basin. This contrasts sharply with siliciclastic systems which in general do not show such early lithification. In addition, in humid climates, the exposed carbonate sediments or rocks may be dissolved away to form a karstic landscape with cave systems. The dissolved calcium carbonate will move into the groundwater system and its only expression in the stratigraphy will be as calcite cements. Thus, carbonates, and also

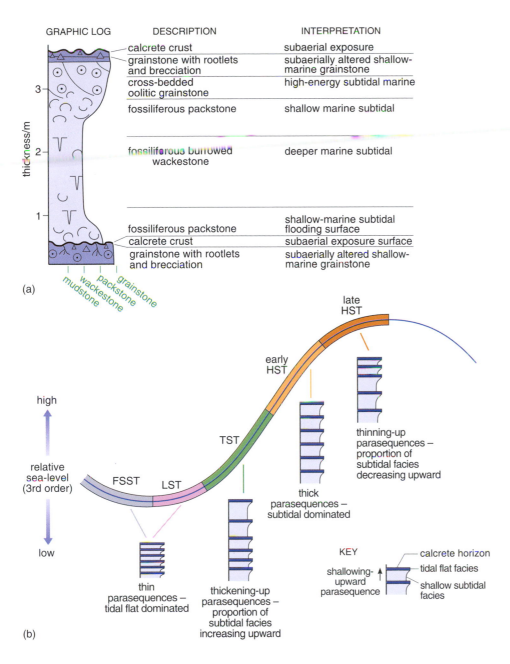

**Figure 12.4** Features and stacking patterns of small-scale carbonate cycles (parasequences). (a) Sketch log of an idealized shallow-marine carbonate parasequence or small-scale cycle. Calcretes are fossil carbonate soils. (b) Idealized stacking patterns of shallow-marine carbonate parasequences on a carbonate platform top. The curve shows a long-term (e.g. third order; Chapter 5) relative sea-level cycle and the mini-graphic logs show the thickness and composition (subtidal vs. intertidal) of the higher frequency fourth or fifth order cycles or parasequences within each systems tract. ((b) Wright and Burchette, 1996 based on Tucker, 1993.)

evaporites, can 'disappear' from the stratigraphical record through processes that have no counterparts in siliciclastic systems. Different carbonate minerals have different solubilities in surface waters. When sea-level changes, these minerals are subjected to different pore waters and significant diagenetic changes such as dissolution, recrystallization and cementation take place. These diagenetic changes have been integrated into our understanding of the sequence stratigraphy of carbonates, as will be shown in Section 12.2.

## 12.1.5  Sediment partitioning and relative sea-level change

In mixed carbonate–siliciclastic and carbonate–evaporite sequences, deposition of the two sediment components is often mutually exclusive, and in some cases one of the components is confined to a basinal or shelf setting during a particular depositional systems tract (Figure 11.3).

Why should mutual exclusivity characterize carbonate–siliciclastic systems? During relatively low sea-level, rivers draining higher land adjacent to carbonate platforms are rejuvenated, and so potentially more silicliclastic sediment is supplied to the basin (Figure 12.5). As the relative sea-level fall may also have caused the emergence of the underlying carbonate HST, rivers will incise into the platform and carry sediment across the eroded top to the platform margin. Siliciclastic sediments will be deposited at the base of the steep edge of the carbonate platform (as can happen in a siliciclastic margin with a shelf break during falling and low relative sea-level). The siliciclastic sands therefore bypass much of the platform top. When relative sea-level rises again to flood the old platform top, rivers are pushed back landward, and carbonate deposition resumes across the platform top. This is exactly what happened along the north-east Australian shelf during the Cainozoic, as illustrated in Figure 12.5.

**Figure 12.5** Sketch sections showing carbonate (in blue and grey) and siliciclastic (orange and beige) sediment partitioning along the north-eastern Australian shelf in the area of the Great Barrier Reef during the Cainozoic. (a) Relative sea-level rise results in carbonate deposition; (b) relative sea-level fall results in siliciclastics being deposited across the carbonate platform. *(Wright and Burchette, 1996 based on Davies et al., 1989.)*

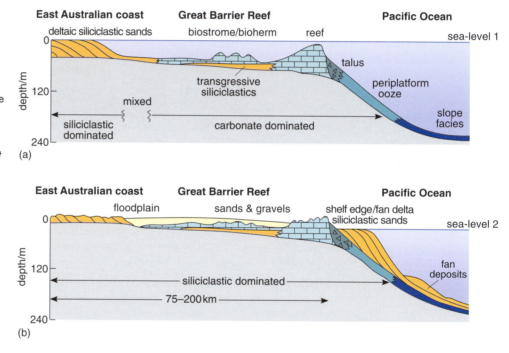

If a basin is situated in an arid climate zone and becomes partially isolated from the oceans, then evaporation will raise salinities and evaporites will be deposited. The partitioning of carbonates and evaporites in FSSTs, LSTs, TSTs and HSTs is summarized in Figure 11.3. During falling and low relative sea-level in an arid, partially isolated basin, evaporites are often precipitated within the basin. The resultant high Mg/Ca ratio of the marine waters may also cause dolomitization of the previously deposited platform carbonates. The evaporitic sediments of the Permian of north-east England and the southern North Sea are just such FSST and LST deposits as are the dolomitized Miocene platforms around the Mediterranean and the associated evaporites (see Section 13.1).

Evaporites may also form as back-barrier lagoonal deposits and sabkhas in TSTs, and as sabkha evaporites in HSTs. In both situations, basinal evaporites do not usually form, as raised sea-level results in more open connections to the ocean.

Next we consider the application of sequence stratigraphical principles and methods to rimmed carbonate shelves and to ramps. In both cases, this application has made a significant advancement to our understanding of the development of carbonate platforms and to basin margin evolution.

## 12.2   Sequence stratigraphy of rimmed carbonate platforms

The significance of how the unique morphology of rimmed carbonate platforms controls the sedimentary response to relative sea-level changes has already been stressed. The steep sides and flat top mean that very specific erosional, depositional and chemical processes are reflected in the facies associations and early diagenesis of flooded and emergent platforms. To discuss these changes, a schematic rimmed shelf platform is examined as it is subjected to the various stages of relative sea-level change (Figure 12.6 overleaf).

### 12.2.1   Transgressive systems tract

During phases of relative sea-level rise, rimmed carbonate platforms may respond in two very different ways depending on the balance between the rate of relative sea-level rise and rate of carbonate production. They may be labelled as backstepping platforms and aggradational–progradational platforms:

*Backstepping platforms.* When rates of increase in accommodation space are greater than rates of carbonate production, then the shoreline will onlap the transgressive surface and facies will retrograde (Figure 12.6a), or the carbonate platform margin may drown, so that offshore facies lie close to, or on top of, the flooded sequence boundary. Reefs will attempt to keep up with relative sea-level rise, producing aggradational geometries with deep lagoons but the reefs will eventually drown if the rate of increase in accommodation space is too high. If reefs have to backstep to a shallower location, this may be located some distance away at the shoreline and not in the topographical low of the lagoon behind the previous reef. The result is the development of a thin TST with an overall deepening-upward vertical succession and retrogradational parasequence sets. Offshore areas will accumulate pelagic facies and sediment starvation may result in submarine cementation and hardground formation. If the rate of relative sea-level rise begins to decrease, then carbonate production in shallow-water areas may start to catch up with the rate of increase in accommodation space; aggradational geometries with shallowing-upward successions may then develop in shoreline areas. These platforms will have a well-marked surface or zone of maximum flooding that marks the top of the TST and base of the overlying HST.

*Aggrading and prograding platforms.* During the development of such platforms, relative sea-level rise provides additional space for shallow-water carbonate production. If production is equal to or faster than the rate of increase in accommodation space, the platform will either aggrade, or prograde into the basin (Figure 12.6b). A good example is the Miocene platform of Mallorca (see Section 13.1). In these cases, the shoreline also progrades and no flooding surfaces are developed during relative sea-level rises. If both the platform margin and the shoreline prograde during relative sea-level rise, then the facies will show a shallowing-upward trend (i.e. be regressive) in vertical section. This creates the unfortunate terminological confusion of regressive successions occurring in a TST. Because these platforms can produce sediment throughout relative sea-level

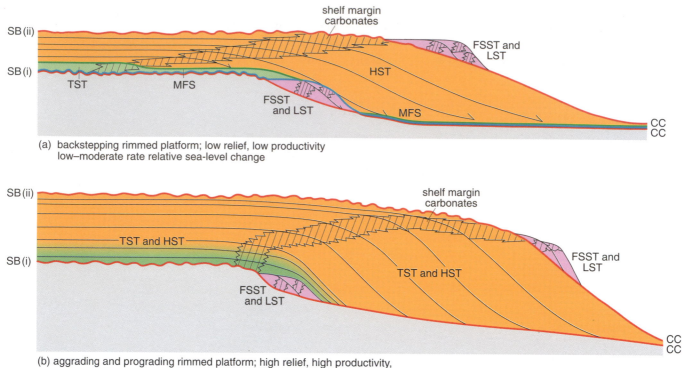

**Figure 12.6** Idealized depositional sequences developed on a rimmed carbonate shelf. Areas with oblique lines represent shelf edge facies (reefs or carbonate sands). FSST and LST are grouped together and both shown in purple. (a) Relatively low productivity platform subjected to low–moderate rates of relative sea-level change. (b) High productivity platform subjected to moderate to high rates of relative sea-level change. (a) Shows pronounced backstepping during deposition of the TST which may leave little, if any, TST on the flat-topped platform. (b) Aggradation and progradation characterize the TST, with sediment being shed basinward to form prograding clinoforms. The high productivity makes the TST and HST hard to separate. CC = correlative conformity. (*Wright and Burchette, 1996.*)

rise, there are large areas available for sediment production and a much thicker TST can be formed than is the case with backstepping platforms. Sediment produced during relative sea-level rise and high sea-level may be shed into the surrounding basin if wave and tidal currents are strong enough to continually sweep the platform top.

## 12.2.2 Highstand systems tract

When the platform top is flooded by shallow-marine waters, there is a very large area for the carbonate factory to occupy. It is at this time, therefore, that platforms are likely to produce the greatest amounts of reefal material, and bioclastic or ooid sand. Many platforms around the world today are approaching this stage, including South Florida and the Great Bahama Bank. As the rate of relative sea-level change is decreasing when maximum sea-level is approached, accommodation space will be rapidly filled so that sediment production will outstrip available space for sediment deposition on the platform top. The net effect of this relationship is that shorelines, back reef/lagoonal banks and islands, platform margin reefs and shoals, and platform slopes will all prograde to infill adjacent accommodation space situated at the platform margin (Figures 11.4, 12.6).

Highstand systems tracts will, therefore, be characterized by prograding clinoforms and regressive successions (Figure 12.6). If the carbonate environments and resultant facies have backstepped during the previous rapid rise in relative sea-level, then a maximum flooding surface may be recognizable (Figure 12.6a). This may coincide with the most open marine or pelagic facies in a vertical section from an offshore setting. If present, this may be used to define the base of the HST. However, as we have just seen, carbonate production commonly exceeds the rate at which accommodation space is created during relative sea-level rise in which case the maximum flooding surface is poorly developed so that the TST/HST boundary cannot easily be identified (e.g. Figure 12.6b).

Aggradational and progradational stratigraphical geometries characterize platform margin and platform slope stratigraphy. The HST comprises a complete range of carbonate facies from peritidal, through lagoonal, back-margin, margin and slope facies.

If the platform backstepped during relative sea-level rise, then HST deposits will prograde over the surface or zone of maximum flooding with slope strata downlapping the open marine, pelagic or condensed (hardground) facies of the TST (Figure 12.6a). This overproduction of carbonate systems during platform flooding causes highstand shedding, as we have seen. The resultant deposits are easily recognized in ancient successions as gravity and density flow deposits with coarse-grained clasts of shallow-water origin (e.g. algae, ooids, coral fragments, large benthic foraminifers). Fine-grained, lagoonal, carbonate muds may be shed considerable distances off the platform to form wedges of periplatform oozes (which become micrites on burial) corresponding to periods of high relative sea-level.

The HST therefore usually contains the maximum amount of potentially porous shelf margin and slope facies that have the potential to become hydrocarbon reservoirs. Not only are these facies more likely to preserve their original interparticle porosity but their shallow-water location means they are likely to be subaerially exposed during any subsequent sea-level fall. In this case, the platform margin and platform top facies will be capped by a sequence boundary. The subaerial exposure will further enhance the reservoir properties of the shallow-water facies through dissolution of calcium carbonate by meteoric waters.

### 12.2.3   Falling stage and lowstand systems tracts

Because of the morphology of rimmed carbonate platforms, falling sea-level will usually lead to the subaerial exposure of large areas of platform top carbonate facies. The new, lower, sea-level position on a steep rimmed margin will be located on the previous submarine slope which will now be a subaerial cliff undergoing mechanical and chemical erosion (Figure 12.6). Basin margin facies may include rock fall deposits, and beaches and fringing reefs may form on the steep slope. The shallow-water area available for carbonate production will be a small zone on the seaward margin of the platform whose size will be controlled by the angle of slope of the platform (Figure 12.1a). This area will decrease as the angle of slope increases. Supply of shallow-water skeletal and precipitated carbonate grains to the adjacent basin will be greatly reduced during falling and low relative sea-level (Figure 12.1). The exposed part of the platform will be affected by diagenetically aggressive ($CO_2$-rich) meteoric waters leading to dissolution and reprecipitation of calcium carbonate in the karst and cave systems of the developing sequence boundary. It is during these stages that large-scale secondary porosity and cementation can form in rimmed platforms. The platform will become lithified in a short space of time (a few tens of thousands of years) so that steep margins may undergo mechanical erosion to generate lowstand breccias.

### 12.2.4  Variations in rimmed platform sequence stratigraphy

The previous discussion is necessarily based on a schematic rimmed platform and many assumptions have had to be made in presenting a single idealized profile. Because of the complex controls (variations in sea-level history, sediment supply etc.), climatic settings and basinal histories, the reality is that each platform is unique in its details. However, the following main variations have been recognized, some of which have already been introduced.

#### Variations in sequence architecture with different subsidence rates

Sequences and sequence sets will develop or stack differently depending on background subsidence rates (Figure 12.3). Assuming constant carbonate production, then retrogradational platforms will form with high rates of subsidence, aggradational platforms with intermediate, and progradational platforms with low rates of subsidence.

#### Fault-block platforms

Faulted carbonate platform margins (Figure 12.7) will force the development of a rimmed carbonate platform, particularly on footwall areas to faults. If rates of fault movement outstrip rates of carbonate production, former shallow-water sites will be lowered beneath the zone of optimum carbonate production to form deep-water, hanging-wall sub-basins. Such steep topography means that progradation off the footwall onto the hanging-wall is limited and aggradational geometries form at the crest of the footwall (Figure 12.7b). Examples of this are seen in Miocene platforms developed on rotated fault blocks along the margin of the Gulf of Suez rift basin.

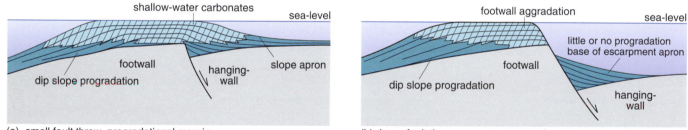
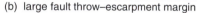

(a)  small fault throw–progradational margin

(b)  large fault throw–escarpment margin

**Figure 12.7**  The effect of normal faulting on carbonate platform development. Areas with oblique lines represent shelf edge facies (reefs or carbonate sands). An isolated platform is illustrated with (a) carbonate production rates greater than rates of fault movement so that the fault is covered by shallow-water carbonates; (b) fault movement greater than rate of carbonate production so that the hanging-wall area drops below the production zone and drowns. Sediment shed from the footwall may accumulate here, but the carbonate platform cannot prograde onto the hanging-wall. *(Wright and Burchette, 1996.)*

#### Isolated platforms and atolls

Carbonate platforms that are not attached to land show significant changes in geometry and facies as oceanographic conditions vary round the margins. Research on the isolated platforms of the Caribbean (including the Bahamas) has shown there is a very strong windward/leeward effect on platform sediment dynamics and preserved facies. Windward margins have low sediment supply as sediment is transported onto the platform and away from the margin (sequence above SB(i) on Figure 12.8). High-energy conditions result in erosion of the windward margin to form steep escarpments with rock fall and grain flow deposits at their base. Leeward margins are characterized by thick, laterally extensive, progradational sequences as material is shed from the platform top and margins (Figure 11.4 and sequence above SB(i) on Figure 12.8). Atolls surrounded by deep oceanic basins have steep slopes and little opportunity for progradation.

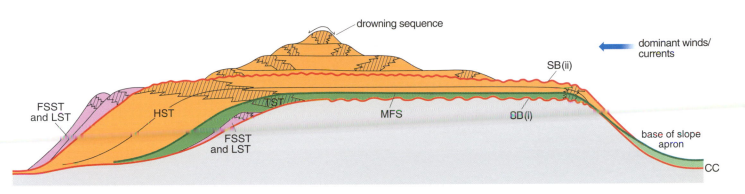

**Figure 12.8**  Sequence development of an isolated platform. The sequence above SB(i) developed during a low rate of relative sea-level rise: dominant winds and currents came from the right, causing a starved margin on the right, and progradation due to highstand shedding on the left. The sequence above SB(ii) developed during a relative sea-level rise that outpaced carbonate production, resulting in retrogradation and eventual drowning (CC = correlative conformity). The FSST and LST are grouped together and both shown in purple. *(Wright and Burchette, 1996.)*

### Drowned platforms

○  Given the rates of carbonate production and changes in accommodation space (Figure 11.2), what conditions are likely to give rise to platform drowning?

●  Platform drowning can result from high subsidence rates or emergence followed by subsidence, and reduction in carbonate production caused by a range of environmental changes such as influx of siliciclastic sediments, low oxygen levels, high or low temperatures, or nutrient excess.

Drowning may be reflected in the later stages of platform stratigraphy as progressively smaller areas of the platform keep pace with sea-level rise, giving a tiered wedding cake pattern (sequence above SB(ii) on Figure 12.8). This results from progressively smaller areas of carbonate production which then, in turn, increase the likelihood of eventual drowning. Drowned platforms are often capped by a small pinnacle that represents the last site to keep pace with relative sea-level rise.

## 12.3  Sequence stratigraphy of carbonate ramps

The different morphology of ramps means that they respond to relative sea-level changes in a different fashion from rimmed shelves. In particular, the lower relief of ramps leads to fewer differences in facies between HSTs, FSSTs and LSTs. This means that a rapidly prograding wedge of carbonate sediment may develop during the late part of the relative sea-level rise, and that progradation may continue at a reduced rate during a relative sea-level fall. During high rates of relative sea-level rise, the low relief of the ramp means it will be rapidly flooded so that the TST is thin and represented in outer ramp settings by condensed sections (Figure 12.9 overleaf).

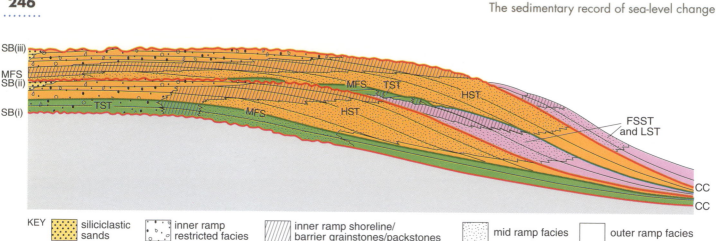

**Figure 12.9** Sequence stratigraphical development of carbonate ramp through two depositional sequences bounded by sequence boundaries (SB(i), SB(ii) and SB(iii)). Colours indicate sequence stratigraphical interpretation except for siliciclastic sands. The lower sequence just above SBi develops in response to relatively high sediment production caused by a steady increase in available accommodation space. Strata therefore aggrade during transgression, but during the highstand, when accommodation space is limited, a large amount of progradation takes place. Note the inner ramp facies tracks relative sea-level rise. Relative sea-level fall results in the formation of an erosive sequence boundary (SB(ii)) in proximal areas, but the boundary passes offshore into a correlative conformity (CC). The falling and lowstand stage prograding wedges of sediment (FSST and LST) develop in response to sediment erosion and redeposition downslope and carbonate production in shallow-water areas during a relative fall and low sea-level. The upper sequence has a relatively lower rate of carbonate sediment production and retrogrades during transgression so that in some places the maximum flooding surface (MFS) is superimposed on the sequence boundary. A large amount of progradation takes place in the highstand and low rates of sediment supply mean that the FSST and LST is a wedge of sediment that is detached from the highstand deposits of the inner ramp. *(Wright and Burchette, 1996.)*

The other main difference between ramps and rimmed platforms is that ramps do not have the highly productive carbonate communities such as fringing and barrier reefs that characterize rimmed carbonate platforms. The few modern-day examples that have been studied suggest that ramps have lower carbonate production rates than rimmed platforms and that there is less of a difference between production in shallow and deep waters.

### 12.3.1 Transgressive systems tract

During relative sea-level rise over the gently sloping topography of a ramp, the shoreline and associated facies belts will migrate large distances landward (Figure 12.9 sequence above SB(ii)). In most cases, the rate of relative sea-level rise outstrips the rate of carbonate production so that a retrogradational parasequence set is deposited. This results in deepening-upward profiles with muddier facies in inner ramp settings, and condensed beds of shelly concentrations, hardgrounds or pelagic facies may be found in mid to outer ramp locations. Inner ramp settings may show subaerial alteration of parasequence tops, depending on the climatic conditions. Inner ramp facies are commonly dominated by retrogradationally stacked barrier or shoreline packstones and grainstones. These will be overlain by marine flooding surfaces.

Some examples are known of vertically aggrading TST on ramps (Figure 12.9 sequence above SB(i)) but most workers have reported backstepping or retrogradation (Figure 12.9 sequence above SB(ii)). There are no reports of ramps prograding during relative sea-level rise. This probably reflects a combination of lower carbonate production rates in ramp systems together with the fact that subsidence is commonly higher in offshore, basinal areas. Such differential subsidence helps maintain the ramp profile through time. A surface or zone of maximum flooding should therefore be recognizable within a ramp stratigraphy and depending on the rate of flooding this may occur close to, or even directly overlie, the underlying sequence boundary in the more distal areas. In the latter case, the TST is only locally developed (Figure 12.9 sequence above SB(ii)).

## 12.3.2   Highstand systems tract

During the decreasing rates of relative sea-level rise characteristic of the period of HST deposition, accommodation space will become limited and will normally be outpaced by sediment production. This results in the development of progradational geometries of inner, mid and outer ramp facies often as parasequences (Figure 12.9). The low angle of slope, and the lack of any confining steep slope on ramps, results in progradation taking place across large distances. This progradation will take place over the surface or zone of maximum flooding to form a low-angle downlap surface. Because of the small angle of depositional slope, this surface may not be recognizable as a change in dip of the beds in the field but may be picked out by the change from the most open-marine facies (maximum flooding) to a shallowing-upward succession or the change from a retrogradational or aggradational parasequence set to progradational parasequence set (Figure 12.9). HST progradation leads to the potential for steepening the ramp slope. Depending on the efficiency of wave and tidal currents to redistribute sediment on the ramp, a margin or slope break may develop.

As accommodation space decreases, the upper surface of the HST in updip sections will be eroded and an unconformity representing the sequence boundary will form (Figure 12.9). Depending on the climatic setting, there may be alteration of the HST deposits by meteoric waters. Potentially, the extensive progradational inner ramp packstones and grainstones will become leached and develop a secondary porosity and/or the sediments will become cemented and lithified. At the more downslope and offshore sections, the unconformity will grade into a correlative conformity marked by evidence of shallowing (e.g. occurrence of clay-rich layers or the formation of submarine hardgrounds) or in very deep waters there may be no evidence of low or falling relative sea-level.

## 12.3.3   Falling stage and lowstand systems tracts

Relative sea-level fall on a ramp will expose updip sections that will develop an unconformity and the locus of sedimentation will shift considerable distances downslope. However, a similar depositional profile will be maintained so that carbonate production and sedimentation, and therefore facies, will be similar to those that occur during other phases of ramp history. This means that downstepping wedges of sediment can easily form on a ramp during a forced regression (Figure 12.9 sequence above SB(ii)). The facies would be expected to be broadly similar to those of the previous HST unless the basin is a barred one in which case restricted facies (i.e. organic-rich mudrocks or evaporites) may form as FSST and LST deposits.

FST and LST comprising breccias or debris flows are not generally found on ramps unless the submarine slope is distally steepened by faulting or an inherited steep slope as discussed below.

## 12.3.4   Variations in carbonate ramp sequence stratigraphy

The main variations on the idealized ramp carbonate sequence stratigraphical model are shown in Figure 12.10 (overleaf) and described below.

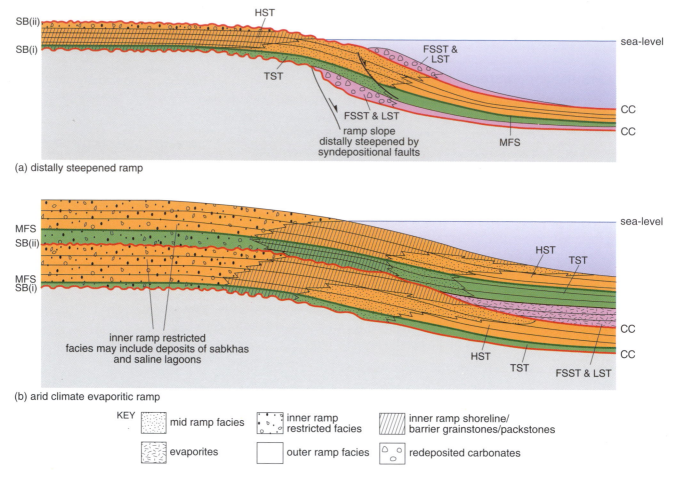

(a) distally steepened ramp

(b) arid climate evaporitic ramp

KEY

mid ramp facies    inner ramp restricted facies    inner ramp shoreline/barrier grainstones/packstones

evaporites    outer ramp facies    redeposited carbonates

**Figure 12.10** Diagrams illustrating variations in the sequence stratigraphy of carbonate ramps: (a) distally steepened ramp; (b) arid climate (evaporitic) ramp.

## Distally steepened ramps

Distally steepened ramps develop where synsedimentary faults cut through to form sea-floor escarpments or when a steep topographic slope is inherited from a previous shelf margin, or when a steep slope forms due to HST progradation (Section 12.3.2). During transgression and relative sea-level highstands, such offshore slopes will become blanketed by outer ramp or pelagic muds unless fault movement is active at this stage and fault escarpment debris flows or breccias form. During low relative sea-level, the steep slope may be affected by higher energy waters or emergence, resulting in localized rock fall or debris flows along the base of slope.

## Arid climate (evaporitic) ramps

In an arid climatic setting, during aggradation and progradation there is the potential for development of extensive back-barrier saline lagoons and mudflats or sabkhas. During periods of lowered sea-level ramps occurring in a semi-barred basin in an arid climate setting such as the Arabian Gulf (Section 11.4), a basinwide evaporite LST could develop (Figure 12.10b). This would be dependent on the balance between the rate of inflow of river waters at the head of the gulf, influx of normal salinity seawater from the ocean, and evaporation rate. The basin could be rapidly infilled with evaporite salts which would also onlap the surrounding carbonates of the previous HST.

### Cool-water carbonate ramps

Possibly because of their relatively low rates of carbonate production, cool-water carbonate factories all have a ramp or distally steepened ramp profiles. Examples studied come mainly from the Cainozoic of the Southern Ocean and the present-day north-east Atlantic margins (Figure 11.1). TSTs and HSTs are predominantly composed of coarse-grained bioclastic sands and gravels affected by wave and storm activity, and associated large-scale dune bedforms are found down to several hundred metres water depth. Glacio-eustatically driven lows, associated with colder climates, can result in the deposition of shell pavements of colder-water faunas or interbedded periglacial deposits such as tills and dropstone facies.

○   What sequence stratigraphical geometries would you expect to develop in cool-water ramps?

●   Sediment geometries should be similar to those illustrated for sequence above SB(ii) in Figure 12.9 with thin progradational FSST and LST, retrograding TST and prograding HST, because of the relatively low rates of carbonate sediment supply. Siliciclastic sediments might also be expected to be deposited during falling or low relative sea-level because of the low rates of production in cool-water carbonates.

## 12.4   Numerical stratigraphical modelling

The numerical modelling of carbonate successions was developed during the 1990s and is proving an exciting new method of predictive stratigraphical analysis. A number of different computer programs are currently in use in the oil industry and academia for predicting subsurface stratigraphy and for analysing the different controls on sediment accumulation. Most of these programs work by the user entering values for the time and spatial framework, and values for the different interpreted sea-level histories for the simulations. Rates for the various sedimentary processes are also entered, such as carbonate production, sediment erosion and transport that are considered to be appropriate for the section being studied (Table 12.1 overleaf). Algorithms within the computer program calculate the sediment production, redistribution and deposition for a series of time steps. The simulated stratigraphy is plotted as a series of sediment surfaces (representing sedimentary geometries) and predictions of facies, based on the depth of deposition of the unit (Figure 12.11a–d, p. 252) or the depositional process that is simulated (Figure 12.11e–h, p. 253). Displays can be as two-dimensional height/length sections through time (Figure 12.11) or three-dimensional. Many programs will also plot out borehole sections for selected parts of the stratigraphy.

By simulating stratigraphical sequences from a number of user-defined variables, whose accuracy will vary on the sections being studied, stratigraphical modelling enables geologists to test different hypotheses concerning the processes that control the accumulation of stratigraphical sequences, test different sequence stratigraphical interpretations, and reconstruct unknown parts of stratigraphical sections in an objective fashion.

New developments include 3-D modelling which is clearly more realistic for carbonate depositional systems such as isolated carbonate platforms and atolls.

A 2-D program is used here that has been developed by David Waltham at the Department of Geology, Royal Holloway University of London (Sedtec2000), parts of which can be accessed at the Royal Holloway website.

The modelling is partly empirical in that data from the study of Holocene carbonate depositional systems are used to obtain the rates for the different parameters controlling platform growth, and partly conceptual in that concepts and algorithms have been developed to simulate, for example, the erosion, transport and redeposition of carbonate sediment. Section 12.4.1 summarizes the main controlling parameters that can be altered in the program and Section 12.4.2 describes how it can be used to simulate the hypothetical stratigraphy of various carbonate settings.

## 12.4.1 Controlling parameters

Rates of *carbonate production* were introduced in Figure 11.2, which showed the differences between warm-water, cool-water and pelagic carbonate factories. If we take the Cainozoic warm-water factory as an example, then sea-floor or benthic production is greatest in shallow waters (Figure 12.1). Production decreases with depth and also with respect to water restriction from the open water shelf margin, to back reef areas and then into lagoons (Table 12.1).

○ Based on information on the South Florida carbonate shelf (Section 11.3.2), recall the main carbonate-sediment producing organisms in order to account for the reduction of production rates from reef to back reef areas.

● Reef communities are dominated by abundant corals and encrusting coralline algae which are both abundant and have high growth rates. Back reef areas are populated by the more sparse, or slower-growing, or more lightly calcified organisms such as molluscs, foraminifers and calcified green algae, and therefore production rates are lower.

The Sedtec2000 program allows different production rates to be entered for platform interior environments that produce finer-grained material and for platform margin environments that produce coarser-grained material. For each time step during the running of the program, the appropriate amount of sediment is added to the sea-floor for shallow-water areas and deep-water areas to receive fall-out of pelagic sediment at the appropriate rates (Table 12.1).

*Erosion rates* are important for the erosion and redistribution of sediment in shallow-water sites and for the subaerial erosion of material during exposure in periods of relative sea-level fall. Subaerial erosion is very variable and depends on climate and soil type. Such rates are obtained by measuring short-term rates in a variety of modern carbonate environments; the range of modern rates are shown in Table 12.1. Submarine erosion takes place in the program down to a user-definable wave-base. Sediment is then redeposited if there is accommodation space available nearby or it is transported basinwards to the nearest available accommodation space. Coarse-grained (platform margin-derived) and fine-grained (platform interior-derived) sediment is transported across different distances according to a user-defined 'transport distance' which is the distance at which half the sediment in the water column is deposited. The total amount of sediment deposited after each user-defined time step is plotted as a black line (Figure 12.11).

**Table 12.1**   Process rates commonly used for modelling carbonate platform stratigraphy. Rates in metres per thousand years (m/ka).

| Process | Average (m/ka) | Minimum (m/ka) | Maximum (m/ka) |
|---|---|---|---|
| Benthic production (max. value decreasing with depth): | | | |
| reef | 2.0 | 0.3 | 6.0 |
| back-reef | 0.3 | 0.1 | 0.5 |
| lagoon | 0.2 | 0.01 | 0.2 |
| Pelagic production | 0.05 | 0.01 | 0.1 |
| Erosion: | | | |
| subaerial | 0.5 | 0.01 | 1.0 |
| submarine (max. value decreasing with depth) | 2.0 | 0.1 | 5.0 |

The time lines are taken as proxies for the stratigraphical geometries within the platform, and they are good proxies to the parasequences. Facies can be plotted either as depth zones based on the water depth at the time of deposition during the program run (Figure 12.11a–d overleaf) or as the main process by which the program generated or deposited the sediment (e.g. 'platform margin carbonates' resulting from *in situ* benthic production, 'pelagic carbonates' and 'reworked platform margin carbonates', Figure 12.11e–h). Platform interior carbonates that have been reworked and redeposited downslope retain their 'platform interior carbonates' colour (green on Figure 12.11e–h).

*Sea-level changes* are either entered as a number of superimposed sinusoidal curves with definable amplitudes and frequencies (Figure 12.11) or a more irregular curve can be entered by the user. Linear rates of rises and falls (Figure 12.11b–c, f–g) can also be entered on their own or superimposed on the curves mentioned above.

### 12.4.2   Modelling carbonate platform stratigraphy

Simulated cross-sections through carbonate platforms generated by Sedtec2000 (Figure 12.11) illustrate some of the sequence stratigraphical principles discussed in this Chapter and demonstrate the effects of varying such controlling parameters as sea-level and carbonate production rate.

○ In Figure 12.11a and e, relative sea-level remains constant but the platform develops from a gently sloping ramp profile to a more steeply sloping rimmed shelf profile. What is the likely cause of this and what stratigraphical geometries are simulated?

● As relative sea-level is constant, the only accommodation space available in shallow-water areas is basinward of the slope. Because more sediment is produced than can be accommodated on this shallow slope, sediment is deposited in the more basinward areas immediately adjacent to the slope and the slope steepens as this space is filled. Sediment from the highly productive shallow-water benthic communities fills all the space available. Therefore the slope will steepen through time to develop a flat-topped rimmed platform. Deep-water areas accumulate equal thicknesses of pelagic sediment produced in upper levels of the open-ocean waters. Progradation is the most obvious geometry and the upper surface of the platform shows toplap and the platform slope clinoforms build out over deep-water pelagic deposits (dark blue areas in Figure 12.11e between the closely spaced black time lines).

**Figure 12.11** Computer-generated profiles of carbonate platforms from Sedtec 2000 showing the stratigraphy resulting from different types of sea-level change. All sediment surfaces are plotted on the same initial surface and run for the same arbitrary time period of 2 to 1.75 Ma. This 0.25 Ma period has been arbitrarily subdivided into 13 equal blocks of time thus the black chronostratigraphical lines (flooding surfaces) on (a) to (h) are every 19.23 ka. Simulations (a)–(d) show sediment deposited with respect to depth of deposition. (a) Sea-level at stillstand throughout run with average values (see Table 12.1) set for production, erosion and sediment deposited with respect to depth of deposition. (b) Sea-level rise of 0.1 m/ka and other values as for (a). (c) Sea-level falling at 0.1 m/ka and other values as for (a). (d) Cyclic sea-level change (10 m amplitude with 100 ka frequency and 5 m amplitude with 20 ka frequency) superimposed on linear rise, other values as in run (a). The sea-level curve used for (d) and (h) is shown in the bottom right-hand corner of each panel. The red time lines are plotted every 19.23 ka and correspond to the black chronostratigraphical lines on (d) and (h). (e)–(h) as for (a)–(d) except that stratigraphy is displayed with respect to depositional processes (see key). No siliciclastics were introduced in these runs. Note that in (e)–(h) reworked platform interior carbonate (green) retains its colour when redeposited downslope from the platform. *(Dan Bosence and Dave Waltham, Royal Holloway University of London.)*

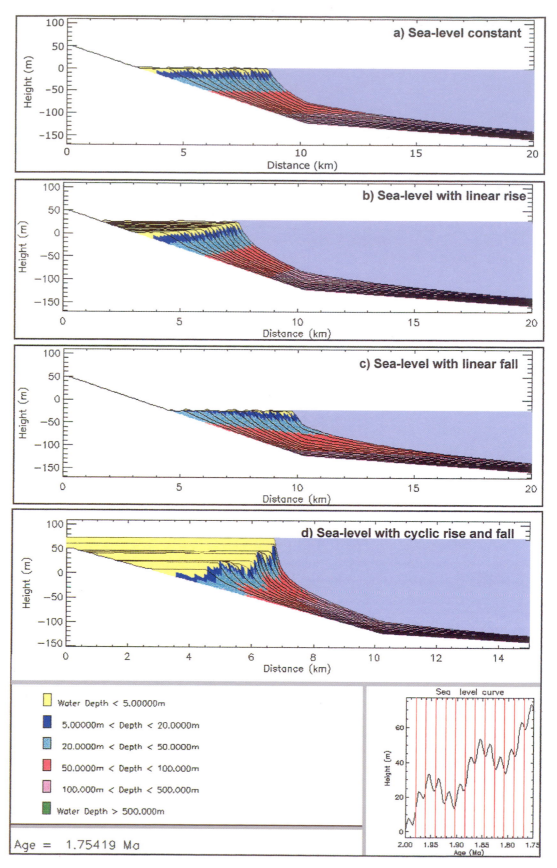

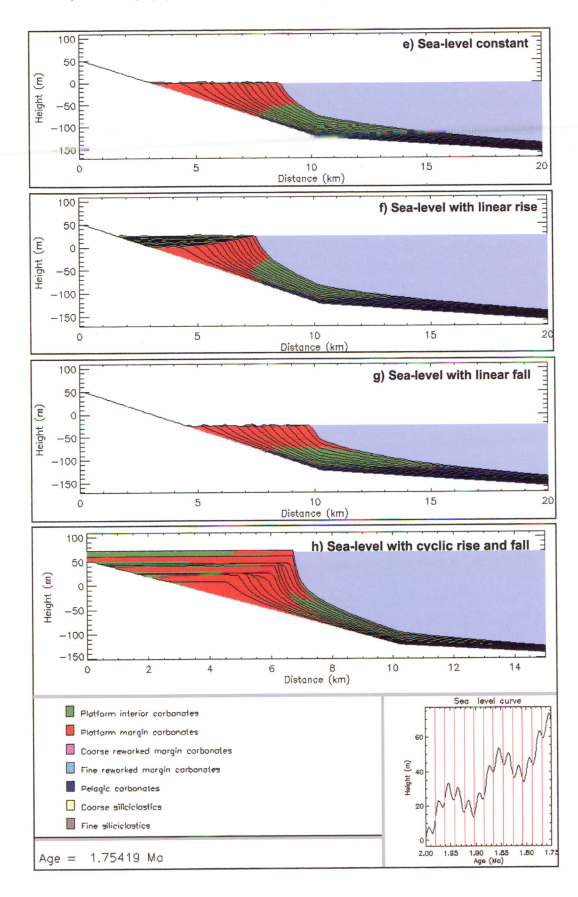

○ The simulation shown in Figure 12.11b and f shows the result of relative sea-level rising at a linear rate, so that the shoreline is transgressing landward up the initial surface. However, a vertical section drilled near the margin of the carbonate platform at 7 km on the scale bar would record a shallowing-upward section. How can this be explained?

● Relative sea-level rise causes onlap of the shoreline along the initial surface but as carbonate production rate is faster than the rate of relative sea-level rise accommodation space is continually filled and the carbonate platform has prograded as well as aggraded. Progradation of the platform margin results in a shallowing-upward or regressive succession whilst the landward shift of the shoreline indicates a transgression.

○ In Figure 12.11c and g, relative sea-level is falling at a linear rate. What stratigraphical geometries are developed and what were the likely fates of the subaerially exposed carbonate?

● The earliest stratigraphical units show erosional truncation of their upper surfaces (cf. Figure 12.11c and g, with 12.11a and b, which have toplap and preserve the shallowest-water facies) and the lower surfaces of the clinoforms build out over pelagic deposits during deposition of this FSST. In addition, all of the carbonate platform facies belts are deposited at progressively lower levels as sea-level falls. The likely fate of the subaerially exposed carbonate is that it has been partially dissolved away and partly physically eroded and redeposited downslope.

Figure 12.11d and h illustrates the stratigraphy that develops when cycles of sea-level change are introduced that are within the Milankovich band (100 ka and 20 ka frequency) and superimposed on a linear relative sea-level rise. The initial TST aggrades and progrades, and the HST progrades in response to the change from rapidly rising sea-level to high sea-level. The subsequent sea-level fall from 1.95–1.9 Ma erodes the previous HST to generate a sequence boundary that passes basinward to a correlative conformity. Subsequent sea-level rise from 1.9–1.85 Ma produces onlap onto the eroded sequence boundary and transgression over the earlier platform to produce a large new platform.

We shall use the numerical modelling in the next Chapter to explore the development of a Miocene rimmed shelf and a Jurassic carbonate ramp.

## 12.5  Summary

• Carbonate production is proportional to the area of flooded platform and so it is usually at its highest during high relative sea-level. If the accommodation space is filled, carbonate factories will shut down. However, space is often maintained on the top of the carbonate platform and production therefore continues because of wave and tide action sweeping sediments off the platform top and onto their margins. This causes the carbonate platform to prograde by 'highstand shedding'.

• During relative sea-level falls and the subsequent lows, carbonate platforms are often exposed to meteoric diagenesis as freshwater from rain and rivers percolates through the previous highstand deposits. This leads to dissolution of exposed carbonate and cementation of underlying units, which means that these deposits are resistant to erosion and little sediment is shed basinwards.

- Sediment partitioning occurs in mixed carbonate–siliciclastic, and carbonate–evaporite depositional systems. In the former, siliciclastic sediment supply is often only voluminous during falling and low relative sea-level whilst carbonates dominate in transgressive and highstand systems tracts. In carbonate–evaporite systems, deep-water basinal evaporites occur in falling stage and lowstand system tracts, and platform top shallow-water and sabkha evaporites occur within transgressive and highstand systems tracts.

- In many carbonate sequences that are not developed on a ramp, transgressive and maximum flooding surfaces cannot be identified because carbonate factories continually infill accommodation space during rising relative sea-level.

- Platform drowning may occur because rates of relative sea-level rise are too high for carbonate factories to keep up, or because environmental changes kill off, or prevent, carbonate factories from becoming established.

- Carbonate sequence development on ramps is more akin to that occurring in siliciclastic depositional systems. Transgressive systems tracts are characterized by retrogradational geometries and downstepping falling stage system tracts are often developed.

- Computer modelling of carbonate sequence development enables the controls responsible for different stratal architectures to be explored and quantified, and different sequence stratigraphical interpretations to be tested.

## 12.6  References

### Further reading

EMERY, D. AND MYERS, K. J. (eds) (1996) (see further reading list for Chapter 4). [Chapter 10 on 'Carbonate Systems' makes good follow-up reading to this book.]

MIALL, A. D. (1997) (see further reading list for Chapter 4).

WALKER, R. G. AND JAMES, N. P. (1992) (see further reading list for Chapter 11).

WRIGHT, V. P. AND BURCHETTE, T. P. (1996) (see further reading list for Chapter 11).

### Other references

BOSENCE, D. W. J. AND WALTHAM, D. A. (1990) 'Computer modeling the internal structure of carbonate platforms', *Geology*, **18**, 26–30. [An early paper dealing with numerical modelling.]

BOYLAN, A. L., WALTHAM, D. A., BOSENCE, D. W. J., BADENAS, B. AND AURELL, M. (2002) 'Digital rocks: Linking forward modelling to carbonate facies', *Basin Research*, **14**, 401–415.

BURCHETTE, T. P. AND WRIGHT, V. P. (1992) 'Carbonate ramp depositional systems', *Sedimentary Geology*, **79**, 3–57. [This was the first paper to develop sequence stratigraphical models for carbonate ramps.]

HARRIS, P. M., SALLER, A. H. AND SIMO, J. A. T. (1999) *Advances in carbonate sequence stratigraphy: application to reservoirs, outcrops and models*, Society of Economic Paleontologists and Mineralogists Special Publication No. 63, 421pp. [A number of detailed studies of different aspects of carbonate sequence stratigraphy. In particular, a discussion of the effects of compaction on sequence development, a number of papers on small-scale cycles and their stacking patterns and the likely controls on these features, and papers on the sequence stratigraphy of slope and basin carbonates.]

KENDALL, C. G. ST. C. AND SCHLAGER, W. (1981) 'Carbonates and relative changes in sea-level', *Marine Geology*, **44**, 181–212. [An early (before the development of sequence stratigraphy) but classic discussion of the responses of carbonate systems to relative sea-level change.]

LOUKS, R. G. AND SARG, J. F. (1993) *Carbonate sequence stratigraphy: recent developments and applications*, American Association of Petroleum Geologists Memoir No. 57, 545pp. [Models for rimmed shelves and ramps are discussed in the introductory article by Louks and Sarg and a number of case studies are presented.]

SCHLAGER, W. (1992) *Sedimentology and Sequence Stratigraphy of Reefs and Carbonate Platforms*, American Association of Petroleum Geologists Continuing Education Course Note Series No. 34, 71pp.

# 13 Application of sequence stratigraphical analysis to ancient carbonate platforms

*Dan W. J. Bosence*

This final Chapter consists of two case studies of ancient carbonate platforms showing contrasting morphologies and scales of development. The case studies illustrate how sequence stratigraphical analysis has significantly improved our understanding of the processes that contributed to carbonate platform development. Finally, the use of numerical modelling is discussed in constraining the rates of processes involved in carbonate platform growth and testing different depositional models.

## 13.1 Case Study 1: Miocene of Mallorca

This case study shows how Luis Pomar from the University of Palma, Mallorca, used sequence stratigraphical concepts to analyse a Miocene succession from south-west Mallorca, Spain, and the results of further collaborative work between Luis Pomar, Dave Waltham, T. Lankester and Dan Bosence using numerical stratigraphical modelling.

### 13.1.1 Setting

During the Late Miocene, reef-rimmed carbonate platforms fringed many of the islands and margins of the Mediterranean Basin. In Mallorca, the Llucmajor Platform prograded some 20 km south-westwards away from its basement in the core of the island (Figure 13.1a overleaf).

Cliff sections near Cap Blanc in the south-west of the island provide near-continuous cross-sections through the outer part of this platform at an orientation more or less parallel to the direction of progradation and normal to the original reef-rimmed margin. The cliff sections (Figure 13.1b) do not show any internal deformation or tilting so that the original depositional slopes of the platform margin are preserved within the stratigraphy.

### 13.1.2 Facies and palaeoenvironments

Four groups of facies occur within these late Tortonian–Messinian aged carbonates. The facies are interpreted to have formed in lagoonal, reef-core, reef-slope and open-shelf environments (Table 13.1, p. 260). The environmental interpretation is considerably assisted by the continuous cliff exposures which make it possible to trace facies and surfaces laterally and along original depositional slopes from one palaeoenvironment to another (Figures 13.1b, 13.2 overleaf). These rocks were originally deposited as limestones but were fairly rapidly transformed to dolomite at the end of the Miocene (late Messinian). This was caused by a relative sea-level fall which closed the Gibraltar Strait and isolated the Mediterranean from the Atlantic. This isolation of the Mediterranean Sea led to evaporation, desiccation and evaporite formation in the basin together with the movement of Mg-rich fluids causing dolomitization of marginal carbonate platforms. This major event is represented in the Cap Blanc stratigraphy by the irregular (sometimes karstic) unconformity cutting into the Miocene platform that is overlain by post-Miocene limestones (Figure 13.1b).

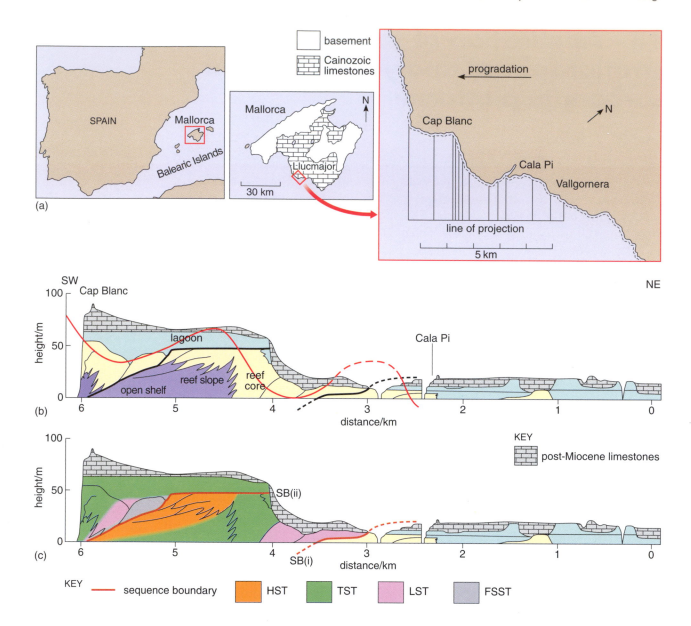

**Figure 13.1**  Miocene carbonate platform in Mallorca: (a) location maps; (b) and (c) composite stratigraphical cross-section constructed from cliff exposures between Vallgornera and Cap Blanc. (b) Facies: these are colour coded the same as Figures 13.3 and 13.4. The sinusoidal-shaped red curve in (b) shows how the height of the reef crest facies (measured or inferred) has varied through the section and therefore through time. This approximates to a relative sea-level curve. (c) Sequence stratigraphical interpretation of the left-hand part of (b); (b) and (c) are vertically exaggerated by ×10. Height is above present-day sea-level. ((a) and (b) Pomar, 1993.)

(a)

(b)

(c)

(d)

(e)

**Figure 13.2** Features of the carbonate platform in Mallorca: (a) cliff sections (*c.* 70 m high) at Cap Blanc showing low angle progradational geometries inclined to the south-west (left); (b) shallow reef-core with preserved *in situ*, dome-shaped coral and surrounding coarse-grained skeletal rudstone (view *c.* 50 cm across); (c) coral reef boundstone formed of stick-like branches of coral *Porites* (view *c.* 50 cm across); (d) lower reef slope with dish-shaped corals (*Porites*) surrounded by reef debris of skeletal rudstone (view *c.* 70 cm across); (e) reef flat grainstone with large (*c.* 10 cm across) rhodolith (coralline algal nodule) with radial growth structure. Compare with Caribbean coral reef zones (Figures 11.11, 11.12). The terms boundstone, rudstone and bafflestone are explained in Box 11.1. (*(a)–(e) Dan Bosence, Royal Holloway University of London.*)

**Table 13.1** Four main facies that occur in the Llucmajor Carbonate Platform, Mallorca. For illustrations of the facies, see Figure 13.2.

| Facies | Subdivisions | Lithologies | Palaeoenvironments |
|---|---|---|---|
| Lagoonal | Outer | Bioturbated, skeletal, grainstone–packstones | Horizontally bedded outer lagoon floor |
|  |  | Coral boundstones | Patch reefs |
|  | Inner | Bioturbated, skeletal, grainstone–packstones and mudstones | Inner lagoon floor, carbonate shoreline |
|  |  | Stromatolites and calcretes | Supratidal–subaerial |
| Reef-core |  | Coral–*Porites* and crustose coralline algae boundstone with coral zones:<br><br>massive & columnar (Fig. 13.2b)<br>branching (Fig.13.2c)<br>dish-shaped (Fig. 13.2d) | Barrier reef with coral depth zonation in reef framework |
|  |  | Coral rudstones; skeletal packstone–grainstone (Fig. 13.2e) | Associated bioclastic reef debris |
| Reef-slope proximal | Proximal | Skeletal and intraclastic grainstone–packstones, rudstones and floatstones | Proximal reef slope talus and debris sloping at <20° |
|  | Distal | Burrowed skeletal packstone–grainstones coralline algal biostromes | Gently sloping distal reef slope debris and coralline algal pavements |
| Open-shelf |  | Fine-grained bioturbated skeletal packstones and wackestones rich in planktonic foraminifers | Flat-lying open marine shelf deposits with pelagic input |

### 13.1.3  Sequence stratigraphy

The facies are generally arranged into sigmoidally shaped packages. Each package is equivalent in scale and origin to parasequences (Chapter 4 and Section 12.1.3) which are interpreted to represent the basic accretionary units of the carbonate platform stratigraphy (Figure 13.3a). Each parasequence comprises a section of reef growth, deposition of lagoonal sediments, shedding of reef debris onto the reef slope and open marine carbonate shelf deposition (Figure 13.3a). These parasequences are metre-scale in thickness and are separated by minor erosion surfaces, and because of their distinctive shape were originally termed 'sigmoids' by Pomar. Each parasequence is interpreted to represent one cycle of accumulation and erosion in response to a high-frequency sea-level change. Figure 13.3b–d shows how the parasequences can be stacked in different patterns in response to superimposed lower-frequency linear and higher-frequency cyclic sea-level changes.

The stratigraphy exposed in the Cap Blanc section shows phases of aggradational and progradational stacking separated by downlapping sections that are associated with sequence boundaries (SB(i) and SB(ii), Figure 13.1b). The sequence boundaries are interpreted to represent times when the platform top was subaerially exposed and eroded whilst reefs re-established themselves at lower levels on a downlap surface, as seen in the lower part of the cliffs at Cap Blanc (Figure 13.1b). Because the reefs preserve much of their original morphology and

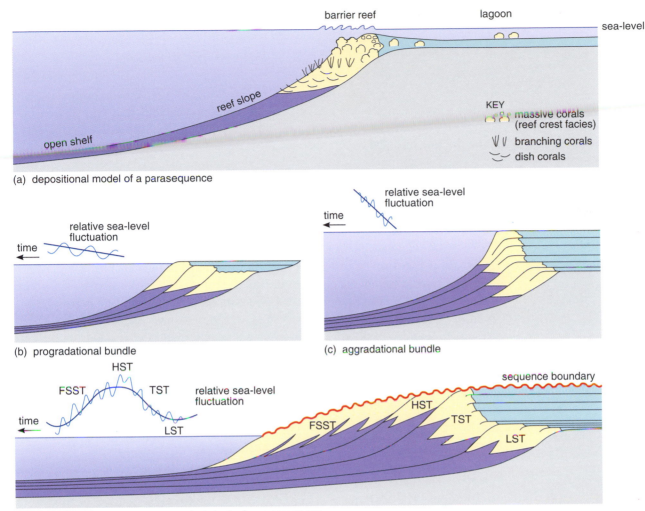

(a) depositional model of a parasequence

(b) progradational bundle

(c) aggradational bundle

(d) depositional sequence composed of stacked parasequences

**Figure 13.3** Miocene platform in Mallorca showing the sigmoidal-shaped units or parasequences, their stacking patterns and sequence stratigraphical interpretation. (a) The parasequence is the basic building block of the accretionary Cap Blanc reef platform. (b) Progradational parasequence sets develop in response to high-frequency relative sea-level fluctuation superimposed on a gradual rise in relative sea-level whilst (c) indicates a progradational to aggradational parasequence set forming in response to high-frequency cycles during a phase of more rapid overall relative sea-level rise. (d) Indicates parasequence sets formed in response to high-frequency cycles superimposed on a lower-frequency cycle of relative sea-level change. Note that the parasequences forming during the phase of falling sea-level have their upper surfaces truncated and are preserved as a wedge-shaped unit rather than a complete sigmoidal shape. (Pomar, 1993.)

zonation, it is possible to use the shallowest coral reef zone (massive *Porites* corals) as a proxy for sea-level. Similarly, if the preserved reef crests are mapped along the cliff sections, then the height variation along this mapped line ('reef crest curve') provides the amplitude of sea-level fluctuation (Figure 13.1b) but not the frequency.

The Cala Pi to Cap Blanc sections provide evidence of two main cycles of reef growth. There is a possible third one to the north-east that is topographically too low to be exposed above present-day sea-level. Because there is a proxy for sea-level preserved within these sections (the reef crest curve), the sequence stratigraphy can be analysed in some detail. Each part of the succession is built up from a series of laterally and vertically stacked parasequences. The sequence boundary (SB(i)) at Cala Pi (Figure 13.1c) represents the base of the well-exposed part of the stratigraphy and our sequence stratigraphical interpretation. This represents a time when the platform top was being eroded and reefs re-established themselves in the subsequent LST several tens of metres lower down this erosion surface. These reefs are exposed just above present-day sea-level at 3.75 km along the scale bar (Figure 13.1b,c). These lowstand reefs prograded for about 0.5 km before they show a phase of aggradational stacking (at 4.25 km) indicating a fluctuating but overall a higher magnitude sea-level rise (cf. Figures 13.1b,c and 13.3c). This aggradational phase represents the TST, but examination of the stratigraphy indicates that the facies show a shallowing-upwards. Therefore, carbonate production from reef growth was outstripping the rate of sea-level rise during deposition of the TST.

Between *c.* 4.5 and 5 km in Figure 13.1b, the reefs show progradational stacking (cf. Figures 13.1b, 13.3b) but the upper parts of the reef have been eroded away by the subsequent sea-level fall and sequence boundary (SB(ii)). This phase of progradation indicates that carbonate production from reef growth was greater than the rate of creation of accommodation space and it is interpreted as the HST. A FSST is recorded by downstepping reefs which re-established themselves fully during the subsequent low sea-level in the lower parts of the cliffs at Cap Blanc (at 6 km on Figure 13.1). Note that in the more offshore slope and open shelf facies, most of the sediment accumulates in the TST, and less in the HST. The FSST and LST are only represented by a thinly bedded condensed horizon in the more distal sections. If it is assumed that most of the sediment was generated in the shallow platform environment, then the shedding of sediment off the platform appears to have been most efficient during the transgressive phase, perhaps when moderate water depths and circulation were maintained. During the highstand of relative sea-level, less sediment was shed off the platform suggesting it had filled to sea-level and was slowing down as an active carbonate factory. During falling and low sea-level, very little sediment was generated and even relatively shallow-water sites are sediment starved. The final phase of platform growth at Cap Blanc is aggradational and progradational, and outer lagoon sediments can be seen to onlap the previous sequence boundary. This represents the last preserved TST of the succession.

This reef-rimmed platform represents an exceptionally clear example of a progradational and aggradational rimmed platform in that there are no retrogradational or successive maximum flooding surfaces developed in the stratigraphy. The Mallorca platform was one of the first platforms to be analysed using high-resolution sequence stratigraphical techniques and this showed that the earlier siliciclastic-based models could not be applied to such sections. It also shows how different orders of sea-level fluctuation may be superimposed to produce distinct stacking patterns in the preserved stratigraphy.

### 13.1.4 Numerical modelling of Cap Blanc section

Simulations of the stratigraphy between Cala Pi and Cap Blanc were undertaken to quantify and constrain the rates at which some of the processes that controlled the development of the stratigraphy operated. This enabled a test to be made about whether the cycles of changing sea-level might relate to high-frequency glacio-eustatic cycles. The modelling (Figure 13.4a,b) was constrained by a partially exposed sequence boundary which was taken as the initial surface for the simulation to start on (SB(i), Figures 13.1c, 13.4c). The position of the lithofacies on the cliff sections and geometries of the prograding and aggrading stratigraphy had to be matched by the simulations. Note that much of the stratigraphy to the north-east and the upper parts of the cliff sections have been removed by Plio-Pleistocene and later erosion (Figures 13.1, 13.4).

The 'reef crest curve' (Figure 13.1b) was used as a proxy for the major cycles of sea-level change seen in the cliff sections. However, there is no precise information on the frequency of these cycles, which are the largest-amplitude cycles used in the modelling. The frequency of these cycles is not known because they cannot be individually dated.

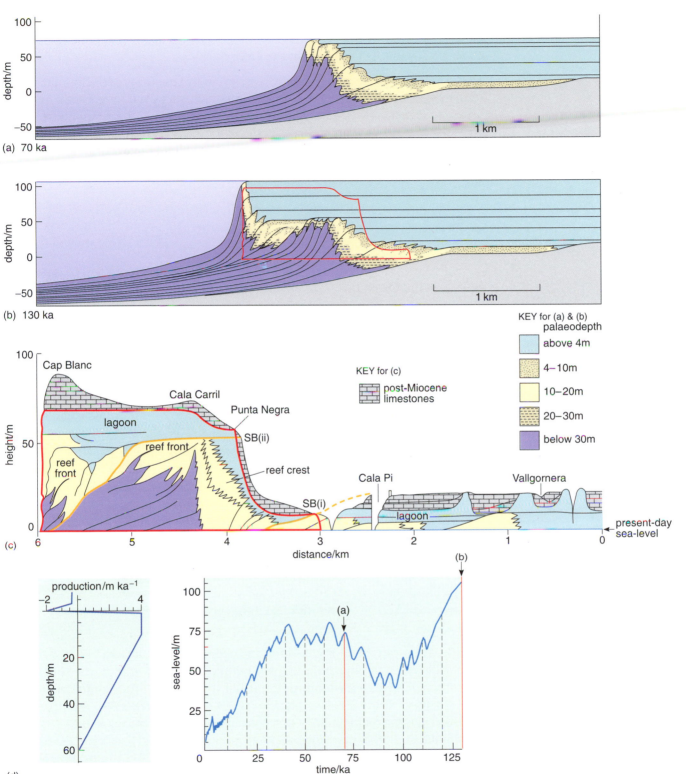

**Figure 13.4** Numerical modelling of Miocene platform in Mallorca: (a) and (b) represent two stages of development of the simulated platform (a) at 70 ka (i.e. 0–70 ka) and (b) at 130 ka (i.e. 0–130 ka) on the sea-level curve in (d). The vertical time lines on the sea-level curve correspond to the sediment surfaces plotted in the simulated stratigraphy in (a) and (b). The facies displayed in the stratigraphy represent the depth of deposition of the sediments and correspond to the different depth-related coral zones on the reef (cf. Figure 13.3a) down to the base of the dish-shaped corals at 30 m. Carbonate production rates used for the modelling are shown on the production/depth graph (d) with negative values used to simulate coastal and subaerial erosion through dissolution. (c) Shows the outcropping Miocene stratigraphy (cf. Figure 13.1). The red line on (b) and (c) encloses the same area of stratigraphy. *(Bosence et al., 1994.)*

Biostratigraphical data correlated with the Miocene radiometric timescale indicate that the whole of the Llucmajor Platform (Figure 13.1a) developed over about 1.9 Ma during the Tortonian and Messinian.

The exposures indicate that the long-term cyclicity (not that controlling the formation of the parasequences) is superimposed on an even longer-term cycle because the present-day altitude of the second of the two troughs on the 'reef crest curve' is about 40 m higher than the first trough. The field evidence for this is the difference in altitude of the reefs at 3.75 km along the horizontal scale compared with the second FSST/LST at 5.75 km along the scale on Figure 13.1b and c.

Superimposed on the main cycles are even higher-frequency cycles revealed by the parasequences discussed and illustrated earlier (Figure 13.3). These represent the highest order cyclicity present in the section and are superimposed on the main sea-level cycles for the modelling runs (Figure 13.4).

Sedimentary process rates (carbonate production, erosion, wave-base etc.) are unknown for the Miocene of Mallorca. Therefore, we used average values obtained from Recent Caribbean examples as shown in Table 12.1. The main unknown factor was the timing of the sea-level fluctuations because of the poor dating of the succession. The modelling showed that a fairly precise match of modelled stratigraphy to actual stratigraphy could be obtained if the main cycles on the 'reef crest curve' (Figure 13.1b) were assigned to *c.* 100 ka eccentricity cycles with higher order cycles superimposed (Figure 13.4d).

Two stages of the model run are shown from time zero to 70 ka (Figure 13.4a) when the first progradational and aggradational phases of reef growth occurred, and the second through to the end of the model run at 130 ka when the reefs developed through falling, low and then rising relative sea-level (Figure 13.4b) resulting in the deposition of the reefs below Cap Blanc (Figure 13.1). Figure 13.4c also shows the outcropping Miocene stratigraphy (cf. Figure 13.1b) so that the simulated and actual stratigraphy can be directly compared.

○ Using Figure 13.4b and c, compare the amounts of progradation, the amount of erosion beneath SB(ii) and the subsequent downstepping of the reef, and describe what the similarities and differences are between the simulated and real stratigraphical sections.

● The major phases of progradation and aggradation of the reef-rimmed margin can be traced through both profiles, but the simulated stratigraphy shows slightly less progradation than the actual stratigraphy. The amount of erosion of the reef during the main phase of sea-level fall above SB(ii) is correct. However, the reef gradually descends over forereef deposits in the simulation, rather than downstepping onto a condensed surface at the base of the cliff as seen at Cap Blanc.

One of the main results of the modelling is that the Holocene rates for carbonate production and erosion can be used to achieve a reasonably good match of real and simulated stratigraphy and that the duration of the main sea-level cycles are constrained to *c.* 100 ka eccentricity cycles. Another possible solution, and generating the same simulated stratigraphy, would be that the main cycles are *c.* 400 ka eccentricity cycles and that the maximum carbonate production rates were 1 m/ka instead of 4 m/ka. This shows how the modelling can be used to test in a consistent and quantitative way different hypotheses about the controlling processes on carbonate platform development.

## 13.2   Case Study 2: an Upper Jurassic ramp from NE Spain

The second case study based on the work of Marc Aurell of the University of Zaragoza describes and interprets a large-scale carbonate ramp from the Jurassic of north-east Spain. This study is also used as an example of how forward stratigraphical modelling can be used to increase our understanding of the origin of carbonate sediment on carbonate ramps.

### 13.2.1   Geological setting, stratigraphy and facies

Upper Jurassic (mainly Kimmeridgian) limestones of the north-eastern margin of the Iberian Basin, Spain are moderately well exposed over a wide geographical area permitting reconstruction of a *c.* 200 km long stratigraphical cross-section extending from inner to outer ramp setting. The Iberian Massif formed the basin margin to the west and the open marine waters of the Tethyan Ocean lay to the east (Figure 13.5a). During periods of high sea-level, there was a marine connection from the Tethyan Ocean through the Iberian Basin and 'Soria Seaway' via a shallow sea to the central Atlantic Ocean. The basin margin sloped gently to the east and carbonate ramps were developed during the Oxfordian and Kimmeridgian, with only local and temporary phases of fault-related subsidence. The area was situated within the tropics and climate models indicate that it was a windward basin margin with both westerly-directed winter storms and summer hurricanes from the south-east.

The Upper Jurassic strata are subdivided into a number of facies that are interpreted to have formed in inner, mid, mid to outer, and outer ramp depositional settings (Table 13.2 overleaf). A major feature of these ramp sediments is the reworking and deposition of sediment in graded storm beds or 'tempestites'. Such beds are coarser-grained than intervening beds that accumulated in fairweather conditions. The storm beds have sharp erosional bases overlain by coarse-grained layers and fining-upward to a wave-reworked or bioturbated top to the bed. The coarse-grained clasts are skeletal grains and ooids, the latter of shallow-water origin, and some siliciclastic grains are derived from the land to the west. Storm processes play an important role in redistributing sediment on shallow, windward-sloping carbonate ramps as storm wave-base will affect a large area of sea-floor.

The correlation between the six logged sections (Figure 13.5c overleaf) is based on ammonites, with deposition spanning from the lowest *platynota* Biozone of the Kimmeridgian through to the *hybonotum* Biozone of the Tithonian (Figure 13.5b), an interval interpreted to represent about 2.8 Ma by comparison with the geological time-scale.

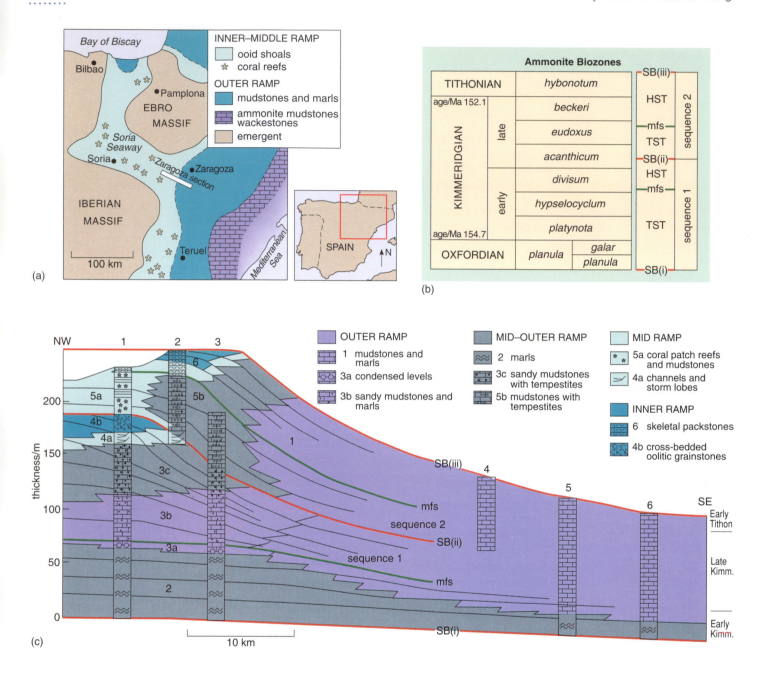

**Figure 13.5**  Upper Jurassic ramp of the Iberian Basin. (a) Location of Iberian Basin, cross-section and palaeogeographic setting of Kimmeridgian ramps. (b) Stratigraphy of the Upper Jurassic of the Iberian Basin. Note that the Upper Jurassic is subdivided and correlated by ammonite biozones and also subdivided by the sequence stratigraphical analysis discussed in the text. (c) Zaragoza cross-section showing location of six measured graphic logs (see numbers at the top of the graphic log columns) through the ramp that have been correlated using ammonites. The facies classification is discussed in Table 13.2 and the sequence stratigraphical interpretation is discussed in the text (Section 13.2.2). *(Aurell et al., 1998.)*

**Table 13.2** The four main depositional settings described from Kimmeridgian ramps, Iberian Basin and their facies and palaeoenvironments (cf. Figure 13.5c).

| Depositional environment | Facies | Processes/Environments |
|---|---|---|
| Inner Ramp | *Skeletal packstones*: 0.5–1 m thick, bioturbated and occasionally cross-bedded | Foraminifers and bioturbation indicate slight restriction in a shallow-marine environment |
| | *Cross-bedded oolitic grainstones* (Fig. 13.6a overleaf): as <8 m cross beds variable amounts of siliciclastic grains | Migrating dunes and megaripples formed by storm-generated currents |
| Mid Ramp | *Coral patch reefs* (Fig. 13.6c): coral and algal framestones and associated skeletal packstones, and *Mudstones*, well bedded with skeletal grains, clastic grains and plant fragments | Patch reef growth is below fairweather wave-base and has associated reef debris. Mid ramp muds derived from shallow waters as suspended sediment from storms |
| | *Channels and storm lobes*: cross-bedded units (0.5–1 m thick) of skeletal grainstones | Storm-scoured sand-filled channels and lobe-shaped, storm-generated bedforms |
| Mid–Outer Ramp | *Mudstones with tempestites*: with ooids, skeletal grains and plant debris in cm-thick graded beds, mm-thick bioturbated muds and locally 0.5–1 m-thick cross-bedded lobes | Mudstones as fairweather accumulations. Tempestites as: proximal tempestites; storm lobes (cross-stratified); distal tempestites |
| | *Sandy mudstones and tempestites*: cm-scale graded and bioturbated beds with skeletal and siliciclastic grains | As above with greater siliciclastic input |
| | *Marl*: blue, laminated and with scarce fossils | Fairweather accumulations of mixed carbonate and siliciclastic muds |
| Outer Ramp | *Mudstones and marls*: rhythmic bedding and bioturbation with scattered benthic skeletal grains and coccoliths | Sub storm wave-base deposition in outer ramp setting with some pelagic input |
| | *Condensed levels* (Fig. 13.6b): ferruginous hardgrounds with shelly concentrations and solitary corals | Breaks in sedimentation allowing matrix and coral colonization |
| | *Sandy mudstones and marls*: thin graded beds within bioturbated marls | Infrequent storm-derived siliciclastics and skeletal debris as distal tempestites |

## 13.2.2 Sequence stratigraphy

The sequence stratigraphical analysis of these rocks used rather different techniques to those applied in Mallorca where stratigraphical surfaces and profiles can be traced as continuous surfaces along the cliff profiles. The 100 km cross-section in Figure 13.5 was constructed using six widely spaced logged sections that were biostratigraphically correlated using ammonites. Aurell used facies analysis and facies interpretations, examination of lateral and vertical facies relations and recognition of preserved stratigraphical geometries (e.g. prograding inner ramp grainstone units) in order to make his sequence stratigraphical interpretation. He recognized two depositional sequences in the proximal to distal ramp section exposed to the south-west of Zaragoza (Figures 13.5c, 13.6a overleaf).

(a)

(b)

(c)

**Figure 13.6**  Exposure photographs of Kimmeridgian limestones from the Iberian Basin, NE Spain. (a) Exposure of the succession shown in graphic log 1 (Figure 13.5c) with darker-coloured mid to outer ramp mudstones, marls and tempestites in the lower slopes that coarsen upwards (note thicker more resistant beds up section). These are overlain by lighter-coloured oolitic grainstones which show progradation to the right (south-east). These are sharply overlain by coral-rich units at the top of the hill (cliff *c.*100 m high). (b) Sequence boundary (SB(i)) at base of Kimmeridgian as expressed in distal setting with an iron (brown) stained bed rich in ammonites, belemnites and crinoids, all indicating sedimentary condensation (view *c.* 50 cm across). (c) Kimmeridgian patch reefs with Marc Aurell and Paul Wright (bottom left) for scale. Nodular weathering massive limestones in centre of photograph are reef limestones with *in situ* preserved corals and microbial crusts. The reefs pass laterally (to the right) into the bedded deposits of intereef bioclastic limestones. *((a)–(c) Dan Bosence, Royal Holloway University of London.)*

### *Sequence 1*

The sequence boundary is placed at the base of these sections where Kimmeridgian strata in the west (proximal sites) lie on eroded and subaerially exposed Oxfordian inner ramp, cross-bedded sandstones. More offshore (eastern) sections show a depositional break in the form of an iron-cemented surface with a concentration of shelly fossils (Figure 13.6b).

The sub-Kimmeridgian sequence boundary (SB(i)) is overlain by the mid-outer ramp marls indicating a relative sea-level rise and that the sequence boundary and transgressive surface are superimposed. The marls are overlain by outer ramp mudstone facies. Therefore, as relative sea-level rose, the rate of increase in accommodation space outstripped the rate of sediment supply so that the stratal geometries are retrogradational. These mid-outer and outer ramp facies are interpreted as the TST. In more proximal areas, condensed levels occur and these are taken to represent the maximum flooding surfaces as they are the most open marine facies recognized in the sections (Figure 13.5c; graphic logs 1–3). In outer ramp locations, the surface of maximum flooding is difficult to recognize and the only evidence may be a zone of thinner beds or slightly iron-stained tops to beds in the outer ramp mudstones and marls. These have been shown to have a pelagic contribution and represent the most offshore facies in the succession.

Following maximum flooding, the lower HST in proximal sites is marked by an upward shallowing in facies. The outer ramp deposits are overlain by the more proximal mid–outer ramp sandy mudstones and tempestites in logged sections 1–3 indicating regression and interpreted progradation of the ramp profile towards the east. This shallowing-upwards continues in these three sections and more proximal, inner ramp, oolitic grainstones are seen to prograde south-eastwards over these deeper-water deposits (Figure 13.6a). To the east, there is no record of this decrease in accommodation space into the highstand, and outer ramp mudstones and marls continued to be deposited in these deeper-water areas.

## Sequence 2

A sharp erosional surface above the oolitic grainstones (Figure 13.5c, graphic log 1), and shift of facies to deeper-water patch reefs (Figure 13.6c) and mudstones is the only evidence of a sequence boundary (SB(ii)) and superimposed transgressive surface in proximal areas. There is no evidence of subaerial exposure at this surface in the available sections and thus this sequence boundary is interpreted to be less well developed than SB(i) in this area. However, the upper surface of a prograding ooid shoal must have formed close to sea-level and the overlying patch reefs and mudstones required additional accommodation space in which to accumulate. There is no evidence of any changes in accommodation space in the deeper-water, outer ramp sections.

The mid ramp coral patch reefs and interbedded mudstones are interpreted to have aggraded in response to relative sea-level rise and can be seen to pass laterally into the sandy mudstones and tempestites to the east between logs 1 and 2 (Figure 13.5c). A single surface of maximum flooding is difficult to pick out in such a section and has been placed at the first indication of shallowing and highstand progradation of the inner ramp skeletal packstones. Again, there is little record of these events in the deeper outer ramp setting. The HST is locally eroded and overlain by a sub-Cretaceous unconformity in this area of the Iberian Basin.

This case study shows many features typical of carbonate ramp sequence stratigraphy: for example, the very large distances of progradation and retrogradation of facies belts on a low angle ramp profile in response to relative sea-level changes, and the retrogradation or aggradation during periods of relative sea-level rise. The case study also highlights the difficulties of recognizing sequence stratigraphical surfaces and systems tracts in deep-water outer ramp locations. Storms are considered to be important in redistributing sediment on the ramp from inshore to offshore areas, and this is clearly shown to have taken place by the mid ramp storm beds containing ooid grains which must have formed originally in an inner ramp environment. This sediment redistribution probably helps maintain the gently sloping ramp profile through time because otherwise higher production rates in shallow-water sites would build up to form a rimmed platform margin (see Figure 12.11a). This aspect is explored more fully in Section 13.2.3.

### 13.2.3  Stratigraphical modelling

Forward numerical modelling has been undertaken on this cross-section through the Upper Jurassic ramps of the Iberian Basin to achieve a better understanding of the dynamics of sediment production, resedimentation and origin of the large amount of carbonate mud in the mid and outer ramp areas. The parameters used in the modelling (Figure 13.7) are given in Table 13.3.

**Table 13.3**  Summary of parameters used in program for simulations shown in Figure 13.7.

| Variable parameters | Run 1 | Run 2 |
|---|---|---|
| Carbonate production (cm/ka) vs. depth (m): | | |
| <1 m | 16 cm/ka | 11 cm/ka |
| 1–10 m | 16 cm/ka | 11 cm/ka |
| 10–70 m | 1 cm/ka | 1 cm/ka |
| >70 m | 0 cm/ka | 0 cm/ka |
| Pelagic sedimentation | 3 cm/ka | 5 cm/ka |
| Erosion above sea-level | 3 cm/ka | 3 cm/ka |
| Erosion at sea-level | 12 cm/ka | 5 cm/ka |
| Wave-base | 10 m | 5 m |
| Offshore transport distance for coarse and fine-grained sediment | 8/50 km | 8/20 km |
| Coarse : fine grain ratio | 10 : 90 | 50 : 50 |

The low production rates compared with those used for modelling the Miocene rimmed platform of Mallorca are based on the fact that continuous reef belts are not present on this ramp, or indeed any other ramps. The values for production are similar to those obtained from the shallow shelf area of Florida Bay (Section 11.3.1) which is the inner part of a carbonate ramp that slopes down into the Gulf of Mexico. The transport distances on the Upper Jurassic ramp are relatively high to account for the large amount of storm-reworked coarse- and fine-grained material in this example. The time-scale is taken from the ammonite biozones assuming that these are of equal duration between the dates for the duration of the Kimmeridgian Stage (Figure 13.5b). The sea-level curve (Figure 13.7) comes initially from the sequence stratigraphical analysis and has been fine-tuned by matching different simulations against the outcrop data. The final relative sea-level curve comprises an overall linear rise of 6 cm/ka with two superimposed sinusoidal curves: one for the overall transgression and regression of the Kimmeridgian (amplitude 28 m, frequency 1.8 Ma) and another slightly higher order cyclicity for the two depositional sequences recognized from the fieldwork (amplitude 15 m, frequency 1 Ma). The stratigraphy is reasonably successfully modelled without any high frequency sea-level changes.

**Figure 13.7 (opposite)**  Actual (a–b) and simulated (c–f) stratigraphy for the Kimmeridgian ramps of the Iberian Basin. (a) shows actual stratigraphy (cf. Figure 13.5c) with facies displayed in terms of depositional setting. (b) Same as (a) but with facies displayed with respect to process of deposition. Vertical lines show position of graphic logs 1–6 shown in Figure 13.5c. (c) and (d) Simulated stratigraphy based on Run 1 displayed as depth of deposition and depositional process respectively. (e) and (f) These show simulated stratigraphy for Run 2 displayed as depth of deposition and dominant depositional process respectively. For details of Runs 1 and 2 and discussion of results, see Table 13.3 and text. (Aurell et al., 1998.)

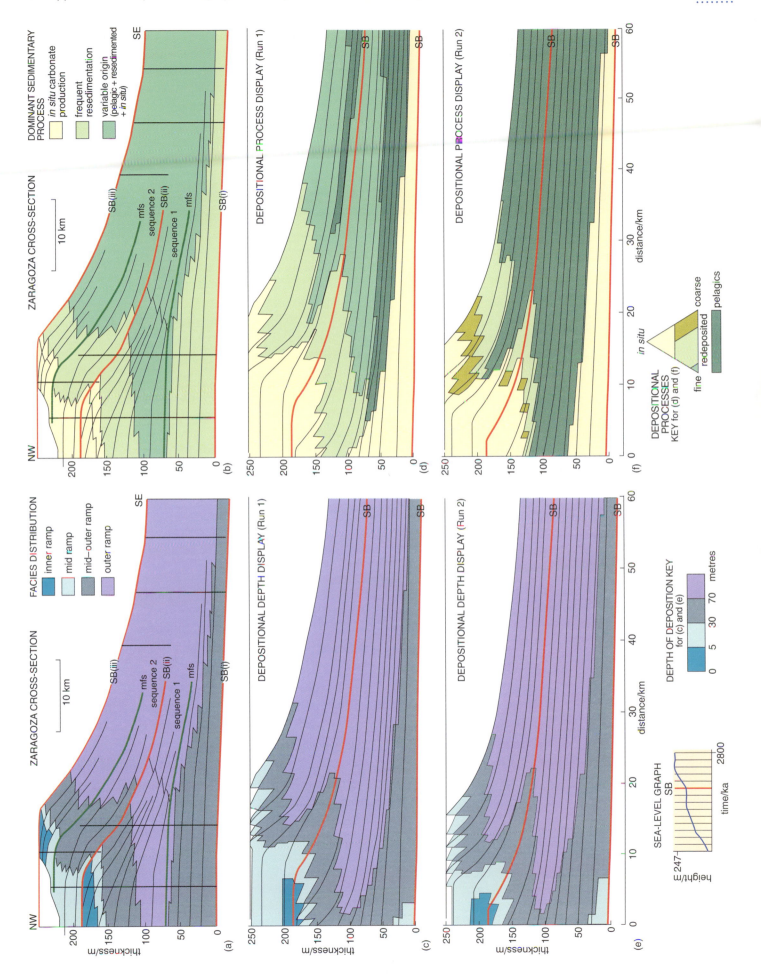

The simulations were matched against the reconstructed cross-section in Figure 13.5c. This is duplicated in Figure 13.7a and b except that in Figure 13.7a the facies are labelled by their depositional setting (i.e. inner, mid, mid-outer and outer ramp) and in Figure 13.7b by the dominant sedimentary process that is thought to have determined the accumulation of the facies (e.g. frequent resedimentation).

Two slightly different solutions were obtained during the modelling and these are referred to as Runs 1 and 2 (Table 13.3 and Figure 13.7). Both of these runs matched the observed stratigraphical thicknesses, locations of the sequence stratigraphical surfaces and systems tracts and the observed stratigraphical geometries. The facies can only be broadly matched because the computer output can only plot either the depth of deposition of the strata (Figure 13.7c,e), which approximates to inner, mid, mid-outer and outer ramp divisions, or it can plot the main process that is used to deposit the sediment, i.e. whether the sediment is deposited more or less where it was produced (e.g. *in situ* deposits such as reefs, ooid shoals, skeletal packstones), or, if it is eroded and redeposited (e.g. tempestites), or whether it is pelagic in origin (Figure 13.7d,f). Comparison of the true stratigraphical cross-section with the depositional depth display and depositional process modelling results shows that there is a reasonably good match and that these two computer runs provide reasonable numerical solutions to this stratigraphy.

The two model runs were undertaken to test two hypotheses concerning the origin of the muddy sediments that could not be resolved from the fieldwork and petrographic analyses. The alternative hypotheses are that the lime mud was produced by pelagic production, or by erosion and redistribution of shallow-water carbonate mud. Run 1 has therefore less offshore pelagic production but increased amounts of inshore mud and larger mud transport distances. Run 2 has higher amounts of offshore pelagic production but decreased amounts of inshore mud and smaller mud transport distances.

When the two model runs are compared, it can be seen that the correct thicknesses of mid to outer ramp carbonate mud are matched in both runs so both are possible solutions for this stratigraphy. However, different outer ramp stratigraphy is predicted by the two model runs. In Run 1, stratigraphical condensation (more closely spaced time lines) is predicted near the surfaces of maximum flooding in the two sections and greater amounts of shallow-water muds are resedimented during highstands and the stillstand period. In Run 2, with the greater proportion of pelagic production, none of these variations are predicted.

○ Given these two different interpretations provided by the modelling, what further work might you carry out to determine which might be the most likely solution?

● The logged sections could be revisited and samples for laboratory analysis collected in order to examine carefully those parts of the sections where facies changes are predicted from the modelling. The main difference in the rock composition predicted by the modelling is in the percentage pelagic component (e.g. coccoliths) to redeposited fine-grained carbonate (which should contain eroded shallow-water components). The offshore carbonate muds could be examined by scanning electron microscopy to assess this difference.

## 13.3  Summary

- The cliff sections of Miocene carbonate rocks in south-west Mallorca, Spain, provide near-continuous outcrops of a prograding reef-rimmed carbonate platform. These are still in their depositional orientation and the facies are very similar to those described from recent reefs in the Caribbean (Section 11.3.2).

- The cliff sections in Mallorca preserve the position of former reef crests which can be used as a proxy for Miocene sea-level positions. When successive reef crests are joined into a 'reef crest line', this gives the amplitude of Miocene sea-level changes, in this case between 30 and 40 m.

- With the sea-level constrained for the Mallorca case study, the depositional geometries can be closely tied to systems tracts. Progradational lowstand systems tract, aggradational transgressive systems tract, progradational highstand systems tract and downstepping falling systems tract are recognized.

- The greatest sediment supply in the Mallorca case study occurs during the transgressive systems tract when the platform top is continually being flooded by shallow-marine waters and the platform aggrades and progrades as relative sea-level rises.

- Numerical modelling constrained by the exposed sediment geometries and facies indicates that reef production rates were similar to those of present-day Caribbean reefs and that the main cycles of sea-level change in the platforms are probably 100 ka eccentricity cycles.

- The case study from the Upper Jurassic (mainly Kimmeridgian) of north-east Spain illustrates the sedimentology and sequence stratigraphy of a large-scale carbonate ramp formed on the margin of the Tethys Ocean.

- Graphic logging of exposed sections in north-east Spain correlated using ammonite biostratigraphy indicates well-preserved sections of inner, mid, mid–outer and outer ramp environments.

- A sequence stratigraphical analysis of the Upper Jurassic ramp carbonates of north-east Spain provides a framework that explains the distribution of the facies as two depositional sequences separated by a sequence boundary.

- The first depositional sequence retrogrades during relative sea-level rise so that outer ramp facies occur over former mid ramp facies in proximal settings. These are later buried by progradation of inner ramp facies (including ooid shoals) in proximal areas. The second depositional sequence starts with reef growth over former shallow-water ooid shoals, and these aggrade during relative sea-level rise (transgressive systems tract) and are subsequently buried by prograding highstand skeletal packstones. No falling stage or lowstand systems tracts are preserved and the deeper-water, outer ramp areas show little evidence of the sea-level changes that affected the inner ramp areas.

- Numerical modelling of these Upper Jurassic ramp carbonates in north-east Spain indicates that the sequences can form from the superposition of two low-order sea-level cycles using production rates from present-day inner ramp and pelagic environments. Two different interpretations of the origin of the large amount of mud in the inner to outer ramp parts of the stratigraphy arise from the modelling:

  1  that the mud comes from a largely pelagic source;

  2  that the mud comes from a largely inner ramp source and is redistributed by down-ramp transport of fines during storms.

2

## 13.4 References

The detailed research on which these two case studies are based are published in the following research papers:

AURELL, M. (1991) 'Identification of systems tracts in low angle carbonate ramps: examples from the Upper Jurassic of the Iberian Chain (Spain)', *Sedimentary Geology,* **73**, 101–115.

AURELL, M., BADENAS, B., BOSENCE, D. W. J. AND WALTHAM, D. A. (1998) 'Carbonate production and offshore transport on a late Jurassic carbonate ramp (Kimmeridgian, Iberian Basin, NE Spain); evidence from outcrops and computer modelling', in WRIGHT, V. P. AND BURCHETTE, T. P. (eds) *Carbonate Ramps,* Geological Society Special Publication No. 149, 137–161.

BOSENCE, D. W. J., POMAR, L., WALTHAM, D. A. AND LANKESTER, T. H. G. (1994) 'Computer modelling a Miocene Carbonate platform Mallorca, Spain', *American Association of Petroleum Geologists Bulletin,* **78**, 247–266.

POMAR, L. (1991) 'Reef geometries, erosion surfaces and high frequency sea-level changes, Upper Miocene Reef Complex, Mallorca, Spain', *Sedimentology,* **38**, 243–269.

POMAR, L. (1993) 'High-resolution sequence stratigraphy in prograding Miocene carbonates: Application to seismic interpretation', in LOUCKS, R. G. AND SARG, J. F. (eds) *Carbonate sequence stratigraphy,* American Association of Petroleum Geologists Memoir No. 57, 389–407.

# Acknowledgements

We would like to thank the following Open University staff for their help during the production of this book: Glynda Easterbrook for commenting on all the draft versions, compiling the index and helping to manage the project; Gerry Bearman for copy editing and his enthusiasm and support; Pam Owen for her patience and expertise in creating most of the diagrams, and Roger Courthold and Jon Owen for completing the remainder; Ruth Drage for her excellent design and layout; Liz Yeomans for her help with the design of this co-published version of the book; and Jo Morris for all her secretarial support.

We are grateful to the following people who provided comments on an earlier version of the book: Stephen Hesselbo (University of Oxford), David Bowler, Derek Gobbett, Fiona Hyden and Steve Killops (Associate Lecturers, Open University) and three referees appointed by Cambridge University Press. Anthony Cohen (Open University) is thanked for providing useful comments and suggestions. Peter Skelton, Iain Gilmour, Rachael James, Mark Sephton, Sandy Smith and Dave Williams (all Open University) for providing valuable references and information. The data and interpretations presented in Part 3 benefited greatly from discussions and work with Keith Adamson, Ciaran O'Byrne, Gary Hampson (who were members of the Stratigraphy Group at Liverpool University and are now at Badley-Ashton, BP Amoco and Imperial College, London respectively), together with Diane Kamola (University of Kansas), Tom McKie (Shell UK), Dave Tabet (Utah Geological Survey), John Van Wagoner (Exxon Production Research), Laine Adair and Mike Glasson (Andelex Resources), and John Mercier (Cyprus Plateau Mining). In addition, the Fry family, Pogue family and people of Price and Green River, Utah, are thanked for field assistance, access to their land and general hospitality. Dave Waltham (Royal Holloway College University of London) is thanked for use of his computer program Sedtec 2000 (unpublished) in Part 4. Thanks also to Giles Clark (Open University), Susan Francis and Sally Thomas (Cambridge University Press) for their support and help with co-publication.

# Figure references for this book

All of the figures in this book that are based on published work have been redrawn and the majority have been modified, some significantly. The sequence stratigraphy figures in Parts 2 and 4 that are based on figures from the literature have been changed to reflect the four system tract sequence stratigraphy models presented in this book. Every effort has been made to trace all copyright owners of the figures, but if any have been inadvertently overlooked, the publishers will be pleased to rectify any omissions in these acknowledgements when the book is reprinted.

AHARON, P. (1983) '140,000-yr. isotope climatic record from raised coral reefs in New Guinea', *Nature*, **304**, 720–723. [**Figure 6.4 (part)**.]

ANDERTON, R., BRIDGES, P. H., LEEDER, M. L. AND SELLWOOD, B. W. (1979) *A Dynamic Stratigraphy of the British Isles*, George Allen and Unwin, London, 301pp. [**Figure 3.7b (part)**.]

ANSELMETTI, F. S., EBERLI, G. P. AND DING, Z.-D. (2000) 'From the Great Bahama Bank into the Straits of Florida: A margin architecture controlled by sea-level fluctuations and ocean currents', *Geological Society of America Bulletin*, **112**, 829–844, The Geological Society of America. [**Figure 11.4**.]

AURELL, M., BADENAS, B., BOSENCE, D. W. J. AND WALTHAM, D. A. (1998) 'Carbonate production and offshore transport on a late Jurassic carbonate ramp (Kimmeridgian, Iberian Basin, NE Spain); evidence from outcrop and computer modelling', in WRIGHT, V. P. AND BURCHETTE, T. P. (eds) *Carbonate Ramps*, Geological Society Special Publication No. 149, 137–161, Geological Society of London. [**Figures 13.5, 13.7**.]

BALLY, A. W. (ed.) (1987) *Atlas of Seismic Stratigraphy*, AAPG Studies in Geology No. 27, Vol. 1, Rice University, Texas, copyright 1987. [**Figure 4.22a**.]

BOND, G. C. AND KOMINZ, M. A. (1984) 'Construction of tectonic subsidence curves for the early Paleozoic miogeocline, southern Canadian Rocky Mountains: Implications for subsidence mechanisms, age of breakup and crustal thinning', *Geological Society of America Bulletin*, **95**, 155–173. [**Figure 5.7**.]

BOSENCE, D. W. J., POMAR, L., WALTHAM, D. A. AND LANKESTER, T. H. G. (1994) 'Computer modelling a Miocene Carbonate platform, Mallorca, Spain', *American Association of Petroleum Geologists Bulletin*, **78**(2), 247–266. [**Figure 13.4** reprinted by permission of the American Association of Petroleum Geologists, whose permission is required for further use.]

CHAPPELL, J. AND SHACKLETON, N. J. (1986) 'Oxygen isotopes and sea level', *Nature*, **324**, 137–140. [**Figure 6.4 (part)**.]

CHURCH, K. D. AND GAWTHORPE, R. L. (1994) 'High resolution sequence stratigraphy of the late Namurian in the Widmerpool Gulf (East Midlands, UK)', *Marine and Petroleum Geology*, **11**, 528–544, Elsevier Science. [**Figures 3.10, 4.20, 5.11, 5.14**.]

COLEMAN, J. M. (1976) *Deltas: Processes of Deposition and Models for Exploration*, Continuing Education Publishing, Champaign, IL. [**Figure 5.8 (part)**.]

COLLINSON, J. D. (1988) 'Controls on Namurian sedimentation in the Central Province basins of northern England', in BESLEY, B. M. AND KELLING, G. (eds), *Sedimentation in a Synorogenic Basin Complex — The Upper Carboniferous of Northwest Europe*, 85–101, Kluwer Academic Publishers. [**Figure 3.7b**.]

CRAIG, G. Y., MCINTYRE, D. B. AND WATERSTON, C. D. (1978) 'James Hutton's Theory of the Earth: The lost drawings', Scottish Academic Press, in association with the Royal Society of Edinburgh and the Geological Society of London, 67pp., and accompanying folio facsimiles. [**Figure 1.2**.]

CROSS, T. A. (1986) 'Tectonic controls of foreland basin subsidence and Laramide style deformation, western United States', in Allen, P. A. and Homewood, P. (eds) *Foreland Basins*, International Association of Sedimentologists Special Publication No. 8, 15–39, Blackwell. [**Figures 7.5, 10.6**.]

DAVIES, P. J., SYMONDS, P. A., FEARY, D. A. AND PIGRAM, C. J. (1989) 'The evolution of the carbonate platforms of Northeast Australia', in CREVELLO, P. D., SARG, J. F., READ, J. F. AND WILSON, J. L. (eds) *Controls on Carbonate Platform and Basin Development*, Special Publication of the Society of Economic Palaeontologists and Mineralogists, **44**, 233–258. [**Figure 12.5**.]

DAVIS, R. A. (1994) *The Evolving Coast*, Scientific American Books, 231pp. © 1994 Scientific American Books, reprinted by permission of Henry Holt & Co. [**Figure 3.5**.]

DUFF, P. MCL. (ed.) (1994) *Holmes' Principles of Physical Geology*, Edward Arnold, London, 791pp. [**Figure 3.3 (right)**.]

DUNHAM, R. J. (1962) 'Classification of carbonate rocks according to depositional texture', in Hann, W. E. *Classification of Carbonate Rocks*, Memoir of the American Association of Petroleum Geologists, **1**, 108–121. [**Figure 11.13 (part)**.]

DUVAL, B., CRAMEZ, C. AND VAIL, P. R. (1992) 'Types and hierarchy of stratigraphic cycles', unpublished conference abstracts from 'Sequence Stratigraphy of European Basins', CNRS–IFP, Dijon, France, 44–45. [**Figure 5.2**.]

EMBRY A. F. AND KOLVAN, J. E. (1971) 'A late Devonian reef tract on north eastern Banks Island, Northwest Territories', *Bulletin of the Canadian Petroleum Geologists*, **19**, 730–781. [**Figure 11.13 (part).**]

EMERY, D. AND MYERS, K. J. (1996) *Sequence Stratigraphy*, Blackwell, 297pp. [**Figure 2.3 (right).**]

ENOS, P. (1977) 'Holocene sediment accumulation of the South Florida shelf margin', in ENOS, P. AND PERKINS, R. D. (eds) *Quaternary Sedimentation in South Florida*, Memoir of the Geological Society of America, **147**, 1–130. [**Figure 11.9f (part).**]

FRAKES, L. A. (1979) *Climates Throughout Geological Time*, Elsevier 310pp. [**Figure 5.4.**]

GALLOWAY, W. E. (1975) 'Process framework for distinguishing the morphologic and stratigraphic evolution of deltaic depositional systems, in BROUSSARD, M. L. (ed.) *Deltas —Models for Exploration*, 87–98, The Houston Geological Society, Houston, Texas. [**Figure 7.7.**]

GALLOWAY, W. E. (1989) 'Genetic stratigraphic sequences in Basin Analysis I: Architecture and Genesis of Flooding — Surface Bounded Depositional Units', *Bulletin of the American Association of Petroleum Geologists*, **73**, 125–142. [**Figure 5.9** by permission of the American Association of Petroleum Geologists, whose permission is required for further use.]

GARCIA-MONDÉJAR, J. AND FERNÁNDEZ-MENDIOLA, P. A. (1993) 'Sequence stratigraphy and systems tracts of a mixed carbonate and siliciclastic platform-basin setting: the Albian of Lunada and Soba, northern Spain', *Bulletin of the American Association of Petroleum Geologists*, **77**, 245–275. [**Figure 12.3.**]

GEBELEIN, C. D. (1974) '*Guidebook of Modern Bahamian Platform Environments*', Geological Society of American Annual Meeting Fieldtrip Guidebook, Boulder, Colorado, 93pp. [**Figure 11.14 (part of a,b).**]

GINSBURG, R. N. (1956) 'Environmental relationships of grain size and constituent particles in some south Florida carbonate sediments', *Bulletin of the American Association of Petroleum Geologists*, **40**, 2384–2427. [**Figure 11.9 d,e and part of b.**]

GINSBURG, R. N. AND JAMES, N. P. (1974) 'Holocene carbonate sediments of continental margins', in: BURKE, C. A. AND DRAKE, C. L., *The Geology of Continental Margins*, 137–155, Springer-Verlag, New York. [**Figure 11.9c (part).**]

GUION, P. D., GUTTERIDGE, P. AND DAVIES, S. J. (2002) 'Carboniferous sedimentation and volcanism on the Laurussian margin', in WOODCOCK, N. AND STRACHAN, R. (eds) *Geological History of Britain and Ireland*, 227–270, Blackwell. [**Figure 3.9.**]

HALLAM, A. (1977) 'Secular changes in marine inundation of USSR and North America through the Phanerozoic', *Nature*, **269**, 762–772.

HAMPSON, G. J., BURGESS, P. AND HOWELL, J. A. (2001) 'Lowstand shoreface deposits constrain relative sea-level history. Examples from the Late Cretaceous strata in the Book Cliffs area, Utah', *Terra Nova*, **13**, 188–196, Blackwell. [**Figures 7.4c, 10.3 (part).**]

HAMPSON, G. J., HOWELL, J. A. AND FLINT, S. S. (1999) 'A sedimentological and sequence stratigraphic re-interpretation of the Upper Cretaceous Prairie Canyon Member and associated strata, Book Cliffs area, Utah, USA', *Journal of Sedimentary Research*, **69**, 414–433, Society for Sedimentary Geology. [**Figure 9.4.**]

HAQ, B. U., HARDENBOL, J. AND VAIL, P. R. (1998) 'Mesozoic and Cenozoic chronostratigraphy and cycles of sea level change', in WILGUS, C. K., HASTINGS, B. S., POSAMENTIER, H., VAN WAGONER, J., ROSS, C. A. AND KENDALL, G. ST. C. (eds) *Sea Level Changes: An Integrated Approach*, Special Publication of the Society of Economic Palaeontologists and Mineralogists No. 42, 71–108. [**Figures 5.6 (part), 10.5 (right part).**]

HARRIS, P. M. AND KOWALIK, W. S. (eds) (1994) *Satellite Images of Carbonate Depositional Settings*, AAPG Methods in Exploration Series, No. 11, 147pp. [**Figure 11.7** by permission of the American Association of Petroleum Geologists, whose permission is required for further use.]

HUTTON, J. (1795) *Theory of the Earth, with proofs and illustrations*, **1–2**, London: Cadell and Davies; Edinburgh: William Creech. [**Figure 1.2.**]

IMBRIE, J., HAYS, J. D., MARTINSON, D. G., MCINTYRE, A. C., MIX, A. C., MORLEY, J. J., PISIAS, N. G., PRELL, W. L. AND SHACKLETON, N. J. (1984) 'The orbital theory of Pleistocene climate: support from a revised chronology of the marine $\delta^{18}O$ record, in BERGER, A., IMBRIE, J., HAYS, J. D., KUKLA, G. AND SALTZMAN, B. (eds) *Milankovitch and Climate*, NATO ASI Series C126, **1**, 269–305, Kluwer Academic Publishers. [**Figure 2.8 (right).**]

JAMES, N. P. AND KENDALL, A. C. (1992) 'Introduction to Carbonate and Evaporite Models', in WALKER, R. G. AND JAMES, N. P. (eds) *Facies Models: Response to Sea Level Change*, 265–275, Geological Association of Canada. [**Figure 11.3 (part).**]

JENKYNS, H. C. (1986) 'Pelagic environments', in Reading, H. G. (ed.) *Sedimentary Environments and Facies* (2nd edn), 343–397, Blackwell. [**Figure 11.1a.**]

JONES, B. AND DESROCHERS, A. (1992) 'Shallow platform carbonates', in WALKER, R. G. AND JAMES, N. P. (eds), *Facies Models — Response to Sea Level Change*, Geological Association of Canada, 277–301. [**Figure 11.8.**]

KAMOLA, D. L. AND VAN WAGONER, J. C. (1995) 'Stratigraphy and Facies Architecture of Parasequences with Examples from the Spring Canyon Member, Blackhawk Formation, Utah', in VAN WAGONER, J. C. AND BERTRAM, G. T. (eds) *Sequence Stratigraphy of Foreland Basin Deposits*, American Association of Petroleum Geologists Memoir No. 64, 27–54. [**Figures 8.7a, 10.1b (part), 10.5 (left, part)** by permission of the American Association of Petroleum Geologists, whose permission is required for further use.]

KOLB, C. R. AND VAN LOPIK, J. R. (1958) *Geology of the Mississippi River Deltaic Plain*, Technical Reports Nos. 3483 and 3484, US Corps of Engineers Waterways Experimental Station. [**Figure 5.8 (part)**.]

LABEYRIE, L. D., DUPLESSY, J. C. AND BLANC, P. L. (1987) 'Variations in mode of formation and temperature of oceanic deep waters over the past 125,000 years', *Nature*, **327**, 477–482. [**Figure 6.4 (part)**.]

LEEDER, M. R. (1999) *Sedimentology and Sedimentary Basins*, Blackwell, 592pp. [**Figure 5.8**.]

LOWE, J. J. AND WALKER, M. J. C. (1997) *Reconstructing Quaternary Environments* (2nd edn), Pearson Education Ltd., 446pp. [**Figure 3.3 (left)**.]

MARTINSON, D. G., PISIAS, N. G., HAYS, J. D., IMBRIE, J., MOORE, T. C. AND SHACKLETON, N. J. (1987) 'Age dating and the orbital theory of the ice-ages: development of a high resolution 0–300,000 year chronostratigraphy', *Quaternary Research*, **27**, 1–29. [**Figure 3.3 (left)**.]

MCARTHUR, J. M., HOWARTH, R. J. AND BAILEY, T. R. (2001) 'Strontium Isotope Stratigraphy: LOWESS Version 3: Best Fit to the Marine Sr-Isotope Curve for 0–509 Ma and Accompanying Look-up Table for Deriving Numerical Age', *Journal of Geology*, **109**, 155–170, University of Chicago Press. [**Figure 2.6**.]

MESTEL, R. (1997) 'Noah's Flood', *New Scientist*, 4 October 1997, 24–27. [**Figure 3.1 (detailed map)**.]

NAISH, T. R. (1997) 'Constraints on the amplitude of late Pliocene eustatic sea-level fluctuations: new evidence from the New Zealand shallow marine sedimentary record', *Geology*, **25**, 1139–1142, The Geological Society of America. [**Figure 3.4**.]

NASA http://earth.jsc.nasa.gov/ [**Figures 7.3a, 7.8a, 7.14g**.]

NICHOLS, G. (1999) *Sedimentology and Stratigraphy*, Blackwell, 355pp. [**Figures 2.3 (left), 2.4, 7.8b**.]

NORRIS, R. D. AND RÖHL, U. (1999) 'Carbon cycling and chronology of climate warming during the Palaeocene/Eocene transition', *Nature*, **401**, 775–778, Macmillan Magazines Ltd. [**Figure 2.9**.]

O'BYRNE, C. J. AND FLINT, S. (1995) 'Sequence, parasequence and intraparasequence architecture of the Grassy Member Blackhawk Formation, Book Cliffs, Utah, USA', in VAN WAGONER, J. C. AND BERTRAM, G. (eds) *Sequence Stratigraphy of Foreland Basin Deposits*, American Association of Petroleum Geologists Memoir No. 64, 225–257. [**Figures 7.1c, 8.3, 10.1b (part)** by permission of the American Association of Petroleum Geologists, whose permission is required for further use.]

OGG, J. G. (1995) 'Magnetic polarity timescale of the Phanerozoic', *Global Earth Physics — A Handbook of Physical Constants*, American Geophysical Union, 240–270. [**Figure 2.5**.]

PATTISON, S. A. J. (1995) 'Sequence Stratigraphic Significance of Sharp-Based Lowstand Shoreface Deposits Kenilworth Member, Book Cliffs, Utah', *Bulletin of the American Association of Petroleum Geologists*, **79**, 444–462. [**Figure 9.3b** by permission of the American Association of Petroleum Geologists, whose permission is required for further use.]

PITMAN, W. C. III (1978) 'Relationship between eustacy and stratigraphic sequences of passive margins', *Geological Society of America Bulletin*, **89**, 1389–1403. [**Figure 5.6 (part)**.]

POMAR, L. (1993) 'High-Resolution Sequence Stratigraphy in Prograding Miocene Carbonates: Application to Seismic Interpretation', in Loucks, R. G. AND SARG, J. F. (eds) *Carbonate Sequence Stratigraphy: Recent Developments and applications*, American Association of Petroleum Geologists Memoir No. 57, 389–407. [**Figures 13.1a,b, 13.3** by permission of the American Association of Petroleum Geologists, whose permission is required for further use.]

POSAMENTIER, H. W. AND MORRIS, W. R. (2000) 'Aspects of stratal architecture of forced regressive deposits', in HUNT, D. AND GAWTHORPE, R. L. (eds) *Sedimentary Responses to Forced Regressions*, Geological Society Special Publication No. 172, 19–46, Geological Society of London. [**Figures 4.13, 8.6b**.]

PURSER, B. H. (ed.) (1973) *The Persian Gulf: Holocene Carbonate Sedimentation and Diagenesis in a Shallow Epicontinental Sea*, Springer Verlag, Berlin, 471pp. [**Figure 11.6 (part of a,c)**.]

RAMOS, A., SOPEÑA, A. AND PEREZ-ARLUCEA, M. (1986) 'Evolution of Buntsandstein Fluvial Sedimentation in the Northwest Iberian Ranges (Central Spain)', *Journal of Sedimentary Petrology*, **56**, 862–879. [**Figure 7.15a**.]

RAMSBOTTOM, W. H. C. (1971) 'Palaeogeography and goniatite distribution in the Namurian and early Westphalian', in *Proceedings of the Sixième Congrès International de Stratigraphie et de Géologie du Carbonifère, Sheffield 11–16 September 1967: compte rendu*, **4**, 1395–1399, Ernest van Aelst, Maastricht. [**Figure 3.7a**.]

REINSON, G. E. (1992) 'Transgressive barrier island and estuarine systems', in WALKER, R. G. AND JAMES, N. P. (eds) *Facies Models: Response to Sea Level Change*, 179–194, Geological Association of Canada. [**Figure 7.13b,c.**]

RUDDIMAN, W. F., RAYMO, M. E., MARTINSON, D. G., CLEMENT, B. M. AND BACKMAN, J. (1989) 'Pleistocene evolution of Northern Hemisphere Climate', *Paleoceanography,* **4**, 353–412. [**Figure 5.3.**]

SELLWOOD, B. W. (1986) 'Shallow-marine carbonate environments', in READING, H. G. (ed.) *Sedimentary Environments and Facies* (2nd edn), Blackwell, 283–342. [**Figure 11.9 (part of b,c,f.**]

SHACKLETON, N. J. (1987) 'Oxygen isotopes, ice volume, and sea level', *Quaternary Science Reviews*, **6**, 183–190. [**Figure 6.4 (part).**]

SOHL, N. F. (1987) 'Presidential address — Cretaceous gastropods: contrasts between Tethys and the temperate provinces', *Journal of Palaeontology*, Palaeontological Society. [**Figure 11.1b.**]

TAYLOR, D. R. AND LOVELL, W. W. (1995) 'Recognition of high frequency sequences in the Kenilworth Member of the Blackhawk Formation, Book Cliffs, Utah', in VAN WAGONER, J. C. AND BERTRAM, G. (eds) *Sequence stratigraphy of foreland basin deposits*, American Association of Petroleum Geologists Memoir No. 64, 257–277. [**Figures 9.2c, 10.1b (part).**]

TUCKER, M. E. (1985) 'Shallow-marine carbonate facies and facies models', in BRENCHLEY, P. J. AND WILLIAMS, B. P. J. (eds) *Sedimentology: Recent Developments and Applied Aspects*, Geological Society Special Publication No. 18, 139–161, Geological Society of London. [**Figure 11.18a (part).**]

TUCKER, M. E. (1991) *Sedimentary Petrology* (2nd edn), Blackwell, 260pp. [**Figures 11.13, 11.18b,d.**]

TUCKER, M. E. (1993) 'Carbonate diagenesis and sequence stratigraphy', in WRIGHT, V. P. (ed.), *Sedimentology Review*, **1**, 51–72, Blackwell, Oxford. [**Figures 5.5 part of b,c, 12.4.**]

TUCKER, M. E. AND WRIGHT, V. P. (1990) *Carbonate Sedimentology*, Blackwell, 482pp. [**Figures 11.6, 11.14 (part of a,b), 11.18a.**]

VAIL, P. R., MITCHUM, R. M. AND THOMPSON, S. (1977) 'Seismic stratigraphy and global changes of sea level', in PAYTON, C. E. (ed.) *Seismic Stratigraphy — Applications to Hydrocarbon Exploration*, American Association of Petroleum Geologists Memoir No. 26, 83–97. [**Figures 5.5 (part), 5.6 (part).**]

VAN WAGONER, J. C. (1995) 'Sequence Stratigraphy and Marine to Nonmarine Facies, Architecture of Foreland Basin Strata, Book Cliffs, Utah, USA', in VAN WAGONER, J. C. AND BERTRAM, G. T. (eds) *Sequence Stratigraphy of Foreland Basin Deposits*, American Association of Petroleum

Geologists Memoir No. 64, 137–224. [**Figure 9.5** by permission of the American Association of Petroleum Geologists, whose permission is required for further use.]

VAN WAGONER, J. C., MITCHUM, R. M., CAMPION, K. M. AND RAHMANIAN, V. D. (1990) *Siliciclastic Sequence Stratigraphy in Well Logs, Cores and Outcrops, American Association of Petroleum Geologists*, Methods in Exploration Series No. 7, 55pp. [**Figures 5.10 (part), 5.13, 8.2a,c, 10.3 (part)** by permission of the American Association of Petroleum Geologists, whose permission is required for further use.]

WANLESS, H. R. AND DRAVIS, J. J. (1989) 'Carbonate environments and sequences of Caicos Platform', *Fieldtrip Guidebook T374: 28th International Geological Congress*, Washington, American Geophysical Union, 75pp. [**Figure 11.8.**]

WEEDON, G. P., JENKYNS, H. C., COE, A. L. AND HESSELBO, S. P. (1999) 'Astronomical calibration of the Jurassic time-scale from cyclostratigraphy in British mudrock formations', *Philosophical Transactions of the Royal Society: Mathematical, Physical and Engineering Sciences*, **357**(1757), 1731–2007, The Royal Society. [**Figure 2.7b–d.**]

WESTERNGECO [**data for Figure 4.22b.**]

WILSON, R. C. L. (1988) 'Sequence stratigraphy: a revolution without a cause?', in BLUNDELL, D. J. AND SCOTT, A. C. (eds) *Lyell: The Past is the Key to the Present*, Geological Society Special Publication No. 143, 303–314, Geological Society of London. [**Figure 5.5.**]

WINN, R. D., JR., ROBERTS, H. H. AND Kohl, B. (1998) 'Upper Quaternary strata of the Upper Continental Slope, north east Gulf of Mexico: Sequence stratigraphical model for a terrigenous shelf edge', *Journal of Sedimentary Research*, **68**, 579–595, Society for Sedimentary Geology. [**Figures 6.1–6.8.**]

WRIGHT, V. P. AND BURCHETTE, T. P. (1996) 'Shallow water carbonate environments,' in READING, H. G. (ed.) *Sedimentary Environments: Processes, Facies and Stratigraphy* (3rd edn), 325–394, Blackwell. [**Figures 11.9, 11.12, 11.14a,b, 11.16a,c, 12.2, 12.3, 12.4b, 12.5–12.9.**]

YOUNG, R. G. (1955) 'Sedimentary facies and intertonguing in the Upper Cretaceous of the Book Cliffs, Utah, Colorado', *Bulletin of the Geological Society of America*, **66**, 177–202, The Geological Society of America. [**Figures 7.4a,b, 10.5 (middle part).**]

ZEIGLER, P. A. (1982) *Geological Atlas of Western and Central Europe,* Shell International Petroleum Maatschappij, B.V. 130pp., plus maps. [**Figure 3.7b (part).**]

# Index

*Note:* page numbers refer to text, figures and tables; **bold** page numbers indicate where a key term or concept is introduced or explained.

# U

*Udotea* 220, 222
unattached (carbonate) platform 215, 231
unconformity 11, 12, 14, 20, 70, 72–74, 77, 80, 85–87, 97, 108, 125, 127, 202, 247, 257, 269
    (angular) 12
uplift (tectonic) 44, 53, 58, 113, 114
upwelling (ocean) 101
Urals 48
uranium isotopes 23
uranium 51, 196

# V

Variscan (highlands) 48
Viosca Knoll (borehole) 118–133
volcanic ash 23
volcanism 43
volcano 46, 215
volumetric partitioning **174**, 177, 189
Vostok (ice core) 40

# W

wacke 225
wackestone 220, **225**, 228, 237, 239
Walther's Law 14, 18–20, 62, 161
warm-water carbonate factory 209–211, 215, 218, 219, 231, 234, 250
Wasatch Plateau 138
washover fan 152, 168
water depth 44, **45**, 54
water table 60, 82, 152, 171, 173, 196
Western Interior Basin 138, 141, 144, 204, 207
Western Interior Seaway 137, 142, 149, 154, 184
Westphalian 50
Wheeler diagram 73 (*see also* chronostratigraphical diagram)
Widmerpool Gulf 50, 117
William Smith 20
Willow Creek 151, 156, 170
Woodside Canyon 137, 149, 151, 189–191

# Y

Yeadonian 49, 50
Younger Dryas 37

# Z

Zagros Mountains 228
Zaragoza 266, 267, 271
zonal boundary 22
zone (fossil) 21, 22, 203, 265, 266